Solid Waste
ENGINEERING

A Global Perspective

Solid Waste
ENGINEERING

A Global Perspective

THIRD EDITION

William A. Worrell • **P. Aarne Vesilind** • **Christian Ludwig**

*San Luis Obispo County
Integrated Waste Management
Authority*

Bucknell University

*École Polytechnique Fédérale
de Lausanne (EPFL)
Paul Scherrer Institute (PSI)*

CENGAGE
Learning®

Australia • Brazil • Mexico • Singapore • United Kingdom • United States

Solid Waste Engineering: A Global Perspective, Third Edition
William A. Worrell, P. Aarne Vesilind, and Christian Ludwig

Product Director, Global Engineering:
 Timothy L. Anderson

Senior Content Developer: Mona Zeftel

Associate Media Content Developer:
 Ashley Kaupert

Product Assistant: Teresa Versaggi

Marketing Manager: Kristin Stine

Director, Content and Media Production:
 Sharon L. Smith

Content Project Manager: D. Jean Buttrom

Production Service: RPK Editorial
 Services, Inc.

Copyeditor: Shelly Gerger-Knechtl

Proofreader: Pat Daly

Indexer: Shelly Gerger-Knechtl

Compositor: MPS Limited

Senior Art Director: Michelle Kunkler

Cover and Internal Designer:
 Imbue Design/Kim Torbeck

Cover Image: James Hardy/PhotoAlto
 Agency RF Collections/Getty Images

Intellectual Property
 Analyst: Christine Myaskovsky
 Project Manager: Sarah Shainwald

Text and Image Permissions Researcher:
 Kristiina Paul

Senior Manufacturing Planner: Doug Wilke

For product information and technology assistance, contact us at
Cengage Learning Academic Resource Center, 1-800-423-0563.
For permission to use material from this text or product, submit all requests online at **www.cengage.com/permissions.**
Further permissions questions can be emailed to
permissionrequest@cengage.com.

Library of Congress Control Number: 2015956726

ISBN: 978-1-305-63520-3

Cengage Learning
20 Channel Center Street
Boston, MA 02210
USA

Cengage Learning is a leading provider of customized learning solutions with employees residing in nearly 40 different countries and sales in more than 125 countries around the world. Find your local representative at **www.cengage.com.**

Cengage Learning products are represented in Canada by Nelson Education Ltd.

To learn more about Cengage Learning Solutions, visit **www.cengage.com/engineering.**

Purchase any of our products at your local college store or at our preferred online store **www.cengagebrain.com.**

Unless otherwise noted, all items © Cengage Learning.

Printed in the United States of America
Print Number: 01   Print Year: 2016

DEDICATION

This book reflects a lifetime of work in the solid waste management field and would not have been possible without the influence of many people during my career. First and foremost would have to be Dr. Aarne Vesilind who provided guidance to a young engineering student at Duke University, hired him to perform a shredder acceptance test, and coauthored a paper that was published by the American Society of Mechanical Engineers. In addition, Frank McAlister inspired me at Duke to finish graduate school. At Brown and Caldwell, Jim Smith and Larry Theisen served as mentors. Joe Ruiz and Tony Sobrino smoothed the transition from the private sector to the public sector at Miami-Dade County. In San Diego County, Rick Anthony and Bow Bowman provided valuable insight. In San Luis Obispo, my staff: Carolyn Goodrich, and Patti Toews, along with Charles Tenborg, Ray Biering, Mike di Milo, Mary Whittlesey, and Ron Munds make my job easy. Finally the San Luis Obispo County Integrated Waste Management Authority Board of Directors, consisting of 13 elected officials, has shown me how elected officials can be dedicated, hard working, and a force for positive change and a real pleasure to work for.

No dedication would be complete without mentioning my family. My children, Hilary, Emily, Michael, and Sarah, who I have driven for two hours around Maui to Hana—not to see waterfalls, but to see the Hana Landfill along with many other landfills, recycling centers, and transfer stations over the years. To my daughter-in-law Lauren, son-in-laws Greg and Ryan, granddaughter Bridget, and grandsons Oliver and Ben, thanks for keeping me young. Finally, to the love of my life, my wife Kathy, who typed my thesis 37 years ago and helped me with this book. When we were just married and I was in graduate school, we went on a field trip to a large wastewater treatment plant. I asked her, wasn't it fascinating to learn what happens when you flush your toilet, and she said there were some things she just didn't care about. Thank goodness she never felt that way about garbage or me.

William A. Worrell

Sometime in the mid-1970s, the threat of nationwide bottle legislation (placing a mandatory deposit on beverage containers to encourage their recycling) prompted some of the large beverage manufacturers like Anheuser-Busch and Pepsi Cola to fund the National Center for Resource Recovery. This organization had a clear agenda—to promote the recovery of waste materials as an alternative to the much-dreaded bottle law. After some years, when the threat of such legislation had abated, the funding for NCRR dried up, and the organization disbanded. While it was operational, however, it produced many fine research publications and provided the funds and ideas for a number of university research programs, including the one at Duke University. Much of the research into the mechanics and policy aspects of materials recovery reported in this book came as a direct result of this support from the NCRR. The president of NCRR was James Abert, and its chief technical officer was Harvey Alter. Their friendship and encouragement was instrumental in my undertaking my own research in municipal solid waste processing, and thus, I would like to dedicate my part of this book to these two gentlemen.

P. Aarne Vesilind

I dedicate my work to all persons who contributed to the success of my professional career and who positively influenced and catalyzed my thinking. The physics and chemistry classes of *Hans-Peter Seiler* lit a flame in me which is still burning today. I started to do chemistry experiments in the basement of the home of my parents, and later I expanded the laboratory at my grandmother's place. I thank my parents and my grandmother who gave me this unique opportunity. *Ueli Roth*, chemistry teacher at the Gymnasium provided chemicals for my home lab, and his excellent courses encouraged me to study chemistry. I shared my hobby with *Silvan Perego* who often joined me at my home lab. Thanks Silvan for this exciting time! Further, I thank my teachers and friends at the University of Berne, *Paul Schindler (†)*, *Geri Furrer* (professor at ETH Zurich) and *Laurent Charlet* (professor at University of Grenoble) who shared their knowledge in geochemistry with me. In this life phase I met my future wife Brigitte. She is not a scientist, but she may know better than anyone else what it means to work as a passionate scientist. I thank her for the patience she had with me over many years.

I am most grateful to *Bill Casey* for his professional and personal support during my postdoc at University California, Davis and his friendship. The most exciting ideas we developed during the sandwich lunches at the weekends and our walks from Hoagland Hall to 3rd street. In a very short period we co-authored several papers and even two in the journal *Nature*.

When I returned to Switzerland I joined the laboratory of *Peter Baccini* at Eawag, where I worked in the research group of *Annette Johnson*. I thank Annette for giving me the chance to perform my first field investigations. *Samuel Stucki* was my superior from 1997 to 2008 at the PSI (Paul Scherrer Institute). Both, Peter and Samuel have strongly influenced my later career. Their spirit and leadership concerning sustainable waste and resources management have enriched and formed my scientific thinking. *Alexander Wokaun*, vice director of PSI, I thank for the confidence placed in me, which helped me to successfully master the challenges that came along with my EPFL-PSI double affiliation. *François Golay*, director of the Environmental Sciences and Engineering Section, and Marilyne Andersen, dean of ENAC, supported the idea of a partial sabbatical which allowed me to contribute to the new edition of Solid Waste Engineering. I thank Bill Worrell and his wife Kathy for the hospitality during my short stay at San Luis Obispo. The authors enjoyed the writing and I hope that you will enjoy the learning.

Christian Ludwig

CONTENTS

CHAPTER 5

Separation Processes 177

A Tale of Two Barges and Birds Dying from Plastics

Not so very long ago, as the coastal cities of the young United States grew to metropolitan regions, the disposal of municipal refuse was expediently achieved by simply loading up large barges, transporting them some distance from shore, and shoveling the garbage into the water. One such scow operated out of New York City during the turn of the 20th century; it is pictured below. Few complained when some of the refuse floated back to the shore. It was simply the way things were done.

(Public Domain)

A different story can be told about another barge, named the *Mobro*, pictured on the next page. The year was 1987. The *Mobro* had been loaded in New York with municipal solid waste and found itself with nowhere to discharge the load, and ocean disposal was now illegal. The barge was towed from port to port, with six states and three countries rejecting the captain's pleas to offload its unwanted cargo.

Mobro garbage barge (AP/World Wide Photos)

The media picked up on this unfortunate incident and trumpeted the "garbage crisis" to anyone who would listen. Reporters honed their finest hyperbole, claiming that the barge could not unload because all of our landfills were full and that the United States would soon be covered by solid waste from coast to coast. Unless we did something soon, they claimed, we could all be buried in garbage.

The difference between the two barges, almost 100 years apart in time, is striking. In 1900, there were few laws restricting refuse disposal, and thus, solid waste disposal practices resulted in severe and permanent detrimental effects to the environment. There is no doubt that much of the refuse being shoveled off the barge 100 years ago is still on the bottom of the New York Bight and will remain there indefinitely as an embarrassment to future generations.

A hundred years later, the public is acutely aware of the problem of solid waste disposal, and today we have achieved a degree of technological sophistication in our management of solid waste. Our landfills are constructed with almost no detrimental environmental effect, our solid waste combustors emit essentially no pollutants, and the public is increasingly participating in recycling programs. We have the problem under control, and yet the public perception is exactly opposite of the reality, as exemplified by the *Mobro* incident.

The story of the hapless *Mobro* is actually a story of an entrepreneurial enterprise gone sour. An Alabama businessman, Lowell Harrelson, wanted to construct a facility for converting municipal refuse to methane gas. He recognized that baled refuse would be the best form of refuse for that purpose. He purchased the bales of municipal solid waste from New York City and was going to find a landfill somewhere on the East Coast or in the Caribbean where he could deposit the bales and start making methane. Unfortunately, he did not get the proper permits for bringing refuse into various municipalities, and the barge was refused permission to offload its cargo. As the journey continued, the press coverage grew, and no local politicians would agree to allow the garbage to enter their ports. Harrelson finally had to burn his investment in a Brooklyn incinerator.

The barge *Mobro* is a poor metaphor for the state of municipal solid waste management, because today we manage the discards of society with engineering skill at reasonable cost and at minimal risk to the public. This is not to say that we cannot do things better. Yes, we have to be concerned about nonreplenishable resources. Yes, we care about the use of land for refuse storage—land that could be used for other purposes. Yes, we can design better packaging for our consumer goods. Yes, we can initiate programs that promote litter-free roads and keep litter out of our oceans. Yes, we can design better devices that more effectively separate the various constituents of refuse. Yes, we can develop solid waste management strategies that promote zero waste. And yes, we can do many good things to improve the solid waste collection, treatment, and disposal process. But we also should be proud of the accomplishments of solid waste engineers in managing the collection, recovery, and disposal of municipal refuse. It is this positive theme that we wish to impress on the readers of this book.

In the first and second editions of this book, most of the focus was on the United States. However solid waste is a global issue as we start to consider our limited resources, impact of pollution on our oceans, and global warming. Solid waste resulting from the tsunami that washed ashore in Fukushima, Japan has been found on the west coast of the United States. E-waste has become a problem of global dimension as toxic wastes are transported between different continents, and an ecological treatment is not always guaranteed. Waste gyres (vortex) exist in the oceans. In 2010 275 million metric tons of plastic waste were generated in 192 coastal countries, with 4.8 to 12.7 million metric tons entering the ocean.[1] Worldwide, many countries, such as Italy, have recurring waste management

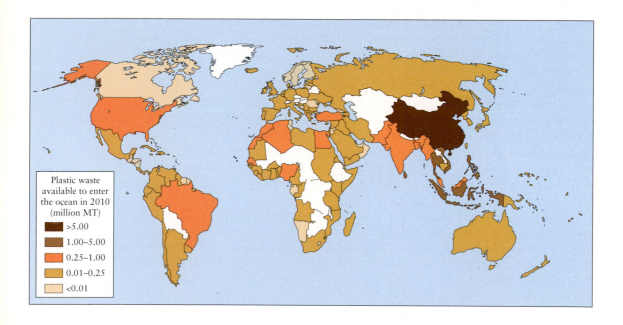

Plastic waste available to enter the ocean in 2010 (million MT)

- >5.00
- 1.00–5.00
- 0.25–1.00
- 0.01–0.25
- <0.01

[1]Jenna R. Jambeck, J.R., Geyer, R., Wilcox, C., Siegler, T.R., Perryman, M., Andrady, A., Narayan, R., Lavender Law, K., 2015. Science, *Plastic waste inputs from land into the ocean.* 347(6223):768–771

Waste in the streets of Napoli, Italy.

problems. In Napoli the army's help was needed in the past to remove the garbage from the streets. At the same time, other countries are also developing advanced new practices for integrated waste management. Because of the significance of this global approach, Dr. Christian Ludwig from Switzerland has been added as a co-author. In addition the third edition has been reorganized to reflect the hierarchy of waste management. For example, landfilling, the last option for waste management, has been moved to the end of the book.

This book is written for the student who wants to learn about solid waste engineering, a subset of environmental engineering. Environmental engineering developed during the last 60 years as a major engineering discipline and is now established as an equal alongside such major engineering fields as civil, chemical, biological, mechanical, and electrical engineering. The emergence of environmental engineering is driven in great part by societal need to control the pollution of our environment. Jobs for environmental engineers continue to increase, and there is no sign that this will slow down.

Using this book as part of a graduate or advanced undergraduate course will help to prepare the student to enter the field. Much of the knowledge in solid waste engineering is gained by actual experience while working with experienced engineers in the field, and it is impossible to include all of this experience in this book. What we hope is that the student, at the conclusion of this course, will be able to enter into meaningful conversations with experienced engineers and eventually put the basic principles learned in this course to beneficial use.

The course, as taught at different universities, usually takes one semester. Some have taught this course by going through the book in sequence, assigning homework problems as appropriate. Others have eschewed homework problems completely and used only a design problem, such as the one found in Appendix A, requiring students to work in groups and individually write weekly chapters. The structure of the course is probably not as important as the education of the students in the fundamentals of solid waste management. It is not enough to train students to solve certain types of problems. It *is* important for them to emerge from this course being able to think reflectively and logically about the problems in and solutions to solid waste engineering.

We believe that the material in this book represents a valuable first course in solid waste engineering. With this revision, we have updated information to reflect current conditions and changes that have occurred in solid waste technology—regulations and practices that have occurred over the past several years. For us, the "proof of the pudding" has been the wide acceptance of our students by the practicing engineering community. We hope that others will be able to use this book to launch exciting and productive careers in solid waste engineering.

Acknowledgments

We would like to thank Debra Reinhart, our co-author from the first edition, for her valuable contribution. In addition, thanks to Barry Shanoff, for his help on the legal issue of flow control, Peter Chromec for assistance with thermal processes and Richard Anthony for this help on zero waste. We would also like to thank the many reviewers of all three editions. Reviewers for this third edition include Clayton J. Clark II, Florida A&M University; Brajesh Kumar Dubey, University of Guelph; R. Ryan Dupont, Utah State University; Joseph F. Malina, The University of Texas at Austin; and Jay N. Meegoda, New Jersey Institute of Technology. Special thanks to the helpful people at Cengage Learning who were very patient with us. We wish to acknowledge and thank our Global Engineering team at Cengage Learning for their dedication to this new book: Timothy Anderson, Product Director; Mona Zeftel, Senior Content Developer; D. Jean Buttrom, Content Project Manager; Kristin Stine, Marketing Manager; Elizabeth Murphy, Engagement Specialist; Ashley Kaupert, Associate Media Content Developer; Teresa Versaggi and Alexander Sham, Product Assistants; and Rose Kernan of RPK Production. They have skillfully guided every aspect of this text's development and production to successful completion.

Finally to all our colleagues around the world whose knowledge and dedication inspire us every day, thank you.

William A. Worrell
P. Aarne Vesilind
Christian Ludwig

WILLIAM A. WORRELL

William A. Worrell received a B.S. and M.S. in Civil Engineering from Duke University in 1976 and 1978, respectively. His Master's Thesis involved evaluating the separation efficiencies of various air classifiers. In 1989, he attended Harvard University's John F. Kennedy School of Government Summer Program for Senior Executives in State and Local Government. Mr. Worrell has published and/or presented 56 professional papers in the United States, England, Switzerland, Japan, Peru, Hong Kong, and China. He is a registered professional engineer in California, Georgia, and Florida.

In 1978, Mr. Worrell joined Brown and Caldwell Consulting Engineers in Atlanta and managed solid and hazardous waste projects throughout the south. Seven years later, he opened Brown and Caldwell's first Florida office. In 1987, he was hired by Miami-Dade County as their chief solid waste engineer. In 1990, he was hired to manage San Diego County's solid waste program, and five years later, he became the first manager of the San Luis Obispo County Integrated Waste Management Authority where he is currently employed. He has also taught the Solid Waste Engineering course at California Polytechnic State University in San Luis Obispo.

During his more than 37 year career, he has had numerous achievements and recognitions. In 1984, he wrote the feasibility study for the Marion County Oregon Waste to Energy Project Bonds—the first American plant to use a dry scrubber and baghouse. In Miami, he was responsible for closing a 640 acre superfund site, retrofitting the largest waste-to-energy plant, and selecting a curbside recycling program for over 200,000 homes. In San Diego, his recycling program was recognized by the National Recycling Coalition as a leading program in the United States. In San Luis Obispo, he implemented the first mandatory retail take-back programs for household batteries, sharps, fluorescent tubes and bulbs, prescription drugs, and latex paint. This household hazardous waste program received the program excellence award from the North American Hazardous Materials Management Association in 2000, 2007, and 2011. In 1998, San Luis Obispo was one of first programs to meet California's 50% waste diversion goal, and by 2014, 67% of the waste was being diverted from landfills. In recognition of his leadership, the California Resource Recovery Association selected Mr. Worrell as the Recycler of the Year.

P. AARNE VESILIND

Following his undergraduate degree in civil engineering from Lehigh University, Vesilind received his PhD in environmental engineering from the University of North Carolina in 1968. He spent a post-doctoral year with the Norwegian Institute for Water Research in Oslo and a year as a research engineer with the Bird Machine Company. He joined the faculty at Duke University in 1970, where he served as chair of the Department of Civil and Environmental Engineering. In 1999, he was appointed to the R. L. Rooke Chair of the Historical and Societal Context of Engineering at Bucknell University. He served in this capacity until his retirement in 2006.

In 1976–1977, he was a Fulbright Fellow at the University of Waikato in Hamilton, New Zealand. He is a former trustee of the American Academy of Environmental Engineers, a past president of the Association of Environmental Engineering Professors, a Fellow of the American Society of Civil Engineers, and a registered Professional Engineer in North Carolina. He is the recipient of the Collingwood Prize awarded by the American Society of Civil Engineers. Other awards include the E. I. Brown Award from the students of the Department of Civil Engineering at Duke University for teaching excellence (four times), and the Tau Beta Pi Teaching Award from the students of the School of Engineering at Duke University.

While at Duke, he headed the Science, Technology, and Human Values program for many years, which was an undergraduate enrichment program that sought to build bridges between the humanities and engineering.

His research into the recovery of materials and energy from municipal solid waste led to many funded projects and numerous articles in professional journals. He has been the primary adviser for 12 PhD students and 46 master's students and has authored over 20 textbooks and other technical and professional books.

CHRISTIAN LUDWIG

Christian Ludwig received his master's degree (1990) and PhD (1993) from the Chemistry Department at the University of Berne, Switzerland. Post-doctoral years were spent at the Department of Land, Air, and Water Resources (LAWR), UC Davis, CA (1994–1995) and at the Swiss Federal Institute for Environmental Science and Technology EAWAG (1995–1997). Since 1997 he has worked at the General Energy Research Department of Paul Scherrer Institute (PSI).

In 2005 he was appointed adjunct professor at the École Polytechnique Fédérale de Lausanne (EPFL) in the field of Solid Waste Treatment. At EPFL he is mainly advising students of the Environmental Sciences and Engineering Section (SIE). He teaches the course Advanced Solid Waste Treatment, makes contributions to the laboratory class Pollutants Analysis in the Environment and the interdisciplinary teaching course Urban Neighborhoods, Infrastructures, and Sustainable Development dedicated to students of architecture, civil and environmental engineering.

Christian Ludwig has chaired or co-chaired several large international conferences focusing on resources and waste management, such as REWAS and World Resources Forum. He had different consulting mandates for industry and governments. Additional landmarks of his occupational career are reflected in his dedication section.

With the third edition of *Solid Waste Engineering*, the authors have decided to expand this university textbook to focus on the worldwide problem of solid waste management. This change is illustrated by the addition of "A Global Perspective" to the title. Given that we are currently using our natural resources at an unsustainable rate, polluting our oceans and land with a variety of waste products, and altering our atmosphere with gases that are causing further global warming, now is the time to educate future engineers with knowledge and tools to address these worldwide problems.

The three co-authors all have significant international experience. For example, Dr. Ludwig is the chair of the World Resources Forum Scientific Committee, and Mr. Worrell is a member. In their official capacity they have interacted with experts from over 80 countries at conferences in Switzerland, China, and Peru. This international knowledge is reflected in the revisions to *Solid Waste Engineering*.

The third edition has been rearranged to follow the hierarchy of solid waste management, reduction, reuse, recycling, and recovery. Thus students will first learn about integrated waste management strategies, an expertise that will support the future engineer to take measures for pollution prevention as well as for resources conservation. In chapter 2 the students are introduced to municipal solid waste characteristics, including the identification of different waste components and materials. Component-specific information is needed for recovery, separation, and recycling of waste materials. The relevance of chemical, physical, and mechanical properties are discussed in more detail as a basis for the chapters that follow. These properties are most helpful in order to identify potentially meaningful recycling pathways as well as to decide about possible technological separation and purification options. The next chapter is dedicated to the collection of municipal solid waste, a key—but many times overlooked—component of integrated waste management. Following collection is mechanical processing, in most cases the necessary first step to the recycling and recovery of municipal solid waste. The students then study mechanical, biological, and thermal processes. For each of these topics the authors have dedicated a separate chapter that introduce the students to the basic principles of these separate disciplines in the context of waste management. Since not all waste streams can be recovered, students move on to residue management by combustion and landfilling. Finally, students are exposed to the current issues in solid waste management and the principles of integrated and sustainable solid waste management. This textbook is an excellent introduction into the field of solid waste engineering and covers most of the relevant topics.

The World Resources Forum envisions the world where influential decision makers, established civil societies, key industrial players, leading scientists and engineers, and the empowered public interact and communicate on setting the agenda and developing solutions on sustainable use of natural resources worldwide, paying close attention to the delicate interplay between the economic, social, and environmental implications of resource use as well as acknowledging the challenges of increasing pressure on the availability of natural resources. Through this interaction of multiple stakeholders, innovative and effective solutions emerge, addressing the issue of efficiency and sufficiency of resource utilization among consumers, producers, and waste management, establishing sustainable practices of resources use worldwide.

Solid waste engineers are needed at the table when setting the agenda and developing solutions on sustainable use of natural resources. *Solid Waste Engineering* is an important tool that reflects the proper approach to solid waste management that students can use in their future endeavors, whether they are working as solid waste engineers for a local municipality or setting worldwide resources policy at the United Nations.

Dr. Xaver Edelmann, President
World Resources Forum (www.wrforum.org)

CHAPTER 1

Integrated Solid Waste Management

All creatures, humans included, constantly make decisions about what to use and what to throw away. A chimpanzee knows that the inside of the banana is good, and that the peel is not, and discards it. A paramecium uses certain high-energy organic molecules and discharges its products after having extracted the energy in the carbon–carbon or carbon–hydrogen bonds. And humans buy a can of soft drink with the full understanding that the can will become waste. Waste is a consequence of everyday life—of all creatures.

The challenge for society is to minimize the waste that is generated and to convert waste into a resource. The goal is to achieve sustainability by reducing the ecological footprint and increasing resources efficiency through the use of strategies such as zero waste. A solid waste engineer must play an important role as society strives to achieve this goal.

1-1   SOLID WASTE IN HISTORY

In this book, we consider a special kind of waste created by humans, the so-called *solid waste*, to distinguish it from the waste we emit into the atmosphere or the waste we discharge into the sewerage system. Humans have been producing solid waste forever as part of life.

When humans abandoned nomadic life at around 10,000 BC, they began to live in communities, resulting in the mass production of solid waste. Waste piled up, and people wallowed in the offal—a characteristic that seems to be unique to the human animal. There were exceptions, of course. In the Indus valley, the city of Mahenjo-Daro had houses with rubbish chutes and probably had waste collection systems. Other towns on the Indian subcontinent—Harappa and Punjab—had toilets and drains, and by 2100 BC, the cities on the island of Crete had trunk sewers connecting homes.[1] The sanitary laws written by Moses in 1600 BC still survive in part. By 800 BC, old Jerusalem had sewers and a primitive water supply. By 200 BC, the cities in China had "sanitary police" whose job it was to enforce waste disposal laws.

But for the most part, people in cities lived among waste and squalor. Only when the social discards became dangerous for defense was action taken. In Athens, in 500 BC, a law was passed to require all waste materials to be deposited more than a mile out of town because the piles of rubbish next to the city walls provided an opportunity for invaders to scale up and over the walls.[2] Rome had similar problems and eventually developed a waste collection program in 14 AD.

The cities in the Middle Ages in Europe were characterized by unimaginable filth. Pigs and other animals roamed the streets, and wastewater was dumped out of windows onto unsuspecting passersby. In 1300, the Black Death, which was to a great degree a result of the filth, reduced the populations in cities and alleviated the waste problems until the Industrial Revolution in the mid-1800s brought people back to the cities.

The living conditions of the working poor in 19th-century European cities have been graphically chronicled by Charles Dickens in novels such as *A Tale of Two Cities* and other writers of that period. Industrial production and the massing of wealth governed society, and human conditions were of secondary importance. Water supply and wastewater disposal were, by modern standards, totally inadequate. For example, Manchester, England had on average one toilet per 200 people. About one-sixth of the people lived in cellars, often with walls oozing human waste from adjacent cesspools. People often lived around small courtyards where human waste was piled and which also served as the children's playground.

The "Great Sanitary Awakening" in the 1840s was spearheaded by a lawyer, Edwin Chadwick (1800–1890), who argued in his 1842 report, "The Sanitary Conditions of the Labouring Population," that there was a connection between disease and filth. The germ theory was not, however, widely accepted until the famous incident with the pump handle on Broad Street in London. The public health physician, John Snow (1813–1858), suspected that the water supply from the Broad Street pump was contaminated and was the cause of the cholera epidemic. He removed the handle and prevented people from drinking the contaminated water, thus stopping the cholera epidemic and ushering in the public health revolution.

In the United States, the conditions in many of the cities were appalling. Waste was disposed of by the judicious method of throwing it into streets where rag pickers would try to salvage what had secondary value. Animals that would devour waste food were considered to be beneficial. For example, in 1834, Charleston, West Virginia enacted a law protecting garbage-eating vultures from hunters.

Recycling in the late 1800s was by individuals who scoured the streets and the trash piles looking for material of value. The first organized municipal recycling program was attempted in 1874 in Baltimore, but for reasons that would not be unfamiliar to today's recycling programs, it did not succeed.[3]

As early as 1657, the residents of New Amsterdam (present-day New York) forbade the throwing of garbage into streets, but the cleanliness of streets was still the responsibility of the individual homeowner.[3] Finally, in 1866, 200 years after the first attempt at cleaning up the streets, the Metropolitan Board of Health in New York declared war on trash, forbidding the throwing of garbage or dead animals into streets. Since municipal waste is rich in nutrients and organics, far into the 20th century it was used as fertilizer and spread untreated onto agricultural land. To protect the people against diseases, the first incineration plant went into operation in 1870 in London, England. Plants in Leeds, Manchester, and Birmingham followed in the years 1876–1877.[18] The first incinerator in Europe was built in 1893 in Hamburg following a cholera epidemic. The first incinerator in the New World was built in 1887 on Governor's Island in New York. The garbage problem finally became a factor in politics, and great effort was made politically to clean up the cities. Municipal collection systems were created, the most famous and best organized one being in New York City, headed by Colonel George Waring. The method of disposal in New York in those days was to carry the waste by barge into the blight area and dump it into the water. Before special barges were built for this purpose, the waste was simply shoveled off the barge by workers. Waring started a comprehensive materials recovery system, using pickers to separate out and then sell recoverable materials. His plan included the collection of separated materials: ashes, garbage, and other material. Public opposition to materials recycling was strong, however, and the program was terminated after several years of operation.[3]

In those days, much of the unwanted material in cities was removed by scavengers, who collected over 2000 yd^3 (1530 m^3) daily in the city of Chicago.[1] Partly because of scavengers and partly because of the lifestyle, in 1916, the municipal collection crews collected only about 0.5 pounds of refuse per capita per day—compared to the current generation rate of over 4 pounds per capita per day today. There are no solid data, however, and the generation rates were no doubt much higher, with creative and clandestine disposal. Open horse-drawn wagons (Figure 1-1) were used to collect the waste, and the horses deposited their own waste on the streets while the refuse was being collected.

The fouling of beaches forced the passage of U.S. federal legislation in 1934 that made it illegal to dump municipal refuse into the sea. Industries and commercial establishments were not covered, however, and continued unabated dumping into offshore waters. The first hole-in-the-ground that was periodically covered with dirt, a precursor of today's modern landfill, was started in 1935 in California. Ironically, the site today is on the U.S. Environmental Protection Agency's (EPA's) Superfund list as containing highly hazardous materials.[3] The American Society of Civil Engineers published the first engineering guide to sanitary landfilling in 1959; the guide discussed the compaction of refuse and the placement of a daily cover to reduce odor and vermin such as rats and buzzards. However, depending on the content and composition of the waste in these old landfills, there can be either a current or future hazard potential. In Europe, sites with hazardous potential are named "Altlasten" (a word with German origin: "alt" = old, "lasten" = burdens), expressing that these sites are potentially burdens from the past, while in the United States these old landfills might become Superfund sites.

Figure 1-1 A typical horse-drawn solid waste collection vehicle, used well into the 1920s in many U.S. cities. (Courtesy NCRR)

How communities respond to these old sites varies. The municipality of Kölliken (Switzerland) decided in 1999 to dig out an entire landfill containing hazardous and municipal wastes in order to properly treat and recycle all toxic and valuable materials contained in the landfill. Such an effort was undertaken because all other possible measures to protect the groundwater from contamination failed. It also turned out that the expected costs for long-term maintenance would exceed the cost of removing the waste. A roof was built over the entire landfill in order to protect the housing nearby from the toxic emissions (Figure 1-2).

The management of municipal refuse has changed over the years, and so has the composition of the waste. Some significant events that changed the characteristics of residential and commercial municipal solid waste include:[3]

1908	Paper cups replace tin cups in vending machines.
1913	Corrugated cardboard becomes popular as packaging.
1924	Kleenex facial tissues are first marketed.
1935	First beer can is manufactured.
1944	Dow Chemical invents Styrofoam.
1953	Swanson introduces the TV dinner.
1960	Pop-top beer cans are invented.

Figure 1-2 Left: The roof constructed over the landfill to protect the neighborhood from emissions during the reconstruction of the landfill. Right: Dust-laden atmosphere inside the hall during operation. (Courtesy Christian Ludwig)

1963	Aluminum beer cans are developed.
1977	PETE soda bottles begin to replace glass.
1982	Plastic grocery bags start to replace paper bags.
1983	First generation LCD TVs become available on the market. Other flat panel and conventional CRT TV technology disappears.
1991	The "producer responsibility principle" is implemented in Germany with the regulation "Verpackungsverordnung," which defines the recovery and recycling of packaging materials for beverages and introduces the trademark "green dot" (Figure 1-3).

Figure 1-3 Green dot symbol.

1998	Stores selling electronic equipment in Switzerland have to take back electronic scrap free of charge (costs are covered by an advanced disposal fee "Vorgezogene Entsorgungsgebühr").[19]
1999	The Council of the European Commission sets targets for systematically reducing biodegradable wastes in landfills.[20]
1999	Landfill Kölliken is the first landfill to be fully dismantled after operation to protect the environment and neighborhood.
2000	Switzerland prohibits landfilling of municipal solid waste and is the first country to do so.[23]
2006	Europe enforces higher recycling rates by law with the Directive 2006/12/EC on waste,[21] which establishes the legislative framework for the handling of waste in the community.

In today's cities, solid waste is removed and is either sent to disposal or reprocessed for subsequent use. This change in thinking from simply getting the stuff out of town to its use for some purpose represents the first paradigm shift in solid waste engineering in nearly 2000 years.

Following rapidly on the move to recover materials is the "Waste Reduction Revolution"—the idea that it is bad to create waste in the first place. These changes have occurred because of both economics and a shift in public attitude.

1-1-1 Economics and Solid Waste

The emergence of the industrial age fostered the science of economics and prompted many leading thinkers to attempt to bring rational order to the seemingly chaotic world around them. The rationalism that resulted led to the common belief that trends could be understood and decisions made best on the basis of numbers. This substitution of the quantitative for the qualitative still pervades modern society and influences our entire set of attitudes toward resources and how they should be distributed.

Adam Smith (1723–1790), through his concept of *the invisible hand*, introduced an element of positive faith and optimism. However, his efforts were overshadowed by a number of pessimistic analysts who predicted continuing misery, poverty, exploitation, and class discrimination. David Ricardo (1772–1823), with his *iron law of wages*, held that wages for the working people would always remain at the poverty level, since any increase in wages would result in a commensurate increase in population, and this would once again drive wages down.

Equally pessimistic was the view held by Thomas Malthus (1766–1834), who in 1798 reasoned that, since population growth is geometric and the increased production of food is arithmetic, a famine must result. This *law of population* was part of the *laissez-faire* school of economic liberalism and was in great part responsible for the earned reputation of economics as a "dismal science." Malthus held that overpopulation can be prevented only by two types of checks: positive and preventive. Numbered among the former are wars, plagues, and similar disasters. Preventive checks include abstention from marriage, limitations on the number of children, and the like. Although the latter is clearly preferable, Malthus had little hope for the world and insisted that the poor were "authors of their own poverty" simply because they failed to use the preventive checks on population growth.

This thesis was widely believed for many years and held as basic economic dogma. But as populations grew, widespread famine and deprivation were avoided, and Malthus's writings fell from favor. Economists began to think of Malthus as an economic anachronism to be studied—but only in the historical context. Technology, the new god, was able to preserve order, avert disaster, and lead us into the promised land.

This optimism was widely shared during the 19th and well into the 20th century with only a few disquieting voices. Henry David Thoreau's (1817–1862) distrust of things technical was tolerated with bemusement as the ramblings of an ungrateful crackpot. During the 1950s and 1960s, a few more voices in the wilderness became audible. Paul Erlich, with his grand overstatements and predictions of doomsday, seemed strangely reminiscent of Malthus. Barry Commoner became known as the first public ecologist and helped promote the feeling of impending disaster. The most respected and well-publicized voice of pessimism came from an interdisciplinary group of scientists at the Massachusetts Institute of Technology. Funded by the Club of Rome (a consortium of concerned industrialists), this group of talented scientists and engineers developed a computer model of the world based on projections of pollution, agricultural production, availability of natural resources, industrial production, and population. The group's ambitious undertaking, led by Dennis Meadows and Jay Forrester, resulted in the publication of a final report (*The Limits to Growth*, 1972) that indicated that even our most optimistic projections will eventually lead to the onset of famine, wars, and the destruction of our economic system.[4] It was, in short, a dismal outlook. Malthus would have been pleased.

The Club of Rome report has been criticized for inaccuracies and misinterpretations, and some of these accusations appear to be valid. Indeed, a revised model has shown an increased chance for world survival,[5] and more accurate data would seem to reduce the level of pessimism. However, in a 30-year update published in 2004, the authors' pessimistic outlook has not changed. The dismal outlook of Malthus is generally reaffirmed by such studies, and we are beginning to realize that our planet is finite and has only limited resources and living space. The scarcity of land and nonrenewable resources could indeed have the ultimate devastating effect envisioned by Malthus and is now (once again) suggested by predictive studies. Many sources are now convinced that our current usage of resources is not sustainable. At the very least, the concern is real, and we should begin to seek alternative life systems in order to have more assurance that these disasters can be avoided.

One (of many) possible potentially beneficial alternatives toward attaining global stability is to eliminate solid wastes generated by our materialistic society that are now deposited on increasingly scarce land. The recovery of these resources from solid waste would be a positive step toward establishing a balanced world system where society is no longer dependent on extraction of scarce natural ores and fuels. Recycling of wastes, however, also has to be evaluated in economic terms. If recycling costs more (energy or dollars) than use and disposal, then we have to recognize this extra cost and balance it in terms of future needs. At the present time, some consumer items, such as aluminum cans, are highly recyclable because the cost of recycled products is less than those manufactured from virgin materials. The objective is to reduce the cost of recycling to the point where many more products can be manufactured economically from recycled materials. While this comparison may seem straightforward, the recycling industry believes it is not a fair fight. One study reports that 15 direct subsidies for virgin resource extraction and waste disposal reduce virgin material costs by an average of $2.6 billion per year or more than $13 billion over five years.[6]

It seems quite clear that society has to adapt, using less technology in some instances and more in others, in order to achieve a balance in materials and energy use. We recognize that at the present time some products and materials cannot be recycled either technically or economically and must be disposed of in the environment, but we view this as a transitional phase in solid waste management. The technology and philosophy necessary for the development of an enlightened solid waste policy for sustained use of the earth's resources is the topic of this text.

1-1-2 Legislation and Regulations

The United States, like other independent countries with British judicial traditions, operates on the concept of *common law*. Common law is derived from the principle of fair play or justice—a purely British invention. Under common law, if a person is wronged, the perpetrator is convicted and sentenced on the basis of *precedence*. That is, if a similar wrongful act occurred previously, then the only right and fair thing to do is to treat the next person in a similar manner. Common law is not written down, except as cases that define the precedent for new cases.

The genius of common law is that it is fair to all, but at the same time, it is able to change as the needs and values of the people change. It is no longer a crime, for example, to be a witch. But common law changes very slowly because most courts are loath to make new law. Common law is also ineffective in protecting the environment and in correcting environmental ills, because under common law, a person (not a forest or a river or any other nonhuman entity) has to be wronged before relief can be sought in the courts. The assumption is that all of nature is owned by humans, and it is only the wealth and welfare of these humans that common law protects. Because of the lack of responsiveness of common law to environmental ills, the Congress of the United States has resorted to passing environmental legislation. Such *legislated law* in effect establishes new common law by setting artificial and immediate precedents.

Prior to the 1960s, the only federal legislation that addressed solid waste was the 1899 *Rivers and Harbors Act*, which prohibited the dumping of large objects into navigable waterways. The federal government was not involved in solid waste matters, except as a major producer of solid waste, much of which was managed by the Department of the Interior. Municipal solid waste was commonly thrown into unlined open dumps, which were intentionally set on fire to reduce volume. In larger communities, solid waste was sent to incinerators, which had minimal air emission controls and did a poor job of reducing the volume of waste. In

Durham, North Carolina, for example, the city incinerator in the 1950s was called the "Durham Toaster" in the newspapers. Apparently the organic matter emerged from the incinerator barely singed.

The first federal legislation intended to assist in the management of solid waste was the 1965 *Solid Waste Disposal Act*, which provided technical assistance to the states through the U.S. Public Health Service. The emphasis in this legislation was the development of more efficient methods of disposal and not the protection of human health.

On January 1, 1970, President Nixon signed the *National Environmental Policy Act* (NEPA), which led to the creation of the Environmental Protection Agency (EPA). The most significant part of this mostly policy statement is Section 102, which requires all federal agencies (except those engaged in national defense work) to write *environmental impact statements* (EISs) whenever there is significant effect on the environment. When this law was enacted, many thought that the EISs would be simple one-paragraph boilerplate statements (much like our present anti-discrimination requirement), which agencies would simply attach to all plans and contracts stating that in their opinion there is no significant effect on the environment. But environmentalists took these first meager environmental impact statements to court, and the courts forced the federal agencies to write meaningful EISs. Since that time, most states have passed environmental impact legislation, and now whenever there is public money involved (such as for all landfills, combustors, and other solid waste management facilities), an environmental impact statement is almost certainly required.

In 1976, the Congress of the United States passed the *Resource Conservation and Recovery Act* (RCRA). With its 1984 amendments, RCRA is a strong piece of legislation that mainly addresses the problem with hazardous waste but also specifies guidelines for nonhazardous solid waste disposal. Subtitle D in this act is the municipal solid waste section, and landfills that fall under these requirements are commonly called *Subtitle D landfills*. In 1991 under Subtitle D, the EPA adopted regulations to establish minimum national landfill criteria for all solid waste landfills. A key component of the standard was to require landfills to install composite liner systems consisting of a plastic liner on top of compacted clay, leachate collection, and groundwater monitoring. Specifications for the composite liner were also included in the regulation. The regulations became effective in 1993.

In response to such hazardous waste disasters as the Love Canal in Niagara, New York, and the Valley of the Drums in Tennessee, Congress passed the 1980 *Comprehensive Environmental Response, Compensation, and Liability Act* (CERCLA), which is commonly known by its nickname—the *Superfund* act. CERCLA created a financial means of cleaning up old hazardous waste sites by tapping into the coffers of present chemical companies, who most likely were not guilty of anything other than association with the real perpetrators.

The combustion of solid waste is controlled by the 1970 *Clean Air Act* (with subsequent amendments). With this legislation began the process of closing burning dumps and uncontrolled incinerators. In every case, the federal agencies involved (mostly the EPA) required the individual states to set up local guidelines that adhere to the federal standards; these guidelines are then approved by the EPA. The actual enforcement of those requirements is then left to the states.

The siting of solid waste facilities is further complicated by local opposition, which often drags out the process. In some states, it now takes over ten years to site a new landfill, even in the absence of significant local opposition. Municipalities that have traditionally managed their solid waste locally are now shipping waste across state borders to remote regional landfills just to avoid having to go through the process of developing a local solution to the problem. Communities are increasingly unwilling to go through the process of siting a new landfill, transfer station, or combustor. In the end, the engineering challenges are minor compared to the regulatory and political challenges of siting new solid waste facilities.

Many people believe that Congress and the EPA have failed to provide any leadership on waste diversion and recycling. As a result, most states have passed strong legislation encouraging and promoting recycling. In the 1990s, over 40 states established recycling goals. For example, in 1989 California mandated that 25% of the waste be diverted from landfills by 1995 and that 50% be diverted by 2000. In 2011 California increased the mandated diversion rate to 75% by 2020. In Pennsylvania, every community with a population of over 5000 is mandated to set up a recycling program. Often local problems in siting new landfills or incinerators drive the recycling effort. If a substantial amount of solid waste can be diverted from the landfill, then either the planned landfill can be smaller in size or it will last longer.

Unlike the United States, the European Union has addressed waste diversion. An April 26, 1999 Landfill Directive stated that municipal waste must be treated prior to being landfilled. The intent was to prevent or reduce the adverse effects of landfills on the environment. This directive was followed by a series of directives addressing the issue of waste, including the one passed in November 2008 (European Union Directive 2008/98/EC), which set up a revised waste management framework based on Directive 2006/12/EC. The goal continued to be the reduction in landfilling by applying the following waste hierarchy:

(a) Prevention

(b) Preparing for reuse

(c) Recycling

(d) Other recovery (e.g., energy or raw materials recovery)

(e) Disposal

This hierarchy directs member states to develop their solid waste management plans by first seeking to prevent the production of solid waste, for example, by encouraging the use of fabric bags in grocery stores (thus eliminating the use of one-time plastic bags). When further elimination is not feasible, the states should consider actions that promote reuse, such as the use of heavy-duty plastic grocery bags that can be reused. Next, the recycling option should be considered, such as the collection and remanufacturing of one-time plastic bags. The fourth option would be to burn the collected bags for energy production. And finally, when all else fails, the bags would be disposed of ["the bags would be disposed of in landfills."] in landfills.

1-2   MATERIALS FLOW

The application of such a hierarchy can be understood by using a model showing the flow of materials in our society (Figure 1-4). This diagram emphasizes the fact that we do not *consume* materials; we merely use them and ultimately return them (often in an altered state) to the environment. The production of useful goods for eventual use by those people called (incorrectly) *consumers* requires an input of materials. These materials originate from one of four sources: raw materials gleaned from the face of the earth and used for the manufacture of products, scrap materials produced in the manufacturing operation, materials separated by users for recycling, and materials recovered after the product has been discarded.

Industrial operations are not totally efficient; they produce some waste that must either be disposed of or used again as raw material. The resulting processed goods are sold to the users of the products, who in turn have three options after use: to discard the product for either disposal or energy recovery, to separate the product for recycling back into the industrial sector, or to reuse the material for the same or a different purpose without remanufacturing.

This is a closed system with only one input and one output; this characteristic emphasizes again the finite nature of our world. At a steady state, the materials injected into the process must equal the materials disposed back into the environment. This process applies to the sum of all materials as well as to certain specific materials. For example, the manufacture of aluminum beverage containers involves the use of a raw material—bauxite ore—which is refined to produce aluminum.

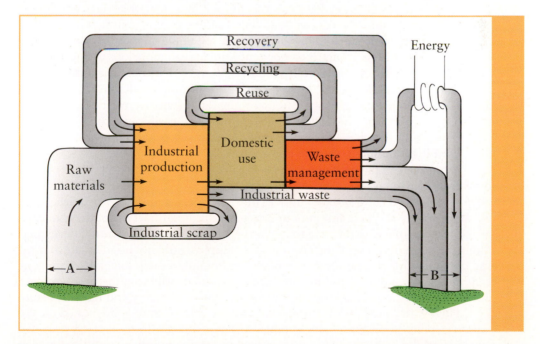

Figure 1-4 Materials flow through society.

The finished products then are sold to consumers. Some of these cans are defective or for other reasons unfit for consumer use and are recycled as industrial scrap. The consumer uses the cans, and the empty containers or other products are disposed of in the usual manner. Some of the aluminum is returned to the industrial sector (for remanufacture), and some of it might be used for other purposes in the home. The aluminum that is recovered and returned to the manufacturing process gets there only by a conscious effort by the community or other organizations that collect and recycle the material through the system.

It is also important to consider the time lag associated with domestic use. For example, an aluminum beer can has a short usage phase and residence time with the user—that is, once the beer is poured out of the can, then the can is available for reuse, recycling, recovery, or disposal. In other cases goods have a long use phase. For example, an aluminum frying pan may last 25 years before the end of its useful life. In this case production and end of life are not synchronized. Within domestic use a stock is built up and may not be available for remanufacturing until after the product has been discontinued. The interaction of the materials flow with the environment is at the input of raw materials and the deposition of wastes. In Figure 1-4, these two interfaces are denoted by the letter **A** for raw materials and by the letter **B** for the materials returned to the environment.

We can argue that both **A** and **B** should be as large as possible, since there are many benefits to be gained by increasing these values. For example, a large quantity of raw materials injected into the manufacturing process represents a high rate of employment in the raw materials industry, and this can have a residual effect of creating cheaper raw materials and reducing the cost of manufacturing. A large **B** component is also beneficial in the sense that the waste disposal industry (which includes people as diverse as the local trash collector and the president of a large firm that manufactures heavy equipment for landfills) has a key interest in the quantity of materials that people dispose of. Thus, a large **B** component would mean more jobs in this industry.

However, large **A** and **B** components also have detrimental effects. A large raw material input means that great quantities of nonreplenishable raw materials are extracted (often using less than environmentally sensitive methods, as exemplified by strip mining). Similarly, large quantities of waste can have a significant detrimental effect, such as land areas used for waste disposal or air pollution created from the burning of waste in combustors.

A high rate of raw material extraction can eventually lead to a problem in the depletion of natural resources. At the present time in the United States, we have already exhausted our domestic supplies of some nonreplenishable materials (such as copper, zinc, and tin) and are importing a substantial fraction of these materials.[7] If the rest of the world were to attain the standard of living that the developed nations have at the present, the raw materials supply would not be adequate to meet the demand. Our present lifestyle is based on obtaining these materials from concentrated sources (ores), and in using them, we are distributing the products over a wide land area. Such a distribution obviously makes recovery and reuse difficult.

Finally, the question of national security for each country is predicated on the nation's ability to obtain reliable supplies of raw materials. In the last 40 years, we experienced the problems that can be created by relying on other countries for such necessities as oil. There is little doubt that cartels will be developed by nations that have large deposits of other nonreplenishable materials, and in the future, the cost of such products as aluminum, tin, and rubber will increase substantially. We have ample justification for reducing the wastes disposed of into the environment to the smallest quantities practical, and we should redesign our economic system to achieve this end.

As shown in Figure 1-2, if the system is in a steady state, the input must equal the output. Hence, a reduction of either **A** or **B** necessarily results in a concomitant reduction in the other. In other words, it is possible to attack the problem in two ways.

Looking first at the **A** component, a reduction in raw materials demand could be achieved by increasing the amount of industrial scrap reprocessed, by decreasing the amount of manufactured goods, or by increasing the amount of recovered materials from the post-consumer waste stream. Increasing industrial scrap would involve increasing either *home scrap* (waste material reused within an industrial plant) or *prompt industrial scrap* (clean, segregated industrial waste material used immediately by another company). But scrap represents inefficiency, and an ultimate goal of industry is to produce as little scrap as possible. Clearly, decreasing the demand for raw materials will require another approach.

One possibility for achieving a low use of raw materials is to decrease the amount of manufactured goods. This *reduction* will necessitate redesigning products in such a way as to use less material. The federal government can legislate a lower rate of material use by placing taxes on excessive packaging, initiating a package charge (e.g., so many cents per pound of packaging), requiring a mandatory longer life of manufactured products, and other options. In addition, it is possible for consumers to buy fewer manufactured goods or simply to buy products that consciously minimize waste.

A second means of reducing waste is to *reuse* the products. Often this is done without much thought, such as using refillable soda and beer bottles, using coffee cans to hold nails, or using paper bags for taking out the garbage. In addition, repairing an item instead of discarding it and buying a replacement is an example of the tradeoff between a labor-intensive society and a consumer-based society.

A third means of reducing the waste destined for disposal is to separate out materials that have some economic value, collect these separately, and use them as a source of raw materials. This process is called *recycling* and involves the active participation of the product user.

The fourth means is to process the solid waste so as to recover useful material from the mixed waste. Some refer to this solid waste as urban ore that can be mined to recover valuable materials. Similar to mining virgin material, the goal is to separate out the valuable resources from the waste materials. *Recovery* can also include the recovery of energy from the solid waste. For example, a waste-to-energy plant or a landfill gas-recovery system will recover the energy value of the solid waste through a transformation process.

In summary, the feasible options for achieving reduced material use and waste generation are known as the four Rs:

1. Reduction
2. Reuse
3. Recycling
4. Recovery

Note that these are similar to the first four options in the solid waste management hierarchy suggested by the European Union, listed previously.

1-2-1 Reduction

Waste reduction can be achieved in three basic ways:

1. Reducing the amount of material used per product without sacrificing the utility of that product,
2. Increasing the lifetime of a product, and
3. Eliminating the need for the product.

Waste reduction in industry is called *pollution prevention*—an attractive concept to industry because in many cases the cost of treating waste is greater than the cost of changing the process so that the waste is not produced in the first place.[7] Any manufacturing operation produces waste. As long as this waste can be readily disposed of, there is little incentive to change the operation. If, however, the cost of waste disposal is great, the company has an incentive to seek improved manufacturing techniques that reduce the amount of waste. Pollution prevention as a corporate concept was pioneered by such companies as 3M and DuPont and has as its driving force the objective of reducing cost (and hence increasing the competitive advantage of the manufactured goods in the market place). For example, automobile manufacturers for years painted new cars using spray enamel paint. The cars were then dried in special ovens that gave them a glossy finish. Unfortunately, such operations produced large amounts of volatile organic compounds (VOCs) that had to be controlled, and control measures were increasingly expensive. The manufacturers then developed a new method of painting, using dry powders applied under great pressure. Not only did this result in better finishes, it all but eliminated the problem with the VOCs. Pollution prevention is the process of changing the operation in such a manner that pollutants are not even emitted.

Reduction of waste on the household level is called *waste reduction* (sometimes referred to as *source reduction* by the EPA).[8,9] Typical alternative actions that result in a reduction of the amount of municipal solid waste being produced include refusing bags at stores, using laundry detergent refills instead of purchasing new containers, bringing one's own bags to grocery stores, stopping junk mail deliveries, and using cloth diapers.[10] Unfortunately, the level of participation in source reduction is low compared to recycling activities. Even though source reduction is the first solid waste alternative for the EPA, and eight states have source reduction goals that range from no net increase of waste per capita to 10% reductions, few people participate in such programs. There is some evidence, however, that where communities have initiated disposal fees based on volume or weight of refuse generated, the amount of refuse is reduced by anywhere between 10% and 30%.[11,12] Public information programs can significantly help in reducing the amount of waste generated. A study of 250 homes in Greensboro, North Carolina found that a 10% waste reduction can be achieved following a public information program.[13] Advice on how to "shop smart" offered by one municipality is shown in Figure 1-5.[14]

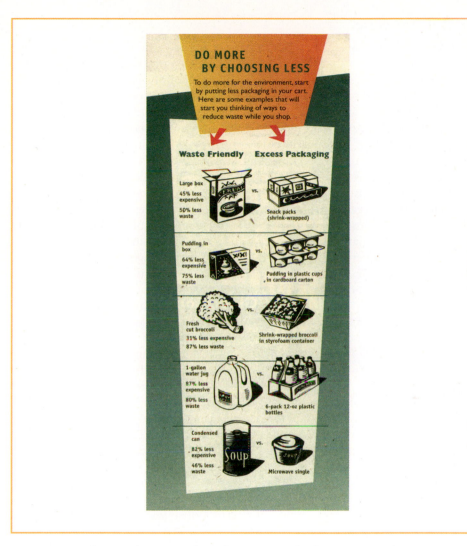

Figure 1-5 Advice on how to reduce the amount of refuse generated. Source: [14]

1-2-2 Reuse

Reuse is an integral part of society, from church rummage sales to passing down children's clothing between siblings. Many of our products are reused without much thought given to ethical considerations. These products simply have utility and value for more than one purpose. For example, paper bags obtained in the supermarket are often used to pack refuse for transport from the house to the trash can or to haul recyclables to the curb for pickup. Old wine barrels are cut in half and used as flower planters. Coffee cans are used to hold bolts and screws. All of these are examples of reuse.

1-2-3 Recycling

The process of recycling requires that the owner of the waste material first separate out the useful items so that they can be collected separately from the rest of the solid waste. Many of the components of municipal solid waste can be recycled for remanufacturing and subsequent use, the most important being paper, steel, aluminum, plastic, glass, and yard waste. In 2012, 34.5% of municipal solid waste was recycled. Nationally, Americans recycled and composted almost 87 million tons of municipal solid waste. This provides an annual benefit of more than 168 million metric tons of carbon dioxide equivalent emissions reduced, comparable to the annual greenhouse gas (GHG) emissions from over 33 million passenger vehicles.

Theoretically, vast amounts of materials can be recycled from refuse, but this is not an easy task, regardless of how it is approached. In recycling, a person about to discard an item must first identify it by some characteristics and then manually segregate it into a separate bin. The separation relies on some readily identifiable characteristic or property of the specific material that distinguishes it from all others. This characteristic is known as a *code*, and this code is used to separate the material from the rest of the mixed refuse using a *switch*.

In recycling, the code may be simple and visual. Anyone can distinguish newspapers from aluminum cans. But sometimes confusion can occur, such as identifying aluminum cans from steel cans, for example, or newsprint from glossy magazines, especially if the glossy magazines are intermingled with the Sunday paper. The most difficult operation in recycling is the identification and separation of plastics. Because mixed plastic has few uses, plastic recycling is more economical if the different types of plastic are separated from each other. Most people, however, cannot distinguish one type of plastic from another. The plastics industry has responded by marking most consumer products with a code that identifies the type of plastic, as shown in Figure 1-6. Plastics that can be recycled are all common products used in everyday life, some of which are listed in Table 1-1.

Theoretically, all a person about to discard an unwanted plastic item has to do is to look at the code and separate the various types of plastic accordingly. In fact, there is almost no chance that a domestic household will have seven different waste receptacles for plastics. Historically each house had three small (10 to 15 gallon) bins for recycling. Residents were asked to place newspapers in one bin, glass bottles in another bin, and plastic and metal bottles in the third bin. The plastic bottles included only the most common types of plastic, PETE (polyethylene terephthalate), the material out of which the 2-liter soft drink bottles are made, and HDPE (high-density polyethylene), the white translucent plastic used for milk bottles. A special truck with three different compartments would then pick up these recyclables.

Figure 1-6 Plastic recycling symbols.

Table 1-1 Common Types of Plastics That May Be Recycled

Code Number	Chemical Name	Abbreviation	Typical Uses
1	Polyethylene terephthalate	PETE	Soft drink bottles
2	High-density polyethylene	HDPE	Milk bottles
3	Polyvinyl chloride	PVC	Food packaging, wire insulation, and pipe
4	Low-density polyethylene	LDPE	Plastic film used for food wrapping, trash bags, grocery bags, and baby diapers
5	Polypropylene	PP	Automobile battery casings and bottle caps
6	Polystyrene	PS	Food packaging, foam cups and plates, and eating utensils
7	Mixed plastic		Fence posts, benches, and pallets

During the past ten years, many communities have adopted a *single-stream recycling system* where all recyclables are all placed into one large bin. This simplified approach for the resident typically results in a larger quantity of material being recycled. However, there are tradeoffs with single stream-stream recycling systems. For example, a more complicated materials recovery facility is need to separate the recyclable materials. In addition, the resulting products will generally have a lower purity.

Taking into account transportation and processing charges, it still appears that the economics for curbside recycling and materials recovery facilities in metropolitan areas (close proximity to refuse and markets) are quite favorable. The proof of this, of course, is the impressive number of materials recovery facilities (MRFs) in operation or under construction. According to the EPA, in 2011 there were 9800 curbside recycling programs and 3090 community composting programs. The success of recycling programs has been in spite of the severe obstacles that our present economic system places on the use of secondary materials. Some of these obstacles are identified here:

Location of wastes. The transportation costs of the waste may prohibit the implementation of recycling and recovery. Secondary materials have to be shipped to market, and if the source is too far away, the cost of the transport can be prohibitive.

Low value of material. The reason that an item is considered waste is that the material (even when pure) has little value. For example, the price paid for secondary polystyrene has fluctuated from a positive payment of $100 per ton to a negative payment of $300 per ton.

Uncertainty of supply. The production of solid waste depends on the willingness of collectors to transport it, the cooperation of consumers to throw things away according to a predictable pattern, and the economics of marketing and product substitution, which may significantly influence the availability of a material. Conversion from aluminum to plastic beverage containers—whether by legislation, marketing options, or consumer preferences—will significantly change the available aluminum in solid waste. The replacement of a high-value material (aluminum) by a low-value material (plastic) will adversely affect recycling. Potential solid waste processors thus have little control over their raw materials.

Administrative and institutional constraints. Some communities are unwilling to pay the additional cost to implement curbside recycling programs. The costs of these programs typically are in the $1 to $4 per month per household range. Other cities may have labor or contractual restrictions preventing the implementation of resource recovery projects. For example, the city of San Diego is prohibited from charging its residents for solid waste service, thus making it difficult to fund the implementation of curbside recycling.

Legal restrictions. Recycling of certain materials may be prohibited. For example, many yard waste composting facilities are prohibited by land-use ordinances from accepting sludge or food waste, both of which may increase the value of the compost.

Uncertain markets. Recovery facilities must depend on the willingness of customers to purchase the end products—materials or energy. Often such markets are fickle, being either small, fragile operations or large, vertically integrated corporations that purchase the products on margin so as to satisfy unusually heavy short-duration demand.

Unclear definitions. The public is confused by terms used to describe recyclable and non-recyclable waste. In Figure 1-7 two waste cans at an airport in China are shown. One can is for recyclable and the second one next to it for non-recyclable wastes. From the photos of the contents of the cans it is difficult to distinguish between the content in the two cans. This may have different causes. Some people may not care about the difference; others may interpret the terms recyclables and non-recyclables in a different way. It may also be a result of globalization, reflecting people with different cultural background and education. However, we can conclude that a common definition is important to improve the quality of the separately collected waste fractions.

1-2-4 Recovery

Recovery is defined as the process in which the refuse is collected without prior separation and when the recyclable materials in the refuse are separated from the non-recyclable materials at a central facility. A typical mixed-waste *materials recovery facility* (*MRF*) is shown in Figure 1-8.

The various recovery operations in an MRF have a chance of succeeding if the material presented for separation is clearly identified by a code and if the switch is then sensitive to that code. Currently, no such technology exists. It is impossible, for example, to mechanically identify and separate all of the PETE soft drink bottles from refuse. Most recovery operations today employ *pickers*, human beings who identify the most readily separable materials—such as corrugated cardboard and HDPE milk bottles—before the refuse is processed mechanically.

Most items in refuse are not made of a single material, and to use mechanical separation, these items must be separated into discrete pieces consisting of a single material. A common "tin can," for example, contains steel in its body, zinc on the seam, a paper wrapper on the outside, and perhaps an aluminum top. Other common items in refuse provide equally challenging problems in separation.

One means of producing single-material pieces (and thereby assisting in the separation process) is to decrease the particle size of refuse by grinding up the

Figure 1-7 Left: Waste cans at an airport of a large city in Asia, one for recyclables and one for non-recyclables. Right: the result of the effort to separate these two fractions from two locations—upper pair (recyclable and non-recyclable) and lower pair (recyclable and non-recyclable). (Courtesy Christian Ludwig)

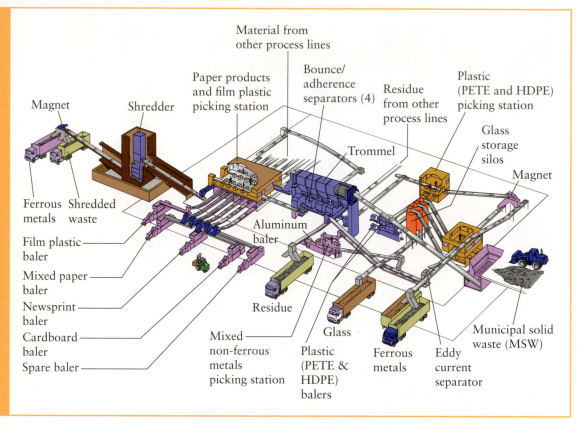

Figure 1-8 Typical mixed-waste materials recovery facility (MRF).

larger pieces. Grinding will increase the number of particles and achieve many clean (single-material) particles. The size-reduction step, although not strictly materials separation, is employed in some materials recovery facilities, especially if refuse-derived fuel is produced. Size reduction is followed by various other processes, such as air classification (which separates the light paper and plastics) and magnetic separation (for the iron and steel) (see Chapter 5).

Paper industry companies are *vertically integrated*, meaning that the company owns and operates all of the steps in the papermaking process. The company owns the lands on which the forests are grown, it does its own logging, and it takes the logs to its own paper mill. Finally, the company markets the finished paper to the public. This is schematically shown in Figure 1-9.

Suppose a paper company finds that it has an annual base demand of 100 million tons of paper. It then adjusts the logging and pulp and paper operations to meet this demand. Now suppose there is a short-term fluctuation of 5 million tons that has to be met. There is no way the paper company can plant the trees necessary to meet this immediate demand; nor is it able to increase the capacity of the pulp and paper mills on such short notice. What the company does then is go to the secondary paper market and purchase the secondary fiber to meet the incremental demand. If several large paper companies find that they

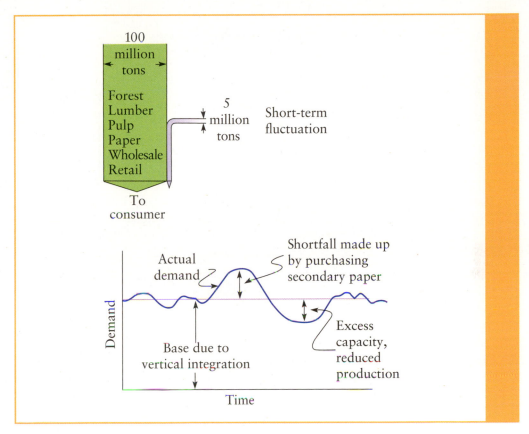

Figure 1-9 Vertical integration of the paper industry.

have an increased demand, they will all try to purchase the secondary paper, and suddenly the demand will shoot the price of waste paper up. When either the demand decreases or the paper company has been able to expand its capacity, it no longer needs the secondary paper, and the price of waste paper plummets. Because paper companies purchase waste paper *on the margin*, secondary paper dealers are always in either a boom or bust situation, and the price is highly variable. When paper companies that use only secondary paper to produce consumer products increase their production (due to the demand for recycled or recovered paper), these extreme fluctuations are dampened out. Price fluctuation is not limited to just old newsprint but includes all recovered materials. For example, Figure 1-10 illustrates how the value of recycled materials has varied over the past 28 years. Such spikes in the cost curve play havoc with long-term planning for secondary materials production and use.

1-2-5 Disposal of Solid Waste in Landfills

The "disposal" of solid wastes is a misnomer. Our present practices amount to nothing more than hiding waste well enough so it cannot be found readily. The only two realistic options for storing waste on a long-term basis are in the oceans (or other large bodies of water) and on land. The former is forbidden by federal

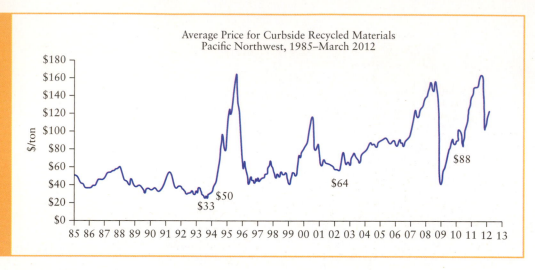

Figure 1-10 Price fluctuations of recycled materials. Source: Zerowaste in the 21st Century, Sound Resource Management Group Inc., Olympia, Washington

law and is similarly illegal in most other developed nations. Little else needs to be said of ocean disposal, except perhaps that its use was a less than glorious chapter in the annals of public health and environmental engineering. Unfortunately the impact of past dumping and litter going into the oceans is still with us. Since 1999, the Algalita Marine Research Foundation and others have been mapping a plastic garbage patch in the Pacific Ocean that is larger than Texas. They have found that in the Central North Pacific Gyre, pieces of plastic outweigh surface zooplankton by a factor of 6 to 1.

The placement of solid waste on land is called a *dump* in the USA and a *tip* in Great Britain (as in *tipping*). The open dump is by far the least expensive means of solid waste disposal, and thus, it was the original method of choice for almost all inland communities. The operation of a dump is simple and involves nothing more than making sure that the trucks empty at the proper spot. Volume was often reduced by setting the dumps on fire, thus prolonging the dump lifespan.

Rodents, odor, air pollution, and insects at the dump, however, can result in serious public health and aesthetic problems, and alternate methods for disposal were necessary. Larger communities can afford to use a combustor for volume reduction, but smaller towns cannot afford such capital investment. This has led to the development of the *sanitary landfill*. The sanitary landfill differs markedly from open dumps in that the latter were simply places to dump wastes, while sanitary landfills are engineered operations, designed and operated according to acceptable standards. The basic principle of a landfill operation is to prepare a site with liners to deter pollution of groundwater, deposit the refuse in the pit, compact it with specially built heavy machinery with huge steel wheels, and cover the material with earth at the conclusion of each day's operation (Figure 1-11). Siting and developing a proper landfill require planning and engineering design skills.

Even though the tipping fees paid for the use of landfills are charged on the basis of weight of refuse accepted, landfill capacity is measured in terms of volume,

Figure 1-11 A typical sanitary landfill. (Courtesy P. Aarne Vesilind)

not weight. Engineers designing the landfills first estimate the total volume available to them and then estimate the density of the refuse as it is deposited and compacted in the landfill. The density of refuse increases markedly as it is first generated in the kitchen and then finally placed into the landfill. The *as generated* bulk density of municipal solid waste (MSW) is perhaps 100 to 300 lb/yd³ (60 to 180 kg/m³), while the compacted waste in a landfill exceeds 1200 lb/yd³ (700 kg/m³). Landfills are required to have daily dirt covers, and the more dirt that is placed on the refuse, the less volume is available for the refuse itself. Commonly, engineers estimate that the volume occupied by the cover dirt is one-fourth of the total landfill volume.

EXAMPLE 1-1

Imagine a town where 10,000 households each fill one 80-gallon container of refuse per week. What volume would this refuse occupy in a landfill? Assume that 10% of the volume is occupied by the cover dirt.

SOLUTION

This problem is solved using a mass balance. Imagine the landfill as a black box, and the refuse goes from the households to the landfill.

(mass out) = (mass in)

$$V_L D_L = V_P D_P$$

where V and D are the volume and density of the refuse, and subscripts L and P denote loose and packed refuse. Assume the density of the refuse when collected is 200 lb/yd³ and is 1200 lb/yd³ in the landfill.

((10,000 households)(80 gal/household)(0.00495 yd³/gal))(200 lb/yd³)

= (1200 lb/yd³) V_P

V_P = 660 yd³

If 10% of the total volume occupied is taken up by the cover dirt, then the total landfill volume necessary to dispose of this waste is

$T = 660 + 0.10T$

where T is the total volume. Thus, $T = 733$ cubic yards.

Sanitary landfills are not inert. The buried organic material decomposes anaerobically, producing various gases (such as methane and carbon dioxide) and liquids that have extremely high pollution capacity when they enter the groundwater. Liners made of impervious clay and/or synthetic materials (such as plastic) are used to try to prevent the movement of this liquid, called *leachate*, into the groundwater. Figure 1-12 shows how a synthetic landfill liner is installed in a prepared pit. The seams have to be carefully sealed, and a layer of soil must be [soil must be placed on] placed on the liner to prevent landfill vehicles from puncturing it.

Figure 1-12 Placement of a synthetic liner in a modern landfill. (Courtesy P. Aarne Vesilind)

Synthetic landfill liners are useful in capturing most of the leachate, but they are never perfect. No landfill is sufficiently tight that groundwater contamination by leachate is totally avoided. Wells have to be drilled around the landfill to check for groundwater contamination from leaking liners, and if such contamination is found, remedial action is necessary. And, of course, the landfill never disappears—it will be there for many years to come, limiting the use of the land for other purposes.

Modern landfills also require the gases generated by the decomposition of the organic materials to be collected and either burned or vented to the atmosphere. The gases are about 50% carbon dioxide and 50% methane—both of which are greenhouse gases. In the past, when gas control in landfills was not practiced, the gases have been known to cause problems with odor, soil productivity, and even explosions. Larger landfills use the gases for running turbines for the production of electricity for sale to the power company, while smaller landfills simply vent the gases to the atmosphere. The fact that landfills produce methane gas has been known for a long time, and many accidental explosions have occurred when the gas has seeped into basements and other enclosed areas where it could form explosive mixtures with oxygen. Modern landfills are required to collect the gases produced in a landfill and either flare them or collect them for subsequent beneficial use.

The first system for collecting and using landfill gas was placed into operation in California after a number of gas-extraction wells were driven around a deep landfill to prevent the lateral migration of gas. These wells burned off about 1000 ft^3/min (44 m^3/min) of gas without the need for auxiliary fuel.[16] Based on this experience, the capture and use of this gas instead of just wastefully burning it off seemed a reasonable alternative. For example, the recently closed Puente Hills

Landfill in the Los Angeles basin has received over 130 million tons of solid waste and produces over 40 mw of electricity. Even though the landfill is closed, it will continue to produce gas for many years as the waste continues to decompose.

1-2-6 Energy Conversion

An alternative to allowing refuse to biodegrade and form a useful fuel is the combustion of refuse and energy recovered as heat. The potential for energy recovery from solid waste is significant. Of the 250 million tons of waste generated annually in the United States in 2008, over 80% is combustible, yielding a heat value equivalent to about 1.15 million barrels of oil per day. The use of refuse as a source of energy clearly has tremendous potential.

Refuse can be burned as is (the so-called *mass burn combustors*) or processed to produce a *refuse-derived fuel*. A picture of a waste-to-energy facility is shown in Figure 1-13. The refuse is dumped from the collection trucks into a pit that serves to mix and equalize the flow over the 24-hour period since such facilities must operate around the clock. A crane lifts the refuse from the pit and places it in a chute that feeds the furnace. The grate mechanism moves the refuse, tumbles it, and forces in air from the bottom as well as the top as the combustion takes place. The hot gases produced from the burning refuse are cooled by a bank of tubes filled with water. The hot water goes to a boiler where steam is produced. This steam can be used for heating and cooling or for producing electricity in a turbine. The cooled gases then are cleaned by dry scrubbers, baghouses, electrostatic precipitators, and/or other control devices and discharged through a stack.

Figure 1-13 Typical waste-to-energy facility for combusting municipal solid waste. (Courtesy Hitachi Zosen Inova)

The more the solid waste is processed prior to its combustion, the better is its heat value and usefulness as a substitute for a fossil fuel. Such processing removes much of the noncombustible materials (such as glass and metals) and reduces the size of the paper and plastic particles so they burn more evenly. Refuse that has been so processed is called a *refuse-derived fuel* (RDF). The simplest form of RDF is shredding the solid waste to produce a more homogeneous fuel without removing any of the metals or other noncombustibles. If the metals and glass are removed, the fuel is improved in its heat value and handling.

One reason waste-to-energy facilities have not found greater favor is the concern with the emissions, but this concern seems to be misplaced. Studies have shown that the risks of MSW combustion facilities on human health are minimal, and with modern air pollution control equipment, there should be minimal measurable effect of either the gaseous or particulate emissions.

The refuse combustion option is, however, complicated by the problem of ash disposal. Although the volume of the refuse is reduced by over 90% in waste-to-energy facilities, the remaining 10% still has to be disposed of. Many special wastes (such as old refrigerators) cannot be incinerated and either must be processed to recover the metals or be landfilled. A landfill is therefore necessary even if the refuse is combusted, and a waste-to-energy plant is not an ultimate disposal facility. Refuse combustion ash concentrates the inorganic hazardous materials in refuse, and the combustion process may create other chemicals that may be toxic. Most solid waste ash is placed in municipal landfills, but other facilities have constructed their own landfills exclusively for ash disposal (called *monofills*). Ash has been increasingly used as a raw material for the production of useful aggregate for producing building materials, as an aggregate for road construction, or in other applications where a porous and predictable (and inexpensive) material is useful.[17]

As an alternative to mass burn or RDF combustion units, municipalities are considering conversion technology such as gasification. The reported advantages of this technology are lower air emissions and less ash, but these advantages are offset by higher costs and lower energy output.

1-3   THE NEED FOR INTEGRATED SOLID WASTE MANAGEMENT

Solid waste professionals recognize that issues related to managing solid waste must be addressed using a holistic approach. For example, if more waste is recycled, this can have a negative financial impact on the landfill because less refuse is landfilled. Since many landfill costs are fixed (there is a cost regardless of whether any refuse is landfilled), a drop in the incoming refuse can have severe economic ramifications. The various methods of solid waste management are therefore interlocking and interdependent.

Recognizing this fact, the EPA has developed a national strategy for the management of solid waste, called the *Integrated Solid Waste Management* (ISWM), which resembles the hierarchy for solid waste management described by the European Union directive discussed previously. The intent of the EPA plan, which reflects the purpose of the proposed EU hierarchy, is to assist local communities in their decision making by encouraging those strategies that are the most environmentally acceptable. The EPA ISWM strategy suggests that the list (from most to least desirable) of solid waste management strategies should be:

1. Reducing the quantity of waste generated
2. Reusing the materials
3. Recycling and recovering materials
4. Combusting for energy recovery
5. Landfilling

Figure 1-14 illustrates this approach. That is, when an integrated solid waste management plan is implemented for a community, the first consideration should be reducing the waste at the source. Waste prevention minimizes the impact on natural resources and energy reserves.

Reuse is the next most desirable activity because it has a minimal impact on natural resources and energy. Recycling is the third option and should be undertaken when most of the waste reduction and reuse options have been implemented.

The EPA confuses recycling with recovery and groups them together as meaning any technique that results in the diversion of waste. As previously defined, recycling is the collection and processing of the separated waste, ending up as new consumer product. Recovery is the separation of mixed waste, also with the end result of producing new raw materials for industry.

During the past 20 years, there has been dramatic growth in recycling/recovery. For example, the State of California was diverting about 10% of residential and commercial solid waste in 1989, and by 2013, it was diverting 65% of the waste. In the United States, the recycling rate has increased from 16% in 1990 to 34.5% in 2012.

The fourth level of the ISWM plan is solid waste combustion, which really should include all methods of treatment. The idea is to take the solid waste stream and transform it into a nonpolluting product. This conversion may be by

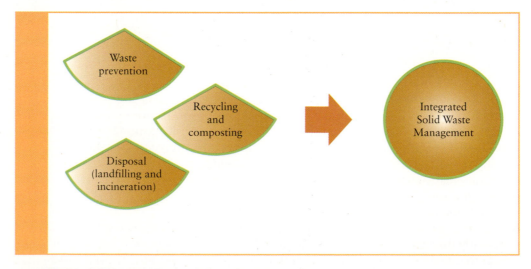

Figure 1-14 Integrated Solid Waste Management approach.

combustion, but other thermal and chemical treatment methods may eventually prove just as effective. In the past 30 years, the combustion of waste for energy recovery has remained fairly constant in the United States but continues to grow worldwide.

Finally, if all of the foregoing techniques have been implemented and/or considered and there is still waste left over (which there will be), the final solution is landfilling. At this time, there really is no alternative to landfilling (except disposal in deep water—which is now illegal), and therefore, every community must develop some landfilling alternative.

While this ISWM strategy is useful, it can lead to problems if taken literally. Communities must balance the various strategies to fit their local needs, and this is where engineering judgment comes into play and where the solid waste engineer really earns his/her salary. All of the options have to be juggled and the special conditions integrated into the decision. The economics, history, politics, and aspirations of the community are important in developing the recommendations.

1-4   SPECIAL WASTES

In addition to the usual residential and commercial solid waste, municipal engineers have to deal with special or unusual waste. These wastes may include items such as inert material (materials that do not degrade, such as rocks), agriculture waste, sludge from both water and wastewater treatment facilities, tires, household hazardous waste, and medical waste. Each of these special wastes must be managed in a way that protects human health and welfare and complies with applicable regulations. While these issues are beyond the scope of this book, solid waste engineers must be prepared to manage these wastes.

1-5   FINAL THOUGHTS

As Kermit the Frog laments, "It's not easy being green." A solid waste engineer might modify this to "It's not easy being a green engineer." Often the engineer is placed in a role of the bad guy who, by doing the right thing, receives severe criticism from some segments of society. Sometimes the engineer is accused of appeasing the establishment and destroying our environment.

One alternative used by some engineers confronted with such harsh criticism is to withdraw into their professional shells. If they are only "hired guns," then what do they care what the outcome of any public decision is? They simply do what the client wants and do not voice their own opinions about societal or environmental values. Engineers can argue that the job is value free, and they are not making any value decisions. Technology, they argue, is value free.

But this is a classical cop-out. The fact is that engineers *do* decide policy in many ways. On a local and mundane level, engineers make seemingly minor decisions daily that affect all of us. For example, the routing of a local highway across a stream that may be a rare habitat for the endangered Venus fly trap, or prohibiting the use of asphalt containing rubber from old tires, are both engineering decisions that will get no press whatsoever. And on a national level, the mere initiation of a major project will result in a momentum that will be difficult to reverse. Thus, it is not enough simply to defer engineering decisions to public comment or environmental oversight. The engineers need to introduce environmental concerns into the planning of projects before the plans become public documents. The decisions of the engineers do make a difference.

Private and governmental clients, if they are to get things done, must use engineering know-how. If a community needs to construct a landfill, it *has* to use engineers to do it. Other professionals—such as geohydrologists, sociologists, philosophers, epidemiologists, or even planners—are useful, but at the end of the day, the engineers are the ones who actually *design* the landfill. Thus, the engineer has the responsibility of viewing his/her role in the broadest sense—to introduce alternatives, concepts, and values that the client may otherwise never consider. For example, if a city hires an engineering firm to design a landfill, it is possible for the engineer to ignore all of the aspects of the job except the determination of the best (safest and least expensive) design. What we are suggesting is that the engineer, recognizing that there are many conflicting values involved in such a project and that it is unlikely that the municipality will seek expert assistance in the resolution of such conflicts, has a responsibility to introduce value (ethical) questions into the project.

A truly professional engineer will infuse ethics into his/her decision making, and with the increasing pressure on the natural environment, a growing population, and accelerated technological development, environmental ethics will play an ever-increasing role in the engineer's professional responsibilities to society. The engineer *can*, and *should*, make a difference.

References

1. Melosi, M. V. 1981. *Garbage in the Cities.* College Station, Texas: Texas A&M Press.
2. Mumford, L. 1961. *The City in History.* New York: Harcourt, Brace & World.
3. "Trash Timeline: 1,000 Years of Waste." *Waste News,* www.wastenews.com /features.
4. Meadows, D. H., et al. 1972. *The Limits of Growth.* Washington, D.C.: Potomac Associates.
5. Boughey, A. S. 1976. *Strategy for Survival.* Menlo Park, California: W. A. Benjamin, Inc.
6. Grass Roots Recycling Network. 1999. *Welfare for Waste, How Federal Taxpayer Subsidies Waste Resources and Discourage Recycling.*
7. Cross, J. F. 1992. "Pollution Prevention and Sustainable Development." *Renewable Resources Journal* 5:13–17.
8. Conn, W. D. 1995. "Reducing Municipal Solid Waste Generation: Lessons from the Seventies." *Journal of Resource Management and Technology* 16:24–27.
9. Sherman, S. 1991. "Local Government Approaches to Source Reduction." *Resource Recycling* 9:112 and 119.
10. Lober, D. 1996. "Municipal Solid Waste Policy and Public Participation in Household Source Reduction." *Waste Management and Research* 14:129–145.
11. Goldberg, D. 1990. "The Magic of Volume Reduction." *Waste Age* 21:98–104.
12. Skumatz, L. A. 1991. "Variable Rates for Solid Waste Can Be Your Most Effective

Recycling Program." *Journal of Resource Management and Technology* 19:1–12.

13. City of Greensboro. 1992. *Waste Stream Characterization Project*. HDR Engineering (quoted in Lober, 1996).

14. *Shop Earth Smart: A Consumer Guide to Buying Products That Help Conserve Money and Reduce Waste*. San Luis Obispo County Integrated Waste Management Authority. 1999. The City and County of San Luis Obispo, California.

15. Franklin Associates. 1999. *Characterization of Municipal Solid Waste in the United States: 1998 Update*. EPA Report No. EPA 530.

16. Dair, F. R. and R. E. Schwegler. 1974. "Energy Recovery from Landfills." *Waste Age* 5:6.

17. Brown, H. 1997. "Ash Use on the Rise in the United States." *World Waste*: p. 16.

18. Lemann, M. 1997. "Fundamentals of Waste Technology," published by C. HERRMANN CONSULTING, Kilchberg, Switzerland.

19. FOEN, 1998 Ordinance on the return, take back and disposal of electrical and electronic equipment Swiss Federal Office of the Environment.

20. EU-Lex, 1999. Directive 1999/31/EC on the landfill of waste.

21. EU-Lex, 2006. Directive 2006/12/EC on waste.

22. ITU, 2015. International Telecommunication Union, May 2015, via mobiThinking (http://mobiforge.com)

23. Denis Meadows, 2011. At World Resources Forum in Davos, September 19–21.

24. United States Environmental Protection Agency "What is Integrated Solid Waste Management?" EPA530-F-02-026a (5306W)

Abbreviations Used in This Chapter

CERCLA = Comprehensive Environmental Response, Compensation, and Liability Act

EIS = environmental impact statement

EPA = Environmental Protection Agency

EU = European Union

HDPE = high-density polyethylene

ISWM = Integrated Solid Waste Management

LDPE = low-density polyethylene

MRF = materials recovery facility

MSW = municipal solid waste

NEPA = National Environmental Policy Act

PETE = polyethylene terephthalate

PP = polypropylene

PS = polystyrene

PVC = polyvinyl chloride

RCRA = Resource Conservation and Recovery Act

RDF = refuse-derived fuel

VOC = volatile organic compound

Problems

1-1. The objective of this assignment is to evaluate and report on the solid waste you personally generate. For one week (selected at random), collect all of the waste you would have normally discarded. This includes food, newspapers, beverage containers, etc. Using a scale, weigh your solid waste and report it as follows:

Component	Weight	Percent of Total Weight
Paper		
Plastics		
Aluminum		
Steel		
Glass		
Food		
Other	_____	_____
Total		100

Your report should include a data sheet and a discussion. Answer the following questions.

a. How do your percentage and total generation compare to national averages?

b. What in your refuse might have been reusable (as distinct from recoverable), and if you had reused it, how much would this have reduced the refuse?

c. What in your refuse is recoverable? How might this be done?

1-2. One of the least studied aspects of municipal refuse collection is the movement of the refuse from the household to the truck. Suggest a method by which this can be improved. Originality counts heavily here, practicality far less.

1-3. Plastic bags at food stores have become ubiquitous. Often recycling advocates point to the plastic bags as the prototype of wastage and pollution, as stuff that clogs up our landfills and becomes litter that pollutes the land and ocean. In retaliation, plastic bag manufacturers have begun a public relations campaign to promote their product. On one of the flyers (printed on paper) they say:

> The (plastic) bag does not emit toxic fumes when properly incinerated. When burned in waste-to-energy plants, the resulting by-products from combustion are carbon dioxide and water vapor, the very same by-products that you and I produce when we breathe. The bag is inert in landfills where it does not contribute to leaching bacterial or explosive gas problems. The bag photodegrades in sunlight to the point that normal environmental factors of wind and rain will cause it to break into very small pieces, thereby addressing the unsightly litter problem.

Critique this statement. Is all of it true? If not, what part is not? Is anything misleading? Do you agree with their evaluation? Write a one-page response.

1-4. The siting of landfills is a major problem for many communities. This is often an exasperating job for engineers because the public is so intimately involved. A prominent environmental engineer is quoted as follows:

> Environmental matters are on the front page today because we, the environmental industry, are not meeting people's expectations. They're telling us that accountability and quality are not open questions that ought to be considered. It's sometimes difficult to grasp in the face of all the misinformation out there. Ultimately, we're responsible for accommodating the public's point of view, not the other way around.[19]

Do you agree with him? Should the engineer "accommodate the public's point of view," or should the engineer impose his/her own point of view on the public, since the engineer has a much better understanding of the problem? Write a one-page summary of what you believe the engineer's role should be in the siting of a landfill for a community.

1-5. Suppose you are the engineer employed by a small community, and the town council tells you to design a "recycling" program that will achieve at least 50% diversion from the landfill. What would be your response to the town council? If you agree to try to achieve such a diversion, how would you do it? For this problem, you are to prepare a formal response to the town council, including a plan of action and what would be required for its success. Numbers are mandatory. Make up data as needed. This response would include a cover letter addressed to the council and a report of several pages, all bound in a report format with title page and cover.

1-6. Same scenario as Problem 1-5, but now you are asked to develop a zero waste program. That is, the community is to produce no solid waste whatsoever. Respond with a letter to the town council.

1-7. How might you use the principles of waste reduction in doing your food shopping at the grocery store? Name at least three specific ways in which you might be able to reduce waste in the purchase of your groceries.

1-8. We often see packaging labeled "Made from 100% recycled materials" or "Made with recyclable materials." The objective is, of course, to make you (the consumer) believe that the company is environmentally conscious and caring and, thus, to make you buy more of its products.

a. Why are such statements as the ones quoted above potentially misleading? What questions would you want to ask the company to determine if it truly is helping with materials recycling?

b. All things considered, if the statements are true, why ought you buy the company's products in preference to products with no recycled material? Be specific.

c. If you were working for the company, how would you write the statements to be more accurate?

1-9. Give three examples of waste reduction that you might be able to implement in your everyday life.

1-10. Consider the meaning and purpose of the various recycling symbols shown in Figure 1-15.

a. Which are used solely as marketing tools, and which tell you what you want to know if your intent is to buy recycled materials?

Figure 1-15 Various recycling symbols.

b. Find and cut out a recycling symbol from some product and attach it to your homework paper. Discuss its purpose and integrity.

1-11. If you were responsible for marketing the paper collected by your city's curbside collection program, what would you tell the purchasing department if it asked you whether the city should buy copy paper that was

a. 75% recycled content (55% pre-consumer and 20% post-consumer)

b. 50% recycled content paper (10% pre-consumer and 40% post-consumer) Why?

1-12. What is the content of recycled fiber in the paper used by your school/college/department? If recycled paper costs more than virgin fiber paper, should your school purchase it?

1-13. Of the six types of plastic shown in Figure 1-4, how many can be recycled in your community? If a type of plastic is not recyclable, do you think it is misleading (ethical) for the plastic manufacturers to place the chasing arrow design on the container?

1-14. Redraw Figure 1-4 to reflect a society that has achieved its goal of zero waste.

1-15. Which of the following items can be placed in the recycling bins at your campus and in the community that you live: glass bottles, aluminum cans, clear plastic bags, junk mail, envelopes with plastic windows, light bulbs, newspaper, plastic containers #1 and #2, plastic containers #3 to #7, telephone books, metal food trays, and Nike shoes? If the items are different between the campus and where you live, please explain why.

1-16. Describe the integrated waste management system in your community. What is the diversion goal and what is the current diversion rate?

1-17. In which devices that you use daily do you find Europium (Eu)?

CHAPTER 2

Municipal Solid Waste Characteristics and Quantities

A great deal of confusion exists in the definitions related to solid waste, and this leads directly to disagreements on quantities and composition. There are both gross and subtle differences in the types and sources of waste material, including the basic question as to what is and is not a solid.

2-1 DEFINITIONS

There are different definitions and categories of wastes used throughout the world. For example, the following definitions and categories of wastes are listed in the 2006 European Union Directive on waste:[28]

1. Production or consumption residues not otherwise specified below
2. Off-specification products
3. Products whose date for appropriate use has expired
4. Materials spilled, lost or having undergone other mishap, including any materials, equipment, etc., contaminated as a result of the mishap
5. Materials contaminated or soiled as a result of planned actions (e.g., residues from cleaning operations, packing materials, containers, etc.)
6. Unusable parts (e.g., rejected batteries, exhausted catalysts, etc.)
7. Substances which no longer perform satisfactorily (e.g., contaminated acids, contaminated solvents, exhausted tempering salts, etc.)
8. Residues of industrial processes (e.g., slags, still bottoms, etc.)
9. Residues from pollution abatement processes (e.g., scrubber sludges, baghouse dusts, spent filters, etc.)
10. Machining/finishing residues (e.g., lathe turnings, mill scales, etc.)
11. Residues from raw materials extraction and processing (e.g., mining residues, oil field slops, etc.)

12. Adulterated materials (e.g., oils contaminated with PCBs, etc.)
13. Any materials, substances, or products the use of which has been banned by law
14. Products for which the holder has no further use (e.g., agricultural, household, office, commercial and shop discards, etc.)
15. Contaminated materials, substances, or products resulting from remedial action with respect to land
16. Any materials, substances, or products which are not contained in the abovementioned categories

However, this book is concerned only with solid wastes produced by communities—*municipal solid waste* (MSW). Most of the waste categories described above can be part of MSW. Some of them are not included in MSW but may nevertheless be co-treated with MSW.

This book is concerned only with solid wastes produced by communities—municipal solid waste or MSW, which can be further defined as having the following components:

- Mixed household waste

- Litter and waste from community trash cans

- Recyclables, such as
    Newspapers
    Cardboards used for packaging, such as corrugated cardboard
    Beverage cartons
    Aluminum cans
    Steel cans
    Plastic drink bottles of different plastic classes

- Household hazardous waste, such as
    Batteries
    Refrigerators
    Electronic equipment
    Lamps, such fluorescent tubes and bulbs
    Syringes
    Paints

- Yard (or green) waste

- Bulky items (furniture, mattresses, rugs)

- Construction and demolition waste

- Other material collected by the community

Often these wastes are defined by the way they are collected. Commonly, the mixed household wastes are collected door-to-door by solid waste collectors, and the material is loaded on trucks specially built for that purpose. The recyclables are collected either in the same truck with the mixed waste in a separate compartment or by a separate truck. Yard waste, which may include food waste and paper contaminated by food, may be collected with the household waste

Figure 2-1 Household hazardous waste collection facility. (Courtesy William A. Worrell)

or placed separately in a dedicated truck. Commercial wastes use large containers that are emptied into specially built trucks. Construction and demolition wastes are collected in roll-off stationary containers that remain on the job site until full or the job is completed. Bulky items are commonly collected on an as-needed basis by larger trucks capable of handling large items. Household hazardous wastes are collected periodically by the community, or are taken to (hazardous) waste collection centers of the community by the homeowners (Figure 2-1).

For many reasons it is convenient to define *refuse* as

- Solid waste generated by households, commercial businesses, institutions such as schools, and government
- Recyclables (whether or not they are collected separately)
- Yard (or green) waste
- Litter and community trash, because the material is produced by individuals

By our definition, refuse does *not* include wastes that generally are not produced by private households, commercial businesses, institutions, and government and typically collected in a garbage truck, such as

- Construction and demolition debris
- Water and wastewater treatment plant sludges
- Industrial wastes

- Leaves and other green waste from community streets and parks that are composted on site

- Bulky items such as large appliances, hulks of old cars, tree limbs, and other large objects that often require special handling

This text is devoted mainly to the collection, disposal, and recovery of materials and energy from *refuse*. The other solid wastes—such as construction and demolition (C & D) wastes—are mentioned only in passing. This slight should not be taken as any indication that these wastes are either small in quantities or unimportant in defining their recycling or disposal options. C & D wastes, for example, are highly recyclable, and that small fraction that cannot be recycled may be landfilled in a special landfill that does not require daily cover. Leaves, in some parts of the United States, represent a serious problem to communities during the autumn, and on-site composting is often developed to handle this seasonal load. In the discussions that follow, however, the material of interest is the fraction of MSW previously defined as refuse.

In summary:

$$(MSW) = (refuse) + (C \& D \text{ waste}) + (leaves) + (bulky \text{ items})$$

In some cases refuse can be defined in terms of *generated, collected,* and *diverted* refuse. Some part of the refuse, such as organic matter and yard waste, is composted on premises. The fraction of refuse that is generated but not collected is called *diverted* refuse.

The collected refuse may be as simple as one garbage can that is sent to a landfill for disposal. Other homes and businesses have multiple containers, such as one or more container for recyclables (paper, glass, plastic, metal, etc.) and one or more container for green waste and/or food waste. In this case the household or business sorts the refuse into the many containers that will result in a usable product. The remaining refuse that must be disposed of would be considered residual refuse and placed in a separate container for disposal.

In summary:

$$(generated \text{ refuse}) = (collected \text{ refuse}) + (diverted \text{ refuse})$$

Sometimes diversion is defined on the basis of MSW instead of refuse. When defined in this way, diverted MSW is that fraction of MSW that is generated but does not find its way to the disposal system of the community. The objective is to extend the life of landfills or to reduce the cost of disposal. One major diversion is the collection of recyclables (aluminum cans, newspapers, etc.) that can be sold on the secondary materials market. In addition, diversion can include waste that was never generated by using resource reduction strategies, such as 2-sided coping and electronic emails. The EPA has challenged communities to increase their diversion to 35%—up from the 25% originally suggested in 1988. California has set an even higher goal of 75% diversion by 2020.

The calculation of diversion in terms of either refuse or MSW is important and controversial. Recall that MSW includes refuse plus C & D debris, leaves, and bulky items. When communities are under state mandates to increase the recycling of consumer products, the calculation often is made using the MSW as the denominator instead of the refuse, thereby achieving large diversion rates. This is demonstrated by the following example.

EXAMPLE 2-1

A community produces the following on an annual basis:

Fraction	Tons per Year
Mixed household waste	210
Recyclables	23
Commercial waste	45
Construction and demolition debris	120
Leaves and miscellaneous	36

The recyclables are collected separately and processed at a materials recovery facility. The mixed household waste and the commercial waste go to the landfill. The leaves are composted, and the C & D wastes are processed and used on the next project. Calculate the diversion.

SOLUTION

If the calculation is on the basis of MSW, the total waste generated is 434 tons per year. If everything not going to the landfill is counted as having been diverted, the diversion is calculated as

$$\frac{23 + 120 + 36}{434} \times 100 = 41\%$$

This is an impressive diversion. But if the diversion is calculated as that fraction of the *refuse* (mixed household and commercial waste) that has been kept out of the landfill by the *recycling* program, the diversion is

$$\frac{23}{210 + 23 + 45} \times 100 = 8.3\%$$

This is not nearly as impressive but may be more honest. Therefore, the published or required diversions can be calculated in many ways.

2-2   MUNICIPAL SOLID WASTE GENERATION

The United Nations Environmental Programme estimates solid waste generation at 2.2 kilograms (4.8 pounds) per person per day with the likelihood that the rate will continue to increase. The World Bank, in its 2012 report *What a*

Waste: A Global Review of Solid Waste Management, estimated current MSW generation at 1.3 billion tonnes (1.43 billion tons) per year with a likely increase to 2.2 billion tonnes (2.4 billion tons) per year by 2025. Note that in this text, ton = 2000 lb and tonne = 1000 kg. China has passed the United States to become the leading generator of MSW. For example, the generation of MSW in China grew from 31.3 million tonnes (34.5 million tons) in 1980 to 212 million tonnes (234 million tons) in 2006.[41]

One issue with solid waste generation statistics is the question of how accurate the statistics are. The World Bank Report states, "Solid waste data should be considered with a degree of caution due to the global inconsistencies in definitions, data collection methodologies, and completeness." In the EU waste statistics are governed by Regulation (EC) No 2150/2002. In the United States, each year the EPA provides information on MSW. The latest EPA municipal solid waste report[1] includes data through the year 2012. Table 2-1 shows the amount of MSW generated from 1960 to 2012 and how that waste has been managed. After 47 years of increasing amounts of MSW being generated, in 2008 there was less MSW generated than in 2007. This trend has continued, with generation through 2012 continuing to be less than 2007 generation. In addition, since 1990 the amount of waste being landfilled has decreased while the amount recycled and composted has increased. This trend is encouraging in that it reflects an improved attempt to better manage our limited resources.

In Europe there is a trend from disposal of household wastes in landfills toward treatment in incinerators. In addition, as shown in Figure 2-2, recycling continues to be an important component in the management of MSW in those countries with incineration.

Table 2-1 **Generation and Management of Solid Waste in the United States from 1960 to 2012 (in millions of tons)**

Activity	1960	1970	1980	1990	2000	2005	2010	2012
Generation	88.1	121.1	151.6	208.3	243.5	253.7	250.4	250.9
Recovery for recycling	5.6	8.0	14.5	29.0	53.0	59.2	65.0	65.3
Recovery for composting*	Negligible	Negligible	Negligible	4.2	16.5	20.6	20.2	21.3
Total materials recovery	5.6	8.0	14.5	33.2	69.5	79.8	85.2	86.6
Combustion with energy recovery†	0.0	0.4	2.7	29.7	33.7	31.6	29.3	29.3
Discards to landfill, other disposal‡	82.5	112.6	134.4	145.3	140.3	142.3	136.0	135.0

* Composing of yard trimmings, food scraps, and other MSW organic material. Does not include backyard composting.
† Includes combustion of MSW in mass burn or refuse-derived fuel form, and combustion with energy recovery or source separated materials in MSW (e.g., wood pallets, tire-derived fuel).
‡ Discards after recovery minus combustion with energy recovery. Discards include combustion without energy recovery. Details might not add to totals due to rounding.
Sources: [1]

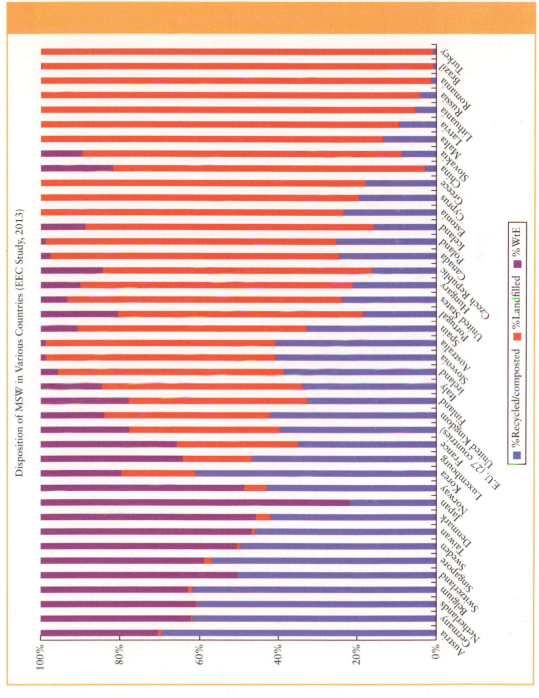

Disposition of MSW in Various Countries (EEC Study, 2013)

Figure 2-2 Treatment of household wastes in different European countries with respect to materials recycling, incineration, or landfilling. Source: Based on Nickolas Themelis, "The Global Waste-to-energy research and Technology council" www.wtert.org. Columbia University, Earth Engineering Center, Page 21.

This compilation refers to the treatment method used for the wastes collected; i.e., this is an "input perspective." Therefore, the amount of ashes landfilled after incineration are not considered in this graph.

Globally, landfilling is still the most common method of waste management. In China in 2007 there were 366 landfills, 17 composting plants, and 66 incinerators.[41] The appendix at the end of this chapter includes a table of MSW disposal methods by country.

On a per capita basis, waste generation in 2012 is lower than the 1990 level as shown in Table 2-2. Note the significant decrease in per capita disposal since 1990 and the corresponding increase in per capita materials recovery.

The generation of refuse in a community also varies throughout the year. Figure 2-3 shows the average generation rate of refuse as measured at five different landfills in Wisconsin[2] and waste generation in New Orleans.[3] In Wisconsin, the very cold months of the winter result in a low generation rate during January and February, while the New Orleans data show little seasonal variation, as can be expected.

The results of a study conducted in Kaunas (Lithuania)[32] are shown in Figure 2-4. Interestingly it turned out that food and green waste increase in the autumn as a result of local activities in harvesting and preserving vegetable food. Not only will this impact collection activities; the seasonal change may lower the heat value of the wastes during this time.

Collection frequency also affects the production of refuse.[4] Generally, the more frequent the collection, the more MSW is produced. Apparently, if the frequency of service is not sufficient, citizens will find other, perhaps less desirable, means of solid waste disposal. On the other hand, this might indicate that more waste diversion activities occur when collection is less frequent.

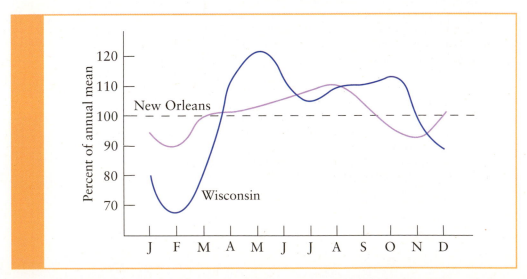

Figure 2-3 Monthly variation in the generation of municipal refuse in Wisconsin and New Orleans. Source: [2, 3]

Table 2-2 **Generation and Management of Solid Waste in the United States from 1960 to 2012 (in pounds per capita per day)**

Activity	1960	1970	1980	1990	2000	2005	2010	2012
Generation	2.68	3.25	3.66	4.57	4.74	4.69	4.44	4.38
Recovery for recycling	0.17	0.22	0.35	0.64	1.03	1.10	1.15	1.14
Recovery for Composting*	Negligible	Negligible	Negligible	0.09	0.32	0.38	0.36	0.37
Total materials recovery	0.17	0.22	0.35	0.73	1.35	1.48	1.51	1.51
Combustion	0.00	0.01	0.07	0.65	0.66	0.58	0.58	0.57
Combustion with energy recovery[†]	0.00	0.01	0.07	0.65	0.66	0.58	0.52	0.51
Discards to landfill, other disposal[‡]	2.51	3.02	3.24	3.19	2.73	2.63	2.41	2.36
Population (millions)	179	203.984	227.255	249.907	281.422	296.410	309.051	313.914

* Composing of yard trimmings, food scraps, and other MSW organic material. Does not include backyard composting.
[†] Includes combustion of MSW in mass burn or refuse-derived fuel form, and combustion with energy recovery of source separated materials in MSW (e.g., wood pallets, tire-derived fuel).
[‡] Discards after recovery minus combustion with energy recovery. Discards include combustion without energy recovery. Details might not add to totals due to rounding.
Sources: [1]

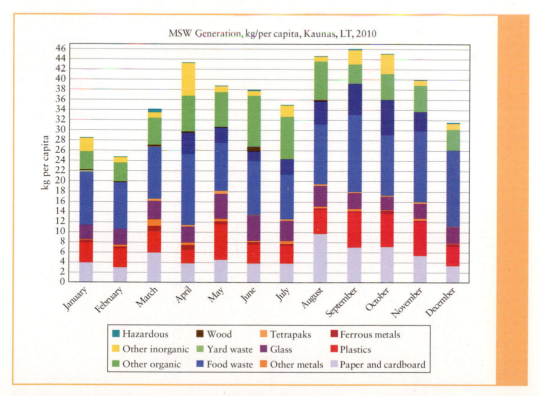

Figure 2-4 Example for seasonal changes in household waste composition. Source: [32]

Table 2-3 Waste Generation by Income Level

| | Current Available Data | | |
| | Total Urban Population (millions) | Urban Waste Generation | |
Region		Per Capita (kg/capita/day)	Total (tons/day)
Lower income	343	0.60	204,802
Lower middle income	1,293	0.78	1,012,321
Upper middle income	572	1.16	665,586
High income	774	2.13	1,649,547
Total	2,982	1.19	3,532,256

On a global level, income and urbanization are highly correlated to waste production. As disposal income and living standards increase, the consumption of goods increases. This leads to the generation of more waste. As shown in Table 2-3, the high income per capita waste generation is over 3 times as great when compared to the low income per capita generation.

The level of urbanization also affects the waste generation rate. Urban residents generate almost twice as much MSW as rural residents.

2-3   MUNICIPAL SOLID WASTE CHARACTERISTICS

As long as the MSW is to be disposed of by landfill, there is little need to analyze the waste much further than to establish the tons of waste generated and perhaps consider the problems of special (hazardous) materials. If, however, the intent is to collect gas from a landfill and put it to some beneficial use, the amount of organic material is important. When recycling is planned, or if materials or energy recovery by combustion is the objective, it becomes necessary to have a better picture of the solid waste. Some of the characteristics of interest are

- Composition by identifiable items (steel cans, office paper, etc.)
- Moisture content
- Particle size
- Chemical composition (carbon, hydrogen, etc.)
- Heat value
- Density
- Mechanical properties
- Biodegradability

2-3-1 Composition by Identifiable Items

The 2012 EPA estimates of national waste composition based on such data are shown in Table 2-4. The last column in the table indicates an annual national solid waste production of 250.9 million tons (227.6 million tonnes)—or on a personal level, a contribution of about 4.38 lb/capita/day (2.0 kg/capita/day).

Figure 2-5 shows the change in quantities of various refuse components generated in the United States. Note the dramatic increase in plastics as a major component in solid waste from 1960 until 2005. Since that time plastic generation has stayed about constant. Figure 2-6 shows the global MSW composition.

On a national level, data from published industry production statistics can be used for estimating waste composition. This method is called the *input method* of estimating solid waste production. For example, the annual production of glass is about 11,570,000 tons (10,496,000 tonnes) annually,[1] and we can safely assume that all of this will end up (sooner or later) in waste that is either disposed of or processed for materials recovery.

The input method of estimating solid waste generation is applicable where the input data can be obtained from specialized agencies that routinely collect and publish industry-wide data. This system also allows for regular updates of waste generation estimates, because gathering the data is expensive. Furthermore, since the data collected by the same institutions include future projections, it is possible to estimate future solid waste generation. The numbers in Table 2-4 were generated using this procedure. National averages are of limited value, however, because they rarely can be used with any degree of precision for local or regional purposes. In addition, with the globalization of production, national numbers may over- or underestimate the actual use of a product in a specific country. On the local level, the only reliable method of estimating refuse composition and production is to use an *output method* of analysis and to perform sampling studies.

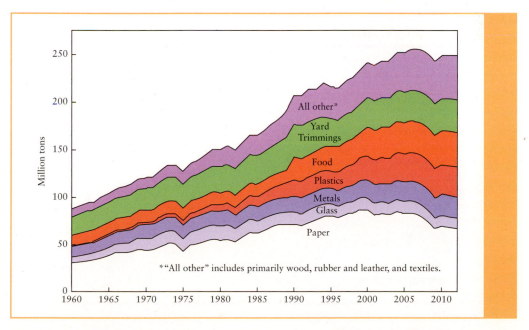

Figure 2-5 Historical trends in municipal solid waste generation and composition in the United States.

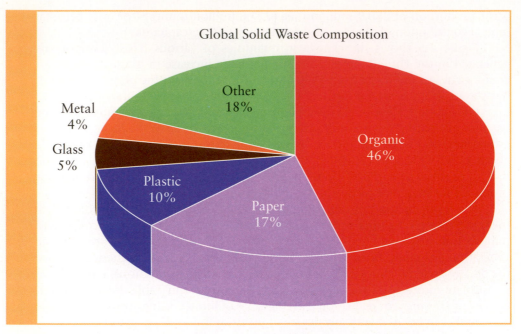

Figure 2-6 Global MSW composition. Source: [42]

Table 2-4 Generation of Municipal Solid Waste Components in the United States, 2012

Item	Weight Generated (millions of tons)	Percent
Paper and paperboard	68.62	27.4
Glass	11.57	4.6
Ferrous metals	16.80	6.7
Aluminum	3.58	1.4
Other nonferrous metals	2.00	0.8
Plastics	31.75	12.7
Rubber and leather	7.53	3.0
Textiles	14.33	5.7
Wood	15.82	6.3
Other materials	4.60	1.8
Food waste	36.43	14.5
Yard trimmings	33.96	13.5
Miscellaneous inorganic	3.90	1.6
Total	250.89	100

Source: [1]

Sampling studies for characterizing refuse must be designed to produce the most useful and accurate data for the least cost and effort. The two variables of importance in designing such a study are sample size and the method of characterizing the refuse.

Measuring the composition of a totally heterogeneous material, such as mixed municipal refuse, is not a simple task, but some determination of its components is necessary if the various fractions are to be separated and recovered. Some authorities, however, suggest that (because a major effort is required to establish the composition with reasonable accuracy) it is often not worth the trouble and expense, and a national average can be used. Composition studies should be used where accurate data are absolutely required for estimating the economics of future solid waste management alternatives.

The sampling plan drives the waste composition study. Even if the sampling procedure is performed adequately, it means nothing unless the plan can produce valid results. First, the waste has to be accurately represented through proper load selection in order to avoid biasing the final analysis. The truckload to be analyzed has to represent (as closely as possible) the average production of refuse in the community.

Once the load has been selected, a methodology for producing a sample small enough to be analyzed but big enough to be a statistical representative of the MSW must be established. The most frequently used methodology for determining the number of samples required in order to achieve statistical validity is the American Society for Testing and Materials (ASTM) *Standard Test for Determination of the Composition of Unprocessed Municipal Solid Waste* (ASTM designation D 5231-92).[12] This method provides a script to follow when conducting a waste composition study, including a statistically based method for determining the number of samples required to characterize the waste. The number of samples required to achieve the desired level of measurement precision is a function of the component(s) under consideration and the desired confidence level. The calculations are an interactive process, beginning with a suggested sample mean and standard deviation for waste components. Typically, a 90% confidence level is adequate for most studies.[13] As a crude first estimate, sorting and analyzing more than 200 lb (90 kg) in each sample would have little statistical advantage.[14] The question is how many of the 200-lb samples are necessary for the testers to feel statistically confident in the results.

To obtain representative 200 lb (90 kg) samples, ASTM recommends *quartering* and *coning*. Quartering is the separation of a truckload of waste into successive quarters after thoroughly mixing the contents with a front-end loader. The samples are then coned again and quartered again until they are about 200 lb (90 kg). The greater the desired precision, the greater will be the number of 200 lb samples analyzed. Figure 2-7, based on the work of Klee and Carruth,[15] represents a good first estimate of the sample sizes required for acceptable precision. For example, if 50 samples of 200 lb each are processed, then the expected precision for the organic fraction would be better than ±5%; the precision for newsprint, aluminum, and ferrous would be about ±15%; and the precision for corrugated cardboard would be only about ±25%. As would be expected, the larger the articles, the more is required to achieve acceptable precision.

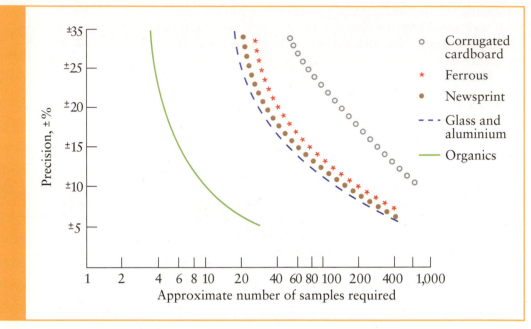

Figure 2-7 Approximate number of 200-lb samples required to achieve desired precision. Source: Klee, A. J., and D. Carruth, 1970. "Sample Weights in Solid Waste Composition Studies." J. Sanit. Eng. Div. ASCE 96, n. SA4: 945. With permission from ASCE.

It is important to decide early in the sampling program what is to be measured—that is, how many categories of waste are to be used. This decision, of course, depends on what the data are to be used for. In one study that was conducted to estimate the possibility of diverting waste from the landfill through prevention and recycling, 43 different categories were established.[16]

- Paper
  Newsprint
  Magazines
  Corrugated cardboard
  Telephone books
  Office/computer paper
  Other mixed paper

- Plastics
  PETE bottles
  HDPE bottles
  PVC containers
  Polypropylene containers
  Polystyrene
  Other containers
  Films/bags

- Organics
    Food waste
    Textiles/rubber/leather
    Fines (unidentifiable small organic particles)
    Other organics

- Ferrous materials
    Ferrous/bimetal cans
    Empty aerosols
    Other ferrous metals

- Nonferrous metals
    Aluminum cans
    Other nonferrous metals

- Electronic components
    Parts and materials from computers
    Printers
    Copy machines

- Glass
    Clear
    Green
    Brown
    Other glass

- Wood
    Lumber
    Pallets
    Other wood

- Inerts
    Asphalt roofing materials
    Concrete/brick/rock
    Sheet rock
    Ceiling tiles
    Dirt/dust/ash and other inerts

- Yard waste
    Grass clippings
    Leaves
    Trimmings

- Hazardous materials
    Lead acid batteries
    Other batteries
    Other hazardous wastes

Such a sampling study obviously would be very cumbersome, time intensive, and expensive, and it is questionable if such detail is really necessary in most cases. Based on the intended use of the study (e.g., construction of a materials recovery facility), a more pragmatic listing of components for a sampling study might be more useful (and certainly less expensive), as follows.

- Paper
  Newsprint
  Corrugated cardboard
  Magazines
  Other paper

- Metal
  Aluminum cans
  Steel cans
  Other aluminum
  Other ferrous
  Other nonferrous

- Glass
  Clear (flint)
  Green
  Brown

- Plastic
  HDPE
  PETE
  Other plastics

- Yard wastes
  Wood (branches and lumber)
  Leaves and clippings

- Food waste

- Other materials that either have little recovery potential or are of low fraction in refuse (such as rubber, ceramics, other glass, bricks, rocks, etc.)

Even careful sampling studies yield imprecise information because of the nature of refuse. Not all items can be readily categorized into the desired components. For example, a tin can with an aluminum top and paper wrapper has four components: steel, tin, aluminum, and paper. Regardless of the final classification of this item, inaccuracies are introduced into the final values.

Common contaminants of waste items include moisture, food, and dirt. Although these materials are normal components of the waste stream, they are of concern when they add significantly to the weight of paper, plastic film, yard waste, and containers. During placement in collection vehicles, the contamination increases when waste is squeezed, causing materials to smear or stick together and forcing moisture from food and other wet wastes into other absorbent items. In addition, contamination can occur during sampling and storing as a result of mixing and/or inclement weather.

After sorting, the samples should be taken to a laboratory where they can be weighed, cleaned of contamination, and air dried. Durable items (such as glass and plastic containers) can be washed prior to air drying, and filled containers can be emptied of their contents. If the contaminant category can be identified (for example, food), each category should be properly adjusted for the weight of contamination. In addition, bulk densities can be determined by knowing the volume of material that is being weighed. Table 2-5 lists the bulk densities of some common refuse components.

Waste composition studies are an important tool for municipal solid waste management. However, because of a lack of consistent procedure and underfunding of studies, the data provided are frequently inaccurate and imprecise. Too often, an insufficient number of samples is obtained, sampling events are not representative of seasonal and economic changes, contamination is not accounted for, and the study is not repeated in response to changes in the community. A poor study may be worse than no study at all if the numbers obtained are to be used for design purposes.

Table 2-5 Bulk Densities of Some Refuse Components

Components	Condition	Bulk Density (lb/yd^3)*
Aluminum cans	Loose	50–74
	Flattened	250
Corrugated cardboard	Loose	350
Fines (dirt, etc.)	Loose	540–1,600
Food waste	Loose	220–810
	Baled	1,000–1,200
Glass bottles	Whole bottles	500–700
	Crushed	1,800–2,700
Magazines	Loose	800
Newsprint	Loose	20–55
	Baled	720–1,000
Office paper	Loose	400
	Baled	700–750
Plastics	Mixed	70–220
	PETE, whole	30–40
	Baled	400–500
	HDPE, loose	24
	Flattened	65
Plastic film and bags	Baled	500–800
	Granulated	700–750
Steel cans	Unflattened	150
	Baled	850
Textiles	Loose	70–170
Yard waste	Mixed, loose	250–500
	Leaves, loose	5–250
	Grass, loose	350–500

* To obtain kg/m^3, multiply by 0.59.

Sources: [18, 19, 20, 21]

2-3-2 Moisture Content

Moisture content continues to change in MSW. While it might be possible to determine the moisture content of a specific item prior to discard, external factors can impact the moisture content after discarding, such as weather, collection location, and method. For example, a transfer of moisture takes place in the garbage can and truck, and thus, the moisture content of various components changes with time. Newsprint has about 7% moisture by weight as it is deposited into the receptacle, but the average moisture content of newsprint coming from a refuse truck often exceeds 20%. In addition, as shown in Figure 2-8, the location of a city and the time of year also impacts moisture content.

The moisture content becomes important when the refuse is processed into fuel or when it is fired directly. The usual expression for calculating moisture content is

$$M = \frac{w - d}{w} \times 100$$

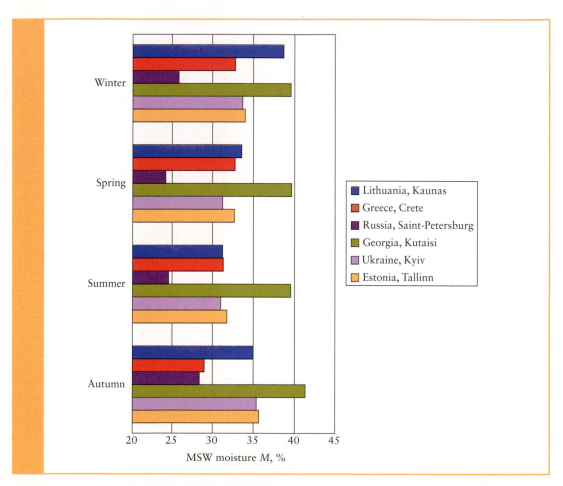

Figure 2-8 Example for change of moisture content in household wastes as function of geographical location and seasonal changes. Source [33]

where

> M = moisture content, wet basis, %
> w = initial (wet) weight of sample
> d = final (dry) weight of sample

Some engineers define moisture content on a dry weight basis as

$$M_d = \frac{w - d}{d} \times 100$$

where M_d = moisture content on a dry basis, %.

This relationship seems at first irrational, because the moisture content can exceed 100%, but in some fields, such as geotechnical engineering, moisture content on a dry basis is useful. In this text, moisture is always expressed on a wet basis unless otherwise indicated.

Drying is usually done in an oven at 77°C (170°F) for 24 h to ensure complete dehydration and yet avoid undue vaporization of volatile material. The moisture content of various refuse components varies widely, as shown in Table 2-6.

The moisture content of any waste can be estimated by knowing the fraction of various components and using either measured values of moisture content or typical values from a list (such as Table 2-6). This calculation is illustrated in Example 2-2.

Table 2-6 Moisture Content of Uncompacted Refuse Components

Component	Moisture Content	
	Range	Typical
Residential		
Aluminum cans	2–4	3
Cardboard	4–8	5
Fines (dirt, etc.)	6–12	8
Food waste	50–80	70
Glass	1–4	2
Grass	40–80	60
Leather	8–12	10
Leaves	20–40	30
Paper	4–10	6
Plastics	1–4	2
Rubber	1–4	2
Steel cans	2–4	3
Textiles	6–15	10
Wood	15–40	20
Yard waste	30–80	60
Commercial		
Food waste	50–80	70
Mixed commercial	10–25	15
Wood crates and pallets	10–30	20
Construction (mixed)	2–15	8

Source: [20 based on 21]

EXAMPLE 2-2

A residential waste has the following components:

Paper	50%
Glass	20%
Food	20%
Yard waste	10%

Estimate its moisture concentration using the typical values in Table 2-6. Assume a wet sample weighing 100 lb. Set up the tabulation:

SOLUTION

Component	Percent	Moisture	Dry Weight (based on 100 lb)
Paper	50	6	47
Glass	20	2	19
Food	20	70	6
Yard waste	10	60	4
			Total: 76 lb dry

The moisture content (wet basis) would then be

$$M = \frac{w - d}{w}(100) = \frac{100 - 76}{100}(100) = 24\%$$

Typically, the moisture content of loose refuse is about 20% if there have not been rainstorms before collection. During rainy weather, the moisture content can go as high as 40%.

In a refuse truck, *moisture transfer* takes place, and the moisture of various components of refuse changes. Paper sops up much of the liquid waste, and its moisture increases substantially. The moisture content of refuse that has been compacted by a collection truck is therefore quite different from the moisture of various components as they are in the can ready for collection.

2-3-3 Particle Size

Any mixture of particles of various sizes is difficult to describe analytically. If these particles are irregularly shaped, the problem is compounded. Municipal refuse is possibly the worst imaginable material for particle-size analysis, and yet much of the MSW processing technology depends on an accurate description of particle size.

No single value can adequately hope to describe a mixture of particles. Probably the best effort in that direction is to describe the mixture by means of

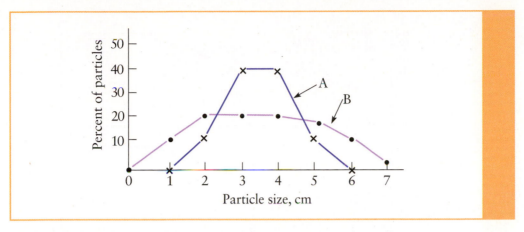

Figure 2-9 Particle-size distribution curves for two mixtures of particles.

a curve showing the percent of particles (by either number or weight) versus the particle size. The curve can be plotted by unit intervals, as shown in Figure 2-9. The two mixtures shown in these curves have very different particle-size distributions. Mixture A has mainly uniformly sized particles, while Mixture B has a wide variability in particle size, and yet the *average particle size* (defined as that diameter where 50% of the particles (by weight) are smaller than—and 50% are larger than—this diameter) is identical for both mixtures.

Particle diameter of nonspherical particles can be defined in any number of ways, and no one definition of diameter is the "correct" one. Only for spherical particles does the term "diameter" have a strict geometrical meaning. Expressions for defining *diameter* for nonspherical particles are discussed further in the appendix to this chapter.

Although the most accurate expression of particle-size distribution is graphical, several mathematical expressions have been suggested. For example, in water engineering, the particle size of filter sand is expressed using the *uniformity coefficient*, defined as

$$UC = \frac{D_{60}}{D_{10}}$$

where UC = uniformity coefficient

D_{60} = particle (sieve) size where 60% of the particles are smaller than that size

D_{10} = particle (sieve) size where 10% of the particles are smaller than that size

Figure 2-10 shows a particle-size distribution based on a sieve analysis for a raw (as collected) refuse for a study conducted at Pompano Beach, Florida.[22] The results of a study performed in two New England towns, illustrated in Figure 2-11, show a similar particle-size distribution by the ability to pass through a sieve.[23]

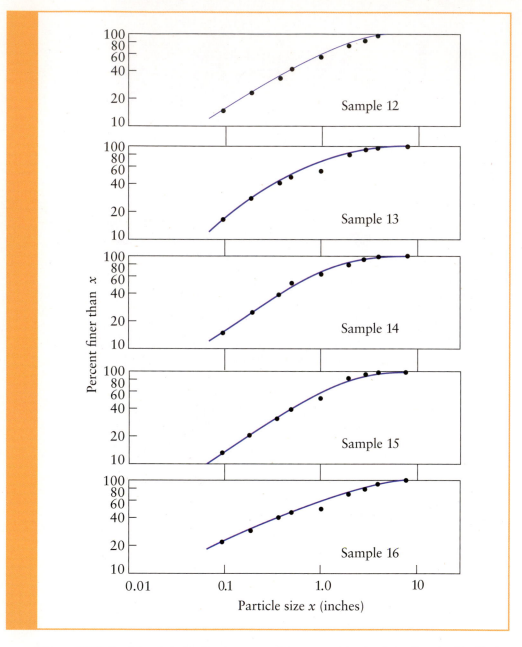

Figure 2-10 Particle-size distribution curve for unprocessed refuse. Source: Vesilind, P. A., A. E. Rimer, and W. A. Worrell. 1980. "Performance Characteristics of a Vertical Hammermill Shredder." *Proceedings 1980 National Waste Processing Conference.* Washington, D.C.: ASME. With permission from ASME.

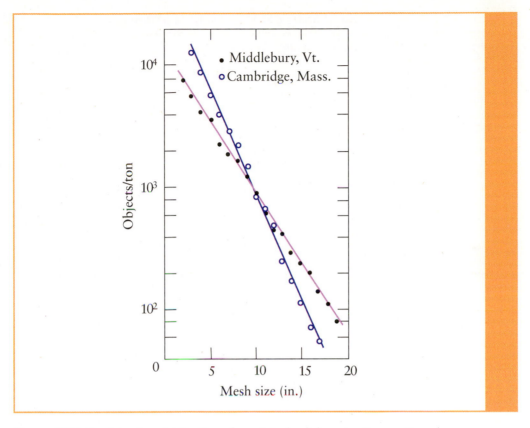

Figure 2-11 Particle-size distribution of municipal solid waste. Source: Based on Rhyner, C. R. 1976, "Domestic Solid Waste and Household Characteristics," *Waste Age* (April): 29–39, 50.

2-3-4 Chemical Composition

The economic recovery of materials and/or energy often depends on the chemical composition of the refuse—the individual chemicals as well as the heat value. Two common means of defining the chemical composition of refuse are the *proximate analysis* and the *ultimate analysis*. Both descriptions were originally developed for solid fuels, especially coal. The proximate analysis is an attempt to define the fraction of volatile organics and fixed carbon in the fuel, while the ultimate analysis is based on elemental compositions. Some data for both proximate and ultimate analysis published by EPA are tabulated in Table 2-7. These data are of limited value for design purposes, because once again, the heterogeneous nature of refuse and its variability with geography and with time results in wide ranges. Accurate information for a specific refuse can be attained only by concerted sampling and analysis.

Table 2-7 **Proximate and Ultimate Chemical Analyses of Refuse**

Proximate Analysis (percent by weight)	
Moisture	15–35
Volatile matter	50–60
Fixed carbon	3–9
Noncombustibles	15–25
Higher heat value (HHV)	3000–6000

Ultimate Analysis (percent by weight)	
Moisture	15–35
Carbon	15–30
Hydrogen	2–5
Oxygen	12–24
Nitrogen	0.2–1.0
Sulfur	0.02–0.1
Total noncombustibles	15–25

Source: [24]

2-3-5 Heat Value

The heat values of refuse are of some importance in resource recovery. Some published values for several fuels are shown in Table 2-8 to illustrate the variability of the fuels according to how they are derived. The production of RDF (refuse-derived fuel) is described further in Chapter 7.

Table 2-8 **Heat Value of Fuels**

Fuel	Heat Value (kJ/kg)	(Btu/lb)	Composition (wt %) S	H	C	N	O	Ash
Natural gas	54,750	23,170	nil	23.5	75.2	1.22	—	nil
Heating oil (no. 2)	45,000	19,400	0.3	12.5	87.2	0.02	nil	nil
Coal, anthracite	29,500	12,700	0.77	3.7	79.4	0.9	3.0	11.2
Coal, bituminous	26,200	11,340	3.22	4.6	40.0	1.0	6.5	9.0
Coal, lignite	19,200	8300	0.4	2.5	32.3	0.4	10.5	4.2
Wood, hardwood	7180*	3090*	—	—	—	—	—	—
Wood, softwood	7950*	18,400*	—	—	—	—	—	—
Shredded refuse[a]	10,846	4675	0.1	—	—	—	—	20.0
RDF[b]	15,962	6880	0.2	—	37.1	0.8	—	22.6
RDF[c]	18,223	7855	0.1	—	45.4	0.3	—	6.0
Unprocessed refuse	10,300	4450	0.1	2.65	25.6	0.64	21.2	20.8
Unprocessed refuse			0.13	4.80	35.6	0.9	29.5	28.9
Paper	24,900	7500	0.1	2.7	20.7	0.13	19.1	2.74

* Lower Heat Value (LHV); all other heat values are Higher Heat Value (HHV)
[a] Shredded, non-air-classified, ferrous removed, not dried; St. Louis RDF facility
[b] Shredded, air-classified, not dried
[c] Same as above, but oversize from a 3/16-in. screen
Source: [18]

In common American engineering language, heat value is expressed as Btu/lb of refuse, while the proper SI designation is kJ/kg. Commonly, the heat values of refuse and other heterogeneous materials are measured with a *calorimeter*, a device in which a sample is combusted and the temperature rise is recorded (see Chapter 7). Knowing the mass of the sample and the heat generated by the combustion, the Btu/lb is calculated (recognizing, of course, that 1 Btu is the heat necessary to raise the temperature of 1 lb of water 1°F).

Refuse can be characterized as being made up of organic materials, inorganic materials, and water. Usually, the heat value is expressed in terms of all three components (the Btu/lb), where the sample weight includes the inorganics and water. But sometimes, the heat value is expressed as *moisture-free*, and the water component is subtracted from the denominator. A third means of defining heat value is to also subtract the inorganics, so the Btu is *moisture- and ash-free*, the ash being defined as the inorganic upon combustion.

An internal report at Paul Scherrer Institute compiled by J. Wochele contains numerous data about major elements, trace elements, as well as high and low heating values of different wastes and substances from different references. Parts of this report are reproduced in Appendix D.

EXAMPLE 2-3

A sample of refuse is analyzed and found to contain 10% water (measured as weight loss on evaporation). The Btu of the entire mixture is measured in a calorimeter and is found to be 4000 Btu/lb. A 1.0 g sample is placed in the calorimeter, and 0.2 g ash remains in the sample cup after combustion. What is the comparable, moisture-free Btu and the moisture- and ash-free heat value?

The one pound of refuse would have 10% moisture, so the moisture-free heat value would be calculated as

$$4000 \text{ Btu/lb} \times \frac{1 \text{ g}}{1 \text{ g} - 0.1 \text{ g water}} = 4444 \text{ Btu/lb}$$

Similarly, the moisture- and ash-free heat value would be

$$4000 \text{ Btu/lb} \times \frac{1 \text{ g}}{1 \text{ g} - 0.1 \text{ g water} - 0.2 \text{ g ash}} = 5714 \text{ Btu/lb}$$

The heat value of various components of refuse is quite different, as shown in Table 2-9. As is the case with moisture content, the heat value of a refuse where the fraction of components is known can be estimated by using such estimated heat values for the various components.

An important aspect of calorimetric heat values is the distinction between *higher heat value* and *lower heat value*. The higher heat value (HHV) is also called the gross calorific energy, while the lower heat value (LHV) is also known as the *net calorific energy*. The distinction is important in design of combustion units. More on this in Chapter 7.

Table 2-9 Heat Values of Some Refuse Components

Component	As Collected	Btu/lb Moisture-Free	Moisture- and Ash-Free
Cardboard	7040	7400	7840
Food waste	1800	6000	7180
Magazines	5250	5480	7160
Newspapers	7980	8480	8610
Paper (mixed)	6800	7570	8050
Plastics (mixed)	14,100	14,390	16,020
HDPE	18,700	18,700	18,900
PS	16,400	16,400	16,400
PVC	9750	9770	9980
Steel cans	0	0	0
Yard waste	2600	6500	6580

Source: Adapted from [20 from 21]

A comparison of the heat values of coal and refuse-derived fuel (RDF) in Table 2-8 suggests implicitly that RDF can be substituted directly for coal by using ratios of the heat values. If the heat value of coal is 10,000 Btu/lb and that of RDF is about 7000 Btu/lb, one might conclude (erroneously) that 10 tons of RDF represents the same energy value as 7 tons of coal. The problems in substituting RDF for coal or oil are immense and often impractical, such as running a car on RDF.

2-3-6 Bulk and Material Density

Municipal solid waste has a highly variable bulk density, depending on the pressure exerted, as shown in Figure 2-12. Loose, as it might be placed into a garbage can by the homeowner, the bulk density of MSW might be between 150 and 250 lb/yd³ (90 and 150 kg/m³); pushed into the can, it might be at 300 lb/yd³ (180 kg/m³). In a collection truck that compacts the refuse, the bulk density is normally between 600 and 700 lb/yd³ (350 and 420 kg/m³). Once deposited in a landfill and compacted with machinery, it can achieve bulk densities of about 1200 lb/yd³ (700 kg/m³). If the covering soil in landfills is included, the total landfilled density can range from about 700 lb/yd³ for a poorly compacted landfill to as high as 1700 lb/yd³ (1000 kg/m³) for a landfill where thin layers of refuse are compacted. Such landfills are quite dense.

For comparison, the density of water is 1690 lb/yd³ (1000 kg/m³). The bulk densities of refuse in various stages of compaction are listed in Table 2-10.

Table 2-10 Refuse Bulk Densities

Condition	Density (lb/yd)³
Loose refuse, no processing or compaction	150–250
In compaction truck	600–900
Baled refuse	1200–1400
Refuse in a compacted landfill (without cover)	750–1250

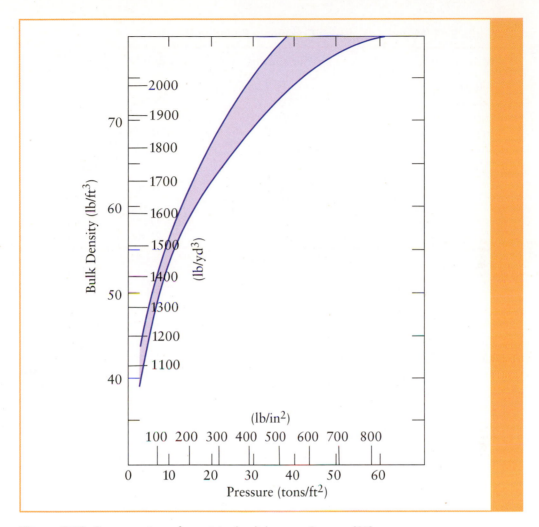

Figure 2-12 Compression of municipal solid waste. Source: [25]

Tables 2-5 and 2-10 show the densities of refuse and its components as *bulk* densities. This is different from *materials* densities, or densities of materials without any void spaces. Table 2-11 shows such material densities for a number of refuse components. For example, for a steel can, the material density is that of steel, or about 7.7 g/cm³. For an uncrushed empty steel can, about 95% of the volume is air, and the bulk density is then only about 0.4 g/cm³.

Because of the highly variable density, MSW quantities are seldom expressed in volumes and are almost always expressed in mass terms as either pounds or tons in the American standard system or kilograms or tonnes in the SI system.

Table 2-11 Material Densities Commonly Found in Refuse

Material	Specific Gravity	lb/yd³
Aluminum	2.70	4536
Steel	7.70	12,960
Glass	2.50	4212
Paper	0.70–1.15	1190–1940
Cardboard	0.69	1161
Wood	0.60	1000
Plastics		
HDPE	0.96	1590
Polypropylene	0.90	1510
Polystyrene	1.05	1755
PVC	1.25	2106

2-3-7 Mechanical Properties

The compressive strength of some typical MSW constituents is shown in Figure 2-13. A wide variation exists in the amount of energy necessary to obtain volume reduction. The curves tend to be mostly linear, indicating that substantial volume reduction can be achieved by expending greater energy in compaction—a fact understood in the use of solid waste balers.

The tensile stress-strain curves for several refuse components are shown in Figure 2-14. As expected, the steel has the greatest ultimate strength, yielding a computed modulus of elasticity (E) of 28.5×10^6 lb/in² (83.8×10^6 kg/cm²), which compares favorably with the usual value of E of 29 to 30×10^6 lb/in² (85 to 88×10^6 kg/cm²) for carbon and low-alloy steels. The E for aluminum was computed as 10×10^6 lb/in² (30×10^6 kg/cm²), while PVC had an E of only 0.2×10^6 lb/in² (0.6×10^6 kg/cm²), as expected.[26]

2-3-8 Biodegradability

All organic material is biodegradable. However, when designing a treatment system, the rate at which a material biodegrades is important. For example, food waste will biodegrade very quickly while wood will take a long time to decrease. From Table 2-4 the fraction of municipal solid waste that is organic can be listed as in Table 2-12.

Table 2-12 shows that only about 55% of MSW will potentially biodegrade rapidly. Treatment techniques (such as composting) must take into account that a large fraction of MSW, which is not biodegradable or which will biodegrade very slowly, must be disposed of by means other than producing useful products using biodegradation.

2-3-9 Greenhouse Gas

MSW can be a significant contributor to greenhouse gases. On a worldwide basis, post-consumer waste accounts for 5% of total greenhouse gas emissions, and methane from landfills represents 12% of total global methane emissions.[42] In the

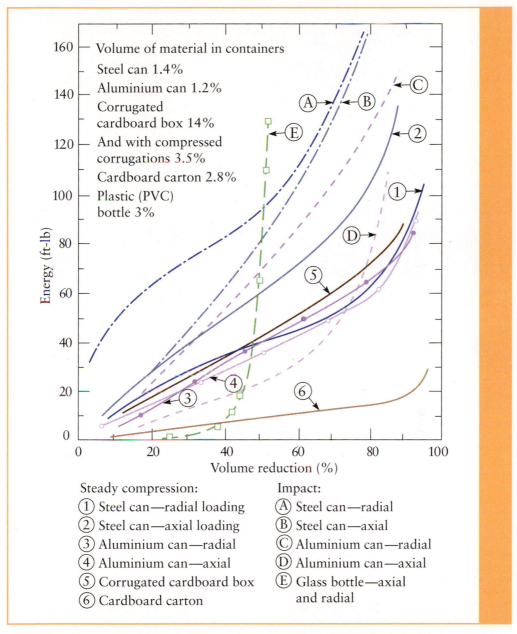

Figure 2-13 Compressive characteristics of some components of solid waste.
Source: [19]

United States, methane is the second most prevalent greenhouse gas emitted from human activities. In 2012, methane accounted for about 9% of all U.S. greenhouse gas emissions from human activities, with landfills being the third largest source. By diverting organic material from landfills to composting facilities, significant reductions in methane generation are achieved. In addition, by recycling, greenhouse gas reductions are possible from the reduced energy consumption and virgin

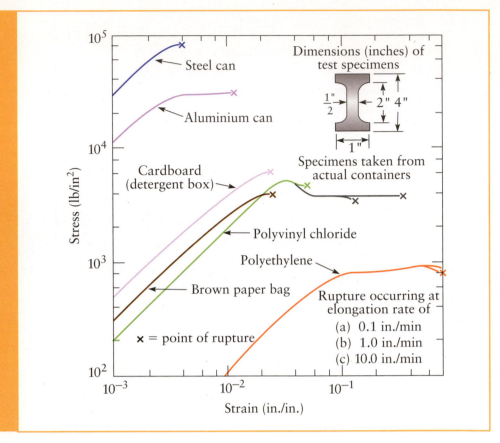

Figure 2-14 Tensile strength of some municipal solid waste components. Source: [26]

Table 2-12 Biodegradable Fraction of MSW

Component	Percent of MSW	Rate of Biodegradable
Paper and paperboard	27.4	rapid
Rubber and leather	3.0	slow
Textiles	5.7	slow
Wood	6.3	slow
Other materials	1.8	varies
Food waste	14.5	rapid
Yard trimmings	13.5	rapid
Miscellaneous inorganic	1.6	varies
Total (organic)	73.8	

Table 2-13 **Greenhouse Gas Benefits Associated with Recovery of Specific Materials, 2012*** **(in millions of tons recovered, MMTCO$_2$E, and in numbers of cars taken off the road per year)**

Material	Weight Recovered (millions of tons)	GHG Benefits MMTCO$_2$E	Numbers of Cars Taken Off the Road per Year
Paper and paperboard	44.4	130.5	27 million
Glass	3.20	1	210 thousand
Metals			
Steel	5.55	9	1.9 million
Aluminum	0.71	6.3	1.3 million
Other nonferrous metals[†]	1.36	5.3	1.1 million
Total metals	**7.62**	**20.6**	**4.3 million**
Plastics	2.80	3.2	670 thousand
Rubber and leather[‡]	1.35	0.7	145 thousand
Textiles	2.25	5.7	1.2 million
Wood	2.41	4.2	900 thousand
Other wastes			
Food, other[^]	1.74	1.4	290 thousand
Yard trimmings	19.6	0.8	170 thousand

* Includes materials from residential, commercial, and institutional sources.

These calculations do not include an additional 1.30 million tons of MSW recovered that could not be addressed in the WARM model. MMTCO$_2$E is million metric tons of carbon dioxide equivalent.

All benefits information that was included in last year's report only took into account the CO$_2$ reduction for recycling of materials. In the report this year, we are accounting for both the recycling of those materials and the CO$_2$ emissions that may occur in the alternative waste management scenarios of landfilling and combustion. This gives us the net overall benefit of recycling these materials.

[†] Includes lead from lead-acid batteries. Other nonferrous metals calculated in WARM as mixed metals.

[‡] Recovery only includes rubber from tires.

[^] Includes recovery of other MSW organics for composting.

Source: WARM model (www.epa.gov/warm), United States Environmental Protection Agency. "Municipal Solid Waste Generation, Recycling, and Disposal in the United States: Facts and Figures for 2012.

feedstock. Using the USEPA WARM model it is possible to calculate the greenhouse gas emission reductions from composting and recycling. Table 2-13 shows the greenhouse gas reductions achieved in 2012.

2-3-10 Toxicity

In this chapter we have mainly focused on the composition of the major fractions of wastes. In our discussion thus far, we have not considered that often trace compounds are those which may harm the environment the most. In Figure 2-15 the composition of a mobile phone in 2001 is compared with the toxicity potential of the single compounds. For example, nickel is only present at about 1% but is responsible for about 20% of the toxicity potential of a mobile phone.[34]

There is a strong tendency toward more complexity of consumer goods. For example the composition of electronic equipment is rapidly changing, sometimes

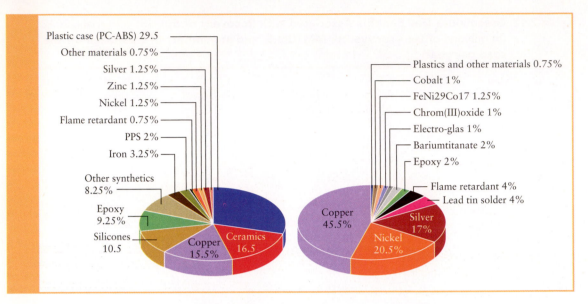

Figure 2-15 Comparison of the content of a mobile phone in 2001 with its toxicity potential according to a study by the Frauenhofer Institute. Source: [34]

using materials which are scarce such as rare earth elements (REE). In this context toxicity, as well as resources shortages for certain elements, needs to be considered. In 1980 a mobile phone contained about 11 elements, while ten years later there were 15. In 2000 one could potentially find 56 different elements.[35] The risk to lose valuable resources increased over the last years with the increase of the diversity of the elements in consumer goods. In Figure 2-16 an estimate of the recycling rates for different elements is shown.[36]

The diversity has not only increased with respect to elements. Materials technology went through substantial developments in the last decade. For example, the application of engineered nanoparticles (ENPs) increased almost exponentially during the last decade. Next to carbon-based ENPs, metal-based ENPs are the most commonly employed. Utilization of such particles for medical applications, remediation of toxic substances, consumer products, new materials, and numerous other fields has been reported already. Despite their extraordinary success and implementation, concerns have been raised about possible negative effects in humans and the environment. Once released, ENPs are easily distributed via the atmosphere and the aquatic environment, where an exposure to organisms can occur. Little about their fate and release to the environment is known today, and there are no appropriate analytical methods available for studying their chemical-physical characteristics in waste treatment processes.[37] It was shown that the physical and chemical transformations of silver nanoparticles in a wastewater treatment plant controls their fate, their transport, and also their toxicity and bioavailability and therefore must be considered in future risk assessments.[38] Little is known about the fate of ENPs in waste treatment facilities. A preliminary risk assessment

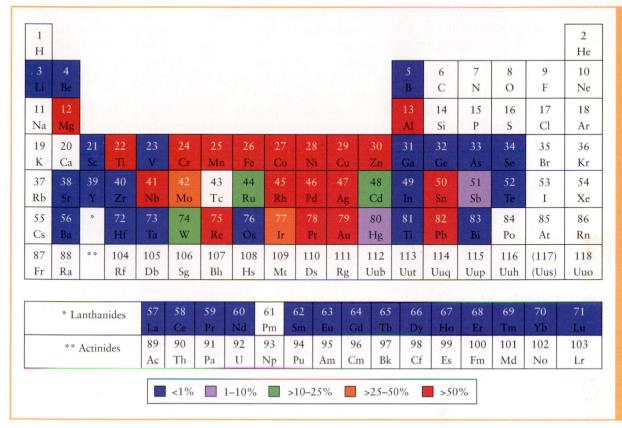

Figure 2-16 Recycling rates for different elements. Source: [36]

has been performed for the fate of a series of ENPs in wastes during incineration. However, the assessment remains difficult and is uncertain for many of the consumer products that are already on the market.[39] But, as demonstrated, ENPs can persist during the treatment of MSW in a conventional incinerator.[40]

2-4   FINAL THOUGHTS

Continued implementation of reduction, reuse, recycling, and resource recovery systems will reduce both the extraction of raw material and the quantities of waste disposed into our environment. Increased recovery would mean less landfilling or incineration (hence, less air and water pollution and other detrimental effects caused by today's common treatment methods).

The rate of the increased use of natural resources for both energy and materials production will be slowed at least by a wider use of secondary materials. It is now commonly accepted that the supply of natural raw materials will be restricted worldwide and that the United States and Europe will feel disproportionately greater pressure as a result of our large consumption and limited domestic supplies. In the high income countries, we recognize that resource recovery is reasonable, logical, prudent, economical, and feasible. However, in low income countries that produce less than one third the amount of waste, resource recovery may not be an important policy issue. Other issues, such as health and raising the standard of living, may take precedence over resource recovery.

Over the course of the development of our basic Western philosophical and ethical framework, environmental concerns have played a minor role. With the exception of Thoreau, St. Francis of Assisi, and a few others, we have not wrestled very long or hard with the problem of environmental ethics. The question posed is thus one that does not yet have a simple answer, nor are there many philosophers even willing to tackle it. Much work remains to be done in this area, as is the case with economics, engineering, and other sciences. The absence of an answer should not, however, deter us from thinking out the "good" in this endeavor. We should still seek understanding, perhaps from a spiritual source.

We respond to nature spontaneously in ways that are not easily accounted for by any other means, and there is evidence that we need nature for psychological health. Of course, we can't just make up a new environmental ethic, but we do have spiritual resources in our cultural traditions, and a study of other traditions can both illuminate our own and provide insights into possibly more common elements of human spiritual needs and what satisfies them.

When Aldo Leopold was a young man working for the Forest Service, one of his first assignments was to help in the eradication of the wolf in New Mexico. One day he and his crew spotted a wolf crossing a creek and instantly opened fire, mortally wounding the animal. Leopold scrambled down to where it lay and arrived in time to "watch the fierce green fire dying in her eyes."[27] This experience was significant in Leopold's recognition of the value of predators in an ecosystem and his eventual acceptance of the intrinsic value of all nature. Did Leopold's spirituality see the fire in the wolf's eyes, or was that spirituality transferred from the wolf to Leopold?

It is unlikely that most of us will ever share such a life-changing experience. But it is also true that most of us already understand, somewhere deep down, the significance and meaning of the spiritual dimension of the environmental ethic.

2-5 APPENDIX: MEASURING PARTICLE SIZE

For nonspherical particles the diameter of a particle may be defined as any of the following:

$$D = l$$
$$D = \frac{h + w + l}{3}$$
$$D = \sqrt[3]{hwl}$$
$$D = \sqrt{lw}$$
$$D = \frac{w + l}{2}$$

where D = particle diameter
$\quad$ l = length
$\quad$ w = width
$\quad$ h = height

When particle size is determined by sieving, the most reasonable definition is

$$D = \sqrt{lw}$$

since only two dimensions must be less than the sieve opening for the particle to fall through. As stated before, the term *particle diameter* has significance only if the shape is circular, and any other geometrical shape is therefore not really describable by diameter.

EXAMPLE 2-4

Consider nonspherical particles that are uniformly sized as length, $l = 2$; width, $w = 0.5$; and height, $h = 0.5$. Calculate the particle diameter by the previous various definitions.

SOLUTION

$$D = l = 2; \quad D = \frac{w + l}{2} = 1.25; \quad D = \frac{h + w + l}{3} = 1.0$$

$$D = \sqrt{lw} = 1; \quad D = \sqrt[3]{hwl} = 2.12$$

Note that the "diameter" varies from 1.0 to 2.12, depending on the definition.

When the mixture of particles is nonuniform, the particle size is often expressed in terms of the *mean particle diameter*. Given an analysis of the various diameters of individual particles (such as by sieving), the mean particle diameter can be expressed in a number of ways, including all of the following:

Arithmetic Mean:

$$D_A = \frac{D_1 + D_2 + D_3 + \cdots D_n}{n}$$

Geometric Mean:

$$D_G = \sqrt[n]{D_1 \times D_2 \times D_3 \times \cdots D_n}$$

Weighted Mean:

$$D_w = \frac{W_1 D_1 + W_2 D_2 + \cdots W_n D_n}{W_1 + W_2 + \cdots W_n}$$

Number Mean:

$$D_N = \frac{M_1 D_1 + M_2 D_2 + \cdots M_n D_n}{M_1 + M_2 + \cdots M_n}$$

Surface Area Mean:

$$D_S = \frac{M_1 D_1^3}{M_1 D_1^2 + M_2 D_2^2 + \cdots M_n D_n^2} + \frac{M_2 D_2^3}{M_1 D_1^2 + M_2 D_2^2 + \cdots M_n D_n^2}$$
$$+ \frac{M_n D_n^3}{M_1 D_1^2 + M_2 D_2^2 + \cdots M_n D_n^2}$$

Volume Mean:

$$D_V = \frac{M_1 D_1^4}{M_1 D_1^3 + M_2 D_2^3 + \cdots M_n D_n^3} + \frac{M_2 D_2^4}{M_1 D_1^3 + M_2 D_2^3 + \cdots M_n D_n^3}$$
$$+ \frac{M_n D_n^4}{M_1 D_1^3 + M_2 D_2^3 + \cdots M_n D_n^3}$$

where n = number of discrete classifications (sieves)
W = weight in each classification
M = number of particles in each classification

EXAMPLE 2-5

Given the following analysis,

Particle diameter, mm, (D)	60	40	20	5
Weight of each fraction, kg (W)	2	10	5	4
Number of particles, (M)	140	300	1000	2000

calculate the arithmetic mean, geometric mean, weighted mean, number mean, surface area mean, and volume mean.

SOLUTION

$$D_A = \frac{(60 + 40 + 20 + 5)}{4} = 31.2 \, mm$$

$$D_{GX} = \sqrt[4]{60 \times 40 \times 20 \times 5} = 21.1 \, mm$$

$$D_W = \frac{(2 \times 60) + (10 \times 40) + (5 \times 20) + (4 \times 5)}{2 + 10 + 5 + 4} = 30.3 \, mm$$

$$D_N = \frac{(140 \times 6) + (300 \times 40) + (1000 \times 20) + (2000 \times 5)}{140 + 300 + 1000 + 2000} = 14.7 \, mm$$

$$D_S = \frac{140 \times 60^3}{140 \times 60^2 + 300 \times 40^2 + 1000 \times 20^2 + 2000 \times 5^2} + \cdots = 40.0 \, mm$$

$$D_V = \frac{140 \times 60^4}{140 \times 60^2 + 300 \times 40^3 + 1000 \times 20^3 + 2000 \times 5^2} + \cdots = 40.0 \, mm$$

MSW Disposal Methods by Country

Country	Dumps (%)	Landfills (%)	Compost (%)	Recycled (%)	WTE (%)	Other (%)
Algeria	96.80	0.20	1.00	2.00	—	—
Antigua and Barbuda		99.00		1.00	—	—
Armenia	—	100.00	—	—	—	—
Australia	—	69.66	—	30.34	—	—
Austria	—	6.75	44.72	26.54	21.10	0.90
Belarus	—	96.00	4.00	—	—	—
Belgium	—	11.57	22.77	31.10	34.32	—
Belize	—	100.00	—	—	—	—
Bulgaria	—	82.90	—	—	—	17.10
Cambodia	100.00	—	—	—	—	—
Cameroon	95.00	—	—	5.00	—	—
Canada	—	—	12.48	26.78	—	60.74
Chile	—	100.00	—	—	—	—
Colombia	54.00	46.00	—	—	—	—
Costa Rica	22.37	71.95	—	0.29	—	5.39
Croatia	—	69.50	0.90	2.40	—	27.20
Cuba	—	100.00	11.10	4.80	—	—
Cyprus	—	87.20	—	—	—	12.80
Czech Republic	—	79.78	3.24	1.27	13.97	1.74
Denmark	—	5.09	15.28	25.57	54.04	0.03
Dominica	—	100.00	—	—	—	—
Greece	—	92	—	8	—	—
Grenada	—	90	—	—	—	10
Guatemala	—	22	—	—	—	78
Guyana	37	59	—	—	—	4
Haiti	24	—	—	—	—	76
Hong Kong, China	—	55	—	45	—	—
Hungary	—	90	1	3	6	0
Iceland	—	72	9	16	9	—
Ireland	—	66	—	34	—	—
Israel	—	90	—	10	—	—
Italy	—	54	33	—	12	—
Jamaica	—	100	—	—	—	—
Japan	—	3	—	17	74	6
Jordan	—	85	—	—	—	15
Korea, South	—	36	—	49	14	—
Kyrgyz Republic	—	100	—	—	—	—
Latvia	60	40	—	—	—	—
Lebanon	37	46	8	8	—	1
Lithuania	—	44	—	4	2	50
Luxembourg	—	19	19	23	39	—
Macao, China	—	21	—	—	—	100
Madagascar	—	97	4	—	—	—
Malta	—	88	—	—	—	13
Marshall Islands	—	—	6	31	—	63
Mauritius	—	91	—	2	—	—
Mexico	—	97	—	3	—	—
Monaco	—	27	—	4	—	132

(Continued)

MSW Disposal Methods by Country (*Continued*)

Country	Dumps (%)	Landfills (%)	Compost (%)	Recycled (%)	WTE (%)	Other (%)
Morocco	95	1	—	4	—	—
Netherlands	—	2	23	25	32	17
New Zealand	—	85	—	15	—	—
Nicaragua	34	28	—	—	—	38
Niger	—	64	—	4	—	32
Norway	—	26	15	34	25	0
Panama	20	56	—	—	—	24
Paraguay	42	44	—	—	—	14
Peru	19	66	—	—	—	15
Poland	—	92	3	4	0	—
Portugal	—	64	6	9	21	—
Romania	—	75	—	—	—	25
Singapore	—	15	—	47	—	49
Slovak Republic	—	78	1	1	12	7
Slovenia	—	86	—	—	—	14
Spain	—	52	33	9	7	—
St. Kitts and Nevis	—	100	—	—	—	—
St Lucia	—	70	—	—	—	30
St. Vincent and the Grenadines	—	78	—	—	—	22
Suriname	100	—	—	—	—	0
Sweden	—	5	10	34	50	1
Switzerland	—	1	16	34	50	—
Syrian Arab Republic	>60	<25	<5	<15	—	—
Thailand	—	—	—	14	—	85
Trinidad and Tobago	6		—	—	—	94
Tunisia	45	50	0	5	—	—
Turkey	66	30	1	—	0	3
Uganda	—	100	—	—	—	—
United Kingdom	—	64	9	17	8	1
United States	—	54	8	24	14	—
Uruguay	32	3	—	—	—	66
Venezuela, RB	59	—	—	—	—	41
West Bank and Gaza	69	30	—	1	—	—

Source: Based on WHAT A WASTE, A Global Review of Solid Waste Management by Daniel Hoornweg and Perinaz Bhada-Tata, March 2012, No. 15, Urban Development & Local Government Unit, World Bank.

References

1. United States Environmental Protection Agency. "Municipal Solid Waste Generation, Recycling, and Disposal in the United States: Facts and Figures for 2012."

2. Rhyner, C. R. 1992. "Monthly Variations in Solid Waste Generation." *Waste Management and Research* 10:67–71.

3. Boyd, G. and M. Hawkins. 1971. *Methods of Predicting Solid Waste Characteristics.* EPA OSWMP, SW–23c, Washington, D.C.

4. Dayal, G., A. Yadav, R. P. Singh, and R. Upadhyay. 1993. "Impact of Climatic Conditions and Socioeconomic Status on Solid Waste

Characteristics: A Case Study." *The Science of Total Environment* 136: 143–153.

5. Davidson, G. R. 1972. *Residential Solid Waste Generation in Low Income Areas.* EPA OSWMP.

6. Grossman, D. J., F. Hudson, and D. H. Marks. 1974. "Waste Generation Models for Solid Waste Collection." *Journal of the Environmental Engineering Division* ASCE 100, EE6:1219–1230.

7. Rathje, W. and C. Murphy. 1992. *Rubbish: The Archeology of Garbage.* New York: Harper-Collins.

8. Ali Khan, M. and F. Buney. 1989. "Forecasting Solid Waste Compositions." *Resources, Conservation and Recycling* 3:1–17.

9. Cailas, M. D., R. Kerzee, R. Swager, and R. Anderson. 1993. *Development and Application of a Comprehensive Approach for Estimating Solid Waste Generation in Illinois.* Urbana-Champaign, Illinois: The Center for Solid Waste Management and Research, University of Illinois.

10. Henricks, S. L. 1994. *Socio-economic Determinants of Solid Waste Generation and Composition in Florida.* MS thesis, Duke University School of the Environment. Durham, North Carolina.

11. Hockett, D., D. J. Lober, and K. Pilgrim. 1995. "Determinants of Per Capita Municipal Solid Waste Generation in the Southeastern United States." *Journal of Environmental Management* 45:205–217.

12. American Society of Testing and Materials. 1992. A *Standard Test Method for Determination of the Composition of Unprocessed Municipal Solid Waste.* D 5231–92. Philadelphia.

13. Reinhart, D., P. M. Bell, B. Ryan, and H. Sfeir. "An Evaluation of Municipal Solid Waste Composition Studies." Orlando, Florida: University of Central Florida.

14. Sfeir, H., D. R. Reinhart, and P. R. McCauley-Bell. 1999. "An Evaluation of Municipal Solid Waste Composition Bias Sources." *Air and Waste Management Association* 49:1096–1102.

15. Klee, A. J. and D. Carruth, 1970. "Sample Weights in Solid Waste Composition Studies." *J. Sanit. Eng. Div.* ASCE 96, no. SA4:945.

16. Miller, C., various articles in *Waste Age.*

17. O'Brien, J. 1977. Unpublished data. Durham, North Carolina: Duke Environmental Center, Duke University.

18. National Center for Resource Recovery. 1978. *Incineration.* Toronto: Lexington Books.

19. *Measuring Recycling: A Guide for State and Local Governments.* 1997. EPA 530-R-97–011. Washington, D.C.

20. Tchobanaglous, G., H. Theisen, and S. Vigil. 1993. *Integrated Solid Waste Management.* New York: McGraw-Hill.

21. Neissen, W. 1977. "Properties of Waste Materials." In D. G. Wilson, ed., *Handbook of Solid Waste Management.* New York: Van Nostrand Reinhold.

22. Vesilind, P. A., A. E. Rimer, and W. A. Worrell. 1980. "Performance Characteristics of a Vertical Hammermill Shredder." *Proceedings* 1980 National Waste Processing Conference. Washington, D.C.: ASME.

23. Rhyner, C. R. 1976. "Domestic Solid Waste and Household Characteristics." *Waste Age* (April): 29–39, 50.

24. *Incinerator Guidelines.* 1969. Washington, D.C.: U.S. Dept. of Health, Education, and Welfare.

25. Ruf, J. A. 1974. *Particle Size Spectrum and Compressibility of Raw and Shredded Municipal Solid Waste.* Ph.D. Diss., University of Florida, Gainesville.

26. Trezek, G., D. Howard, and G. Savage. 1972. "Mechanical Properties of Some Refuse Components." *Compost Science* 13, no. 6.

27. Leopold, A. 1966. *A Sand County Almanac*. New York: Ballantine.

28. Franklin Associates. 1999. *Characterization of Municipal Solid Waste in the United States: 1998 Update*. EPA 530–R-98–007, Washington, D.C.

29. EU-Lex, 2006. Directive 2006/12/EC on waste.

30. EU-Lex, 2000. Directive 2000/53/EC on end-of-life vehicles.

31. Hellweg, S., 2010. in: K. Schenk (Ed.), KVA-Rückstände in der Schweiz. Der Rohstoff mit Mehrwert, Swiss Federal Office of the Environment.

32. Denafas, G., Ruzgas, T., Martuzevičius, D., Shmarin, S., Hoffmann, M., Mykhaylenko, V., Ogorodnik, S., Romanov, M., Neguliaeva, E., Chusov, A., Turkadze, T., Bochoidze, I, Ludwig, C., 2014. *Resources, Conservation and Recycling*, 89:22–30.

33. Denafas, G., Zavarauskas, K., Martuzevičius, D., Vitkauskaitė, L., Ludwig, C., Hoffmann, M., Shmarin, S., Mykhaylenko, V., Chusov, A., Romanov, M., Negulyaeva, E., Lednova, Y., Turkadze, T., Bochoidze, I., Butskhrikidze, B., Karagiannidis, A., Antonopoulos, J., Kriipsalu, M., Horttanainen, M., 2010. Third International Symposium on Energy from Biomass and Waste, Venice, November 8–11, Italy.

34. Griese, H., 2001. IZM aus Mediendienst 1/2001—Thema 5: Grüne Elektronik in Handys, Fraunhofer-Institut für Zuverlässigkeit und Mikrointegration IZM, Berlin.

35. Mc Manus, T., 2006. Intel Corporation, presented by T. Graedel at World Resources Forum 2012, Beijing, October 21–23.

36. Graedel, T. E., Allwood, J., Birat, J.-P., Buchert, M., Hagelüken, C., Reck, B.K., Sibley, S.F., Sonnemann, G., 2011. What do we know about metal recycling?, *J. Ind. Ecol.*, 15:355–366.

37. Hagendorfer, H., 2011. New Analytical Methods for Size Fractionated, Quantitative, and Element Specific Analysis of Metallic Engineered Nanoparticles in Aerosols and Dispersions, PhD Thesis No. 5202, EPFL, Lausanne, Switzerland.

38. Kaegi, R., Voegelin, A., Sinnet, B., Zuleeg, S., Hagendorfer, H., Burkhardt, M., Siegrist, H., 2011. Behavior of Metallic Silver Nanoparticles in a Pilot Wastewater Treatment Plant, *Environ. Sci. Technol.*, 45(9):3902–3908.

39. L. Roes, M.K. Patel, E. Worrell, C. Ludwig, 2012. Preliminary evaluation of risks related to waste incineration of polymer nanocomposites, *Sci. Total Environ.*, 417:76–86.

40. Walser, T., Limbach, L. K., Brogioli, R., Erismann, E., Flamigni, L., Hattendorf, B., Juchli, M., Krumeich, F., Ludwig, C., Prikopsky, K.; et al. Persistence of engineered nanoparticles in a municipal solid-waste incineration plant. *Nat. Nanotechnol.*, 7:520–524.

41. Zhang, DQ., Tan, SK., and Gersberg RM., Municipal Solid Waste Management in China: Status, problems and challenges, *Journal of Environmental Management*, August 2010.

42. World Bank 2012. *What a Waste: A Global Review of Solid Waste Management*. Urban Development & Local Government Unit, World Bank, 1818 H Street, NW Washington, DC 20433 USA

Abbreviations Used in This Chapter

ASTM = American Society for Testing and
    Materials
Btu = British thermal unit
C = chemical symbol for carbon
C & D = construction and demolition
E = modulus of elasticity
ENP = Engineered nanoparticles
EPA = Environmental Protection Agency
H = chemical symbol for hydrogen
HDPE = high-density polyethylene
HHV = higher heat value
IRM = Integrated resources management
ISWM = Integrated solid waste management

IWM = Integrated waste management
LHV = lower heat value
MSW = municipal solid waste
N = chemical symbol for nitrogen
O = chemical symbol for oxygen
PETE = polyethylene terephthalate
RDF = refuse-derived fuel
REE = Rare earth elements
S = chemical symbol for sulfur
SI = System International, the metric system
    of units
UC = uniformity coefficient

Problems

2-1. A large furniture manufacturer produces 1000 tons of industrial solid waste per day. This firm has decided to locate in a small community of 20,000 residents. If the wastes from this plant go to the town landfill, which has an estimated life of 10 years, how much sooner will the town need a new landfill? Show all assumptions.

2-2. Estimate, using input analysis, the quantity of paper of different categories that might be obtained from the building where you work or study.

2-3. A landfill operation uses a tractor to compact the waste. The tractor weighs 8 tons and has two tracks, each 2 ft. × 10 ft. Estimate the maximum compaction attainable in the landfill.

2-4. A politician states "There is enough energy in our garbage to replace all the oil we presently get from Iraq and Iran." Using your knowledge as a solid waste engineer, write a helpful letter to the editor.

2-5. What is meant by *moisture transfer* in refuse management, and why is this important in studies on refuse composition and materials recovery or energy conversion?

2-6. What do you believe is the approximate composition of solid waste produced at your university? Indicate your estimates by checking the appropriate boxes in Table P2–6. Make sure your answers sum to 100%. Submit as your homework a table showing the fraction of each component.

Table P2-6

Component	Percent by weight					
	<1	5	10	20	40	60
Paper						
Garbage						
Glass						
Aluminum						
Steel						
Plastics						
Other						

2-7. Use examples and definitions to describe what is meant by *code* and *switch* in material recovery operations.

2-8. What impact would there be on the generation of plastic if your community banned single use plastic bags?

2-9. The town of Chapel Hill, North Carolina, has about 100,000 people (including college students). The Smith Center has a volume of about 180,000 cubic yards. If the town of Chapel Hill used the Smith Center as a refuse disposal site, how long would it take to fill up? Make all necessary assumptions.

2-10. Describe what is meant by *as-received*, *moisture-free*, and *moisture- and ash-free* heat value. How are these calculated?

2-11. The state of California requires its communities to divert 75% of municipal solid waste from landfills. Discuss both the objective and the calculations used to attain such high diversion rates. How would you revise the mandate to achieve the original intent of the program—to recycle more refuse? Show your calculations.

2-12. Is polystyrene recycled in your community? If it is not recycled, call the community solid waste/recycling office and ask why it is not. How would you respond to those reasons? If it is, call and ask what happens to it.

2-13. Compare the volume occupied by 1000 lb of refuse following baling with the same waste as it would be in a garbage can. Assume the composition shown in Table P2–13. For miscellaneous wastes, assume a loose density of 300 lb/yd³ and a baled density of 1000 lb/yd³. What is the density of the combined waste, loose and baled? (Use the representative values for bulk densities found in Table 2-5.)

Table P2-13

Component	Percent by weight
Newsprint	21
Office paper	15
Cardboard	8
Glass	12
HDPE	3
PETE	3
Steel cans	5
Yard waste	18
Aluminum cans	4
Miscellaneous	11

2-14. Using the moisture contents (wet basis) of the waste components given in Table P2–14, calculate the overall moisture content of the waste having the composition also shown. This calculation can be simplified by assuming 100 lb of waste.

Table P2-14

Component	Moisture content, percent (wet basis)	Percent by weight
Food	70	10
Paper	6	33
Cardboard	5	8
Plastics	2	5
Textile	10	4
Rubber	2	3
Yard waste	60	18
Metals	3	10
Miscellaneous	6	9

2-15. Calculate the moisture content of the waste in Problem 2–14 on both a wet and a dry basis. Why the difference? Which number makes more sense to you, and why?

2-16. Determine the composition of the waste in Problem 2–14 if yard waste is separately collected. How would this affect the moisture content?

2-17. Using the energy content (wet basis) of the waste components given in

Table P2–17, calculate the overall energy content of the waste having the composition shown. This calculation can be simplified by assuming 100 lb of waste.

Table P2-17

Component	Energy, Btu/lb	Percent by weight
Food	2000	10
Paper	7200	33
Cardboard	7000	8
Plastics	14,000	5
Textile	7500	3
Rubber	10,000	4
Yard waste	2800	18
Metals	300	10
Miscellaneous	3000	9

2-18. Determine the composition of the waste in Problem 2–17 if 10% of the wood waste, 50% of the paper, 30% of the metals, and 25% of the plastics are recycled. How would this affect the energy content?

2-19. To achieve a 60% volume reduction of a cardboard carton, how much energy would be needed?

2-20. How has municipal solid waste generation and diversion in the United States changed since 2012?

CHAPTER 3

Collection

Solid waste collection is an exercise in reducing entropy. The pieces that make up the solid waste are scattered far and wide, and the role of the collector is to gather this material together into one container. In most parts of the developed world, solid waste collection systems are invariably person/truck systems. With only a few minor exceptions, the collection of MSW is done by men and women who traverse a town in trucks and then ride with the truck to a site at which the truck is emptied. This may be an intermediate stopover where the refuse is transferred from the small truck into trailers, larger vans, barges, or railway cars for long-distance transport or the final site, such as the landfill, waste to energy plant, compost site, or materials recovery facility.

3-1 REFUSE COLLECTION SYSTEMS

The process of refuse collection should be thought of as a multiphase process, and it is possible to define at least five separate phases, as shown in Figure 3-1. First, the individual homeowner must transfer whatever is considered waste (defined as material having no further value to the occupant) to the refuse can, which may be inside or outside the home. The second phase is the movement of the refuse can or bin to the collection truck, which is usually done by the collection crew and is called *backyard collection*. If the can is moved to the street by the waste generator or the home occupant, the system is called *curbside collection*.

More and more separated materials (commonly called *recyclables*) and yard wastes, which sometimes include food waste, are collected separately, either in separate compartments of the same trucks as the mixed refuse or in separate vehicles. The following analysis applies to any or all of these materials, although most of the discussion is about mixed (nonseparated) waste.

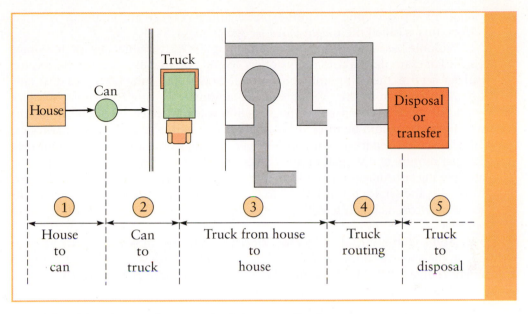

Figure 3-1 Five phases of municipal solid waste collection.

The truck must collect the refuse from many homes in the most efficient way possible, and when it is full (or at the end of the day), it must travel to the materials recovery facility, the point of disposal, or the transfer site. The fifth phase of the collection system involves the location of the final destination (materials recovery facility, disposal site, or transfer station). This is a planning problem, often involving more than one community.

3-1-1 Phase 1: House to Can

The house-to-can phase has received almost no attention or concern by researchers or government because the efficiencies and conveniences gained here are personal and not communal. As discussed later under source separation, one major drawback of collecting separated material is the inconvenience suffered by the individual.

Some communities use tax funds to operate the solid waste collection and disposal system, or they charge for the service just as they charge for water consumption and wastewater disposal. Such a system gives the generators of waste *carte blanche* to generate as much as they please because the cost is the same regardless of how much they contribute. The need to manage and control the amount of waste generated has led communities to try novel ways of funding solid waste programs. Some communities have adopted a ***volume-based fee system*** to pay for solid waste collection and specify the containers that must be used. In a volume-based fee system, residents are offered cans in three sizes—such as 19-, 30-, 60-, or 90-gallon (72-, 110-, 230-, and 340-liter) cans. The fee for refuse service is based on the size of can used. Over 4000 communities have adopted volume-based rates for solid waste collection. The EPA reports typical reduction of 25% to 35% for communities that have gone to volume-based rates.[1] Other communities are taking this approach one step further by weighing every can and charging by actual weight; this is called the *weight-based fee system*.

Volume-based fee systems can either be 100% variable rate or less than 100%. For example, if a community offers 30-, 60-, or 90-gallon waste wheelers for garbage service, the community must decide how to price the service to raise sufficient revenue. In a 100% variable rate community, the 60-gallon waste wheeler would be twice as expensive as the 30-gallon waste wheeler, while the 90-gallon waste wheeler would be three times as expensive as the 30-gallon waste wheeler. Thus 30-gallon service might cost $12 per month, 60-gallon service would cost $24 per month, and 90-gallon service would cost $36 per month. The 100% variable pricing encourages the public to maximize waste reduction. However, the cost to provide the service is not 100% variable. Thus with the exception of the landfill fee, it may cost as much to collect a 30-gallon waste wheeler as a 90-gallon waste wheeler.

Some communities try to more closely tie the cost of service to the garbage fee. In the above examples, instead of charging $12/$24/$36, another community might charge $20/$24/$28. While these rates might more closely match the cost to provide the service, the cost difference between the different size waste wheelers is smaller and thus there is not as large a financial incentive to reduce the amount of waste being generated.

One study looked at 8 similar communities with the same type of service (pay for garbage based on the size of the container and no additional charge for recycling and green-waste service[26]). Four communities had 100% variable rates and 4 had rates that varied by 23 to 88%. The communities with the 100% variable rates had an average garbage service level of 41 gallons. The communities with the less than 100% variable rates had a garbage service level of 51 gallons.

Pay-as-you-throw may have generated renewed interest in the home compactor. This device, originally introduced in the 1950s without much success, sits under the kitchen counter and compresses about 20 lb (9 kg) of refuse into a convenient block within a special bag. The bulk density of the compacted refuse is about 1400 lb/yd^3 (830 kg/m^3), which yields a compaction ratio of about 1:5.

EXAMPLE 3-1

A family of four people generates solid waste at a rate of 2 lb/cap/day, and the bulk density of refuse in a typical garbage can is about 200 lb/yd^3. If collection is once a week, how many 30-gallon garbage cans will the family need, or alternately, how many compacted 20-lb blocks would the family produce if it had a home compactor? How many cans would the family need in that case?

SOLUTION

2 lb/cap/day × 4 persons × 7 days/week = 56 lb refuse
56 lb/200 lb/yd^3 = 0.28 yd^3
0.28 yd^3 × 202 gal/yd^3 = 57 gal

The family will need a 60-gallon can.

If the refuse is compacted into 20-lb blocks, the family would need to produce three such compacted blocks to take care of the week's refuse. If each block of compacted refuse is 1400 lb/yd^3, the necessary volume is

$$\frac{56\ lb}{1400\ lb/yd^3} \times 202\ gal/yd^3 = 8.1\ gal$$

The family would need only one 30-gal can.

3-1-2 Phase 2: Can to Truck

At one time, the most common system of getting the solid waste into the truck was the collectors going to the backyard, emptying the garbage cans into large tote containers, and carrying these to the waiting truck. This system was not only expensive in dollar cost to the community, but it was expensive in terms of the extremely high injury rate to the collectors. At one time, solid waste collectors had the highest injury rate of any vocation—three times higher than the injury rate for coal miners, for example. Even now, with all of the improvements in collection technology, solid waste collection is still one of the most hazardous jobs in America. A survey by the U.S. Department of Labor's Occupational Safety and Health Administration (OSHA) Statistics Department found that fully 40% of solid waste workers had missed time during the preceding year due to various injuries, including strains, bruises, and fractures.[2]

The traditional trucks used for residential and commercial refuse collection are rear-loaded and covered compactors called *packers* and vary in size and design, with 16- and 20-yd^3 (12- and 15-m^3) loads being common (Figure 3-2). The truck size is often limited not by its ability to store refuse but by its wheel weight. Residential streets are not designed to carry large wheel loads, and refuse trucks can easily exceed these limits. Commonly, the refuse is emptied from garbage cans into the back of the packers, where it is scooped up by hydraulically operated compaction mechanisms that compress the refuse from a loose density of about 100 to 200 lb/yd^3 (60 to 120 kg/m^3) to about 600 to 700 lb/yd^3 (360 to 420 kg/m^3). The compaction (packing) mechanism of one manufacturer is shown in Figure 3-3. In addition, many companies are now buying natural gas fired vehicles (Figure 3-4). In combination with biogas production from green and kitchen waste or landfills, these trucks could run on biogas from waste.

Two revolutionary changes are occurring that have had a great impact on both the cost of collection as well as the injury rate of the collectors. The first is wide acceptance of the *can-on-wheels* idea, known as waste wheelers. These containers are provided to the customer by either the garbage company or the local municipality. The resident fills the large plastic container on wheels and then pushes it to the curb for collection. These containers can be used for mixed refuse,

Figure 3-2 A rear-loading packer truck for collecting residential solid waste. (Courtesy William A. Worrell)

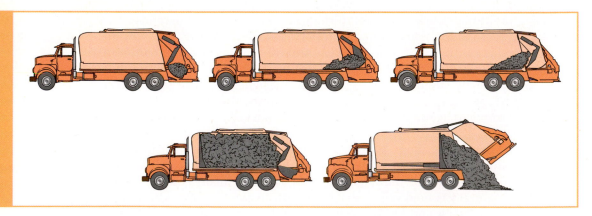

Figure 3-3 Compacting mechanism for a packer truck.

recyclables, and/or yard waste (Figure 3-5). The collection vehicles are equipped with hydraulic lifters that are used to empty the contents into the truck, as shown in Figure 3-6. The collectors do not come into contact with the refuse, thus avoiding dangerous materials that can cut or bruise. In addition, the collectors do not lift the heavy container, thus avoiding back injuries. This system, referred to as

Figure 3-4 Natural gas powered packer truck. (Courtesy William A. Worrell)

Figure 3-5 Separate containers for recyclables, yard waste, and mixed refuse at the curb. (Courtesy William A. Worrell)

Figure 3-6 Blue mixed recycling container being dumped into a truck. (Courtesy William A. Worrell)

semi-automated collection, typically requires a driver and one or more collectors. A further development in solid waste collection technology is the *can snatcher*, a truck equipped with long arms that reach out, grab a can, and lift it into the back of the truck (Figure 3-7). Such systems, called *fully automated collection*, can be operated by a single driver. Communities that have converted from the manual system to the fully automated system have saved at least 50% in collection costs, much of it in reduced crew size and medical costs.[3]

3-1-3 Phase 3: Truck from House to House

Once the refuse is in the truck, it is compacted as the truck moves from house to house. The higher the compaction ratio, the more refuse the truck can carry before it has to make a trip to the landfill.

EXAMPLE 3-2

Assume each household produces 56 lb of refuse per week (as in Example 3-1). How many customers can a 20-yd^3 truck that compacts the refuse to 500 lb/yd^3 collect before it has to make a trip to the landfill?

SOLUTION

20 yd^3 × 500 lb/yd^3 = 10,000 lb

10,000 lb/56 lb/customer = 178 customers

(Note that the refuse weighs 5 tons, and if the truck itself weighs 2 tons, the common 6-ton residential load limit can be exceeded before the truck is full.)

Figure 3-7 Automated garbage truck. Source: Peter Cron/San Luis Obispo County Integrated Waste Management Authority.

The size of the truck crew can range from one to over five people. If backyard pickup is offered, a larger crew size is needed because the crew must service cans that might be at some distance from the collection vehicle. Curbside pickup requires a smaller crew, and, of course, fully automated systems require only one person. Studies have shown that the greatest overall efficiencies can be attained with the smallest possible crews. For curbside refuse collection, three-person crews do not collect three times as much refuse as a one-person crew.

As a rough guideline, for most residential curbside collections, a single truck should be able to service between 700 and 1000 customers per day if the truck does not have to travel to the landfill. In one California community, an automatic collection system allows a truck to average about two and a half loads per day (10-hour shift) at 10 tons per load.[4] Realistically, most trucks can service only about 200 to 300 customers before the truck is full and a trip to the landfill is necessary.

EXAMPLE 3-3

Suppose a crew of two people requires 2 minutes per stop, at which the crew can service four customers. If each customer generates 56 lb of refuse per week, how many customers can crew members service if they did not have to go to the landfill?

SOLUTION

A working day is 8 hours, minus breaks and travel from and to the garage—say 6 productive hours, $6 \times 60 = 360$ minutes. At 2 minutes per stop, a truck should be able to make 180 stops and service $180 \times 4 = 720$ customers.

(Note, however, from Example 3-2, that the truck has to go to the landfill after only 178 customers or fewer still if its wheel loading is exceeded for the streets!)

An organized way of estimating the amount of time the crew actually works in collecting refuse is to enumerate all of the various ways crew members spend time. The total time in a workday can be calculated as

$$Y = a + c(b) + c(d) + e + f + g$$

where

Y = the total time in a workday

a = time from the garage to the route, including the marshaling time or that time needed to get ready to get moving

b = actual time collecting a load of refuse

c = number of loads collected during the working day

d = time to drive the fully loaded truck to the disposal facility, deposit the refuse, and return to the collection route

e = time to take the final (not always full) load to the disposal facility and return to the garage

f = official breaks including time to go to the toilet
g = other lost time such as traffic jams, breakdowns, etc.

All variables, of course, have to be in consistent time units, such as minutes.

Such an analysis, while it may not be very useful for calculations, strikingly demonstrates that a working day is not the same as the time spent collecting solid waste. If the value of b in the equation (the amount of time the crew actually spends collecting refuse) is known, the number of customers served by that truck and crew can be estimated.

If a region is fairly homogeneous, travel times (d in the previous equation) can be estimated by driving representative routes and generalizing the data. Once sufficient travel-time data are available, the data can be regressed against the "crow-fly" distance. One such regression[5] for New Jersey resulted in the expression

$$d = 1.5D - 0.65$$

where

d = actual one-way travel time, min
D = one-way travel time as the crow flies, assuming different truck speeds along the route, min

EXAMPLE 3-4

A truck is found to be able to service customers at a rate of 1.25 customers per minute. If crew members find that the actual time they spend on collection is 4 hours, how many customers can be served per day?

SOLUTION

If the crew can service 1.25 customers in one minute, what can crew members do in 4 × 60 minutes?

$$\frac{1.25}{1} = \frac{X}{4 \times 60}$$

X = 300 customers per day

If the number of customers that a single truck can service during the day is known, the number of collection vehicles needed for a community can be estimated by

$$N = \frac{SF}{XW}$$

where

N = number of collection vehicles needed
S = total number of customers serviced
F = collection frequency, number of collections per week
X = number of customers a single truck can service per day
W = number of workdays per week

EXAMPLE 3-5

Calculate the number of collection vehicles a community would need if it has a total of 5000 services (customers) that are to be collected once per week.

SOLUTION

A single truck can service 300 customers in a single day and still have time to take the full loads to the landfill. The town wants to collect on Mondays, Tuesdays, Thursdays, and Fridays, leaving Wednesdays for special projects and truck maintenance.

$$N = \frac{SF}{XW} = \frac{5000 \times 1}{300 \times 4} = 4.2 \text{ trucks}$$

The community will need five trucks.

3-1-4 Phase 4: Truck Routing

The routing of a vehicle within its assigned collection zone is often called *micro-routing* to distinguish it from the larger-scale problems (phase 5) of routing to the disposal site and establishing the individual route boundaries. The latter problem is commonly known as *macrorouting* or *districting* and is discussed later.

The present question is how to route a truck through a series of one- or two-way streets so that the total distance traveled is minimized. Put another way, the objective is to minimize *deadheading*, which is passing a collection point again after a previous pickup. The assumption is that if a route can be devised that has the least amount of deadheading possible, it is the most efficient collection route.

The problem of designing a route to eliminate all deadheading was actually addressed as early as 1736. The brilliant mathematician Leonhard Euler was asked to design a route for a parade across the seven bridges of Königsberg, a city in eastern Prussia, such that the parade would not cross the same bridge twice but would end at the starting point. The problem is illustrated in Figure 3-8 together with a schematic diagram. The routes are shown by lines called *links*, and the locations are known as *nodes*. The system shown has four nodes and seven links.

Euler not only proved that the assignment was impossible, but he generalized the two conditions that must be fulfilled for any network to make it possible to traverse a route without traveling twice over any road. These two conditions are

1. All points must be connected (one must be able to get from one place to another).
2. The number of links to any node must be of an even number.

The first condition is logical. The second similarly makes sense, in that if one travels to a location such as island **A** or **B** in Figure 3-8, one must be able to get off again—hence two roads. Euler's parade problem had all the nodes with an odd

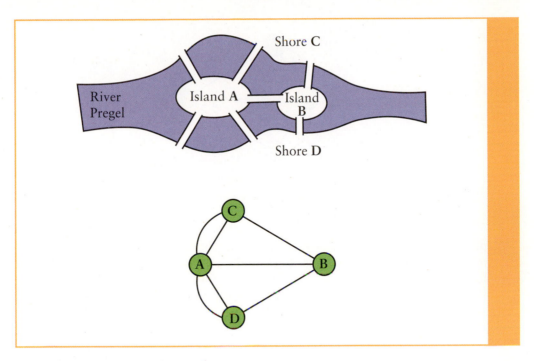

Figure 3-8 The bridges of Königsberg.

number of links—a clearly impossible situation. The number of links connecting a node designates its degree, and the existence of any odd-degree nodes in a system indicates that a route without deadheading is impossible. A system that has all nodes of even degree is known as a *unicoursal network*, and an *Euler's tour* is theoretically possible.[6]

In the real world, one-way streets, dead ends, and other restrictions can often make a practical application of the theoretical analysis difficult. One-way streets can be considered in Euler's theory by recognizing again that one must be able to get to a node exactly as many times as one leaves it. Thus a node with three one-way streets leading to it and a single one-way street leading away from it immediately makes a network nonunicoursal, even though the number of links at that node is even.

The development of a least-cost route involves making a system unicoursal with the least number of added links. For example, the Königsberg bridge problem would require only two additional (deadhead) links to make the system unicoursal (Figure 3-9). With this system, a theoretical Euler's tour exists, and the problem is now one of finding the proper route.

Kwan[7] has provided a means of achieving the most efficient unicoursal network (and also provided the name for this procedure: the *Chinese postman problem*) by observing that networks are really a series of loops where each node appears exactly once. By minimizing the additional connecting links (deadheads) necessary to achieve a unicoursal system, one can in fact achieve an overall optimum system. For example, the unicoursal network of the Königsberg bridge

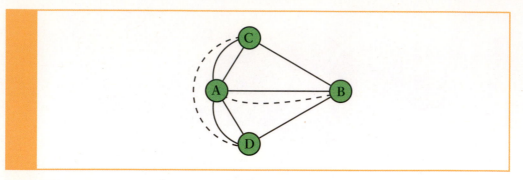

Figure 3-9 One possible route for the king's parade.

problem shown in Figure 3-9 is clearly a poor choice. (A new bridge is required!) It would make much more sense to trade the two deadheads shown in Figure 3-9 for the two in Figure 3-10. The latter is an obviously more efficient solution. The skill of the route planner must come into play in such tradeoffs, since a shorter street with many traffic problems, in fact, may be a more expensive alternative to a longer but clear street.

Once a unicoursal network has been designed, it remains to route the truck through this network. The method of **heuristic** (commonsensical) **routing** has found wide application.[8] The following set of rules apply to microrouting. Some of these are pure commonsensical judgment, and some are useful guidelines for determining overall strategy when planning a network.

1. Routes should not overlap, should be compact, and should not be fragmented.
2. The starting point should be as close to the truck garage as possible.
3. Heavily traveled streets should be avoided during rush hours.
4. One-way streets that cannot be traversed in one line should be looped from the upper end of the street.
5. Dead-end streets should be collected when on the right side of the street.
6. On hills, collection should proceed downhill so that the truck can coast.

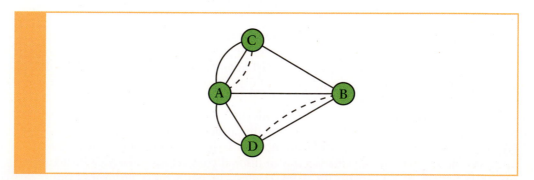

Figure 3-10 An alternative route for the parade.

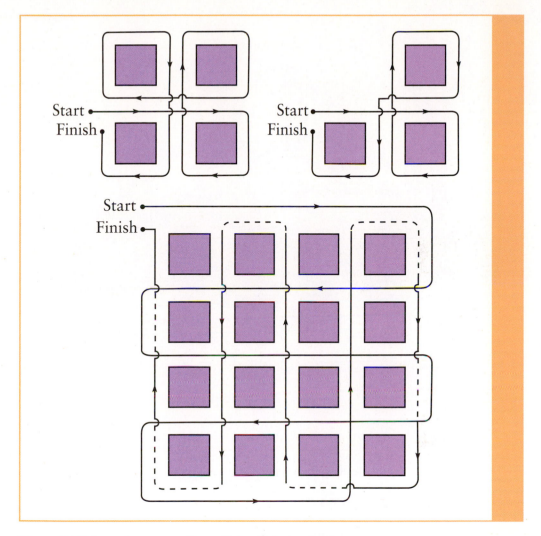

Figure 3-11 Large loops usually result in efficient collection.

7. Clockwise turns around blocks should be used whenever possible.
8. Long, straight paths should be routed before looping clockwise.
9. For certain block patterns, standard paths, as shown in Figure 3-11, should be used.
10. U-turns can be avoided by never leaving one two-way street as the only access and exit to the node.

These rules can be used to develop effective routes with minor deadheading. Figure 3-12 is an example of some of these routing rules applied to a large area. Elegant computer programs have been developed by a number of researchers, but in practice, it has been found that the tours constructed manually are almost always better than those done by mechanical tour-building codes.[7]

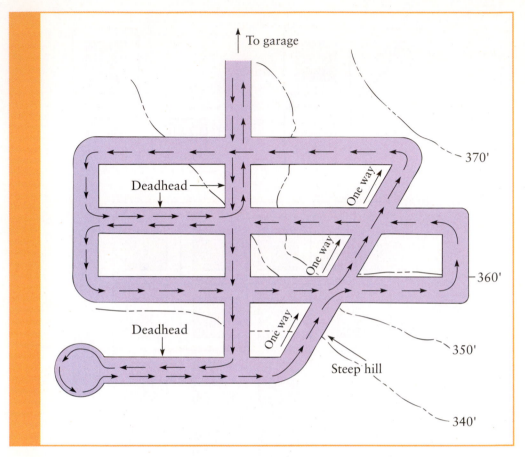

Figure 3-12 A sample routing for a collection truck.

3-1-5 Phase 5: Truck to Disposal

For smaller isolated communities, the macrorouting problem reduces to one of finding the most direct road from the end of the route to the disposal site. For regional systems or large metropolitan areas, however, macrorouting in terms of developing the optimum disposal and transport scheme can be used to great advantage. The available techniques, called *allocation models*, are all based on the concept of minimizing an objective function subject to constraints, with linear programming being the most common technique.

The simplest allocation problem is the assignment of solid waste disposal to more than one disposal site. Often the solution is obvious—the closest sources are allocated first, followed by the next closest, and so on. With more complex systems, however, it becomes necessary to use optimization techniques. The most appropriate one is the transportation algorithm, which is a type of linear programming. This technique is illustrated in Appendix 3-7.

3-2   COMMERCIAL WASTES

Commercial solid waste is almost always collected with *dumpsters*, which are large steel or plastic containers that are commonly lifted overhead by a front loader collection truck (Figure 3-13). Dumpsters range in size from 1.5 to 8 cubic yards, with the most common sizes being 3 or 4 cubic yards. As with the can snatchers, the driver does not have to get out of the truck, which is both an advantage and a disadvantage. In both cases, the driver does not see what has been placed in the container. Hazardous or dangerous materials can be transferred to the truck and could cause dangerous situations in the landfill or combustor. At the landfill, the refuse in the full truck is then pushed out (Figure 3-14).

A recent innovation has been the development of the split container. In many older communities the dumpster is placed inside an enclosure. These

Figure 3-13 A dumpster used for commercial collection.

Figure 3-14 Dumpster collection truck being emptied at a landfill.

Figure 3-15 Split container for recyclables and garbage. (Courtesy William A. Worrell)

enclosures hide the dumpster from public view. Unfortunately the enclosures typically only have room for one dumpster. Thus it is difficult to add recycling service because there is no room for the recycling dumpster. The San Luis Obispo County Integrated Waste Management Authority solved this problem by creating a split dumpster. This dumpster has two separate locking compartments, which can be serviced by a standard front loader truck. When dumping garbage the recycling side is locked, and when dumping recycling the garbage side is locked. Special signs were developed to inform users about the split dumpster (Figure 3-15).

Roll-off containers (Figure 3-16) are commonly used on construction sites. The roll-off containers typically range in size from 10 to 40 cubic yards. The 10-yd^3

Figure 3-16 Roll-off containers. Source: Peter Cron/San Luis Obispo County Integrated Waste Management Authority.

(7.6-m³) units are used for demolition material, while the 40-yd³ (30-m³) units can be used as rural transfer stations for household garbage. The containers are pulled onto trucks that then take the full containers to the landfills. A special truck transports one roll-off container at a time to the landfill.

3-3   TRANSFER STATIONS

When the waste disposal unit is remote to the collection area, a *transfer station* is employed. At a transfer station, waste is transferred from smaller collection vehicles to larger transfer vehicles, such as a tractor and trailer, a barge, or a railroad car.

Transfer stations can be quite simple, or they can be complex facilities. The design of the facility is based on its intended use, with small transfer stations typically relying on a tipping floor where collection vehicles drop their loads. Waste then can be loaded into open-top trailers using a wheeled loader. More complex facilities might employ pits for vehicles to dump into. Transfer vehicles then can be loaded by using a compacting unit. A facility also might have a tunnel for the transfer vehicle to drive into. Chutes or an opening in the floor would allow these units to be loaded by having waste pushed over the edge into them. Some typical transfer stations are shown in Figure 3-17.

Transfer vehicles can be as large as 105 cubic yards (80 cubic meters). The total weight of a transfer vehicle is limited by allowable wheel loads up to 80,000 lb (36,000 kg). In addition, many states also have weight limits on each axle. Some transfer stations employ scales so that trailers can be loaded to their maximum payload weight. The available payload is usually about 40,000 lb (18,000 kg). Transfer vehicles are constructed from either steel or aluminum, which can affect the available payload.

One of four methods is used to unload the transfer trailer:

1. *Live bottom*, or *walking floors*, on the floors of the vehicles. The back-and-forth movement of the longitudinal floor sections causes the refuse to be pushed out of the trailer.
2. *Push blade*, similar to the blade in a packer truck. A telescoping rod pushes a blade from the front of the trailer, which forces the waste out.
3. *Drag chains*. Some vehicles have chains on sprockets that go from the front to the back of a trailer, and by pulling on the chain, the refuse is dragged out of the vehicle.
4. *Tipper*. Some units have no unloading mechanism, and a large tipper at the landfill lifts the entire transfer vehicle up at an angle, causing the door to open and the refuse to slide out.

Transfer stations can also transfer refuse to trains or even barges. In New York City, solid waste was transferred onto barges, which then moved the refuse to Fresh Kills Landfill on Staten Island until the landfill closed in 2001. In Seattle, solid waste is transferred to railroad shipping containers that are then placed on a rail car for shipment to eastern Oregon. Durham, North Carolina, sends all of its refuse by railroad to southern Virginia. Los Angeles is also planning to ship garbage by

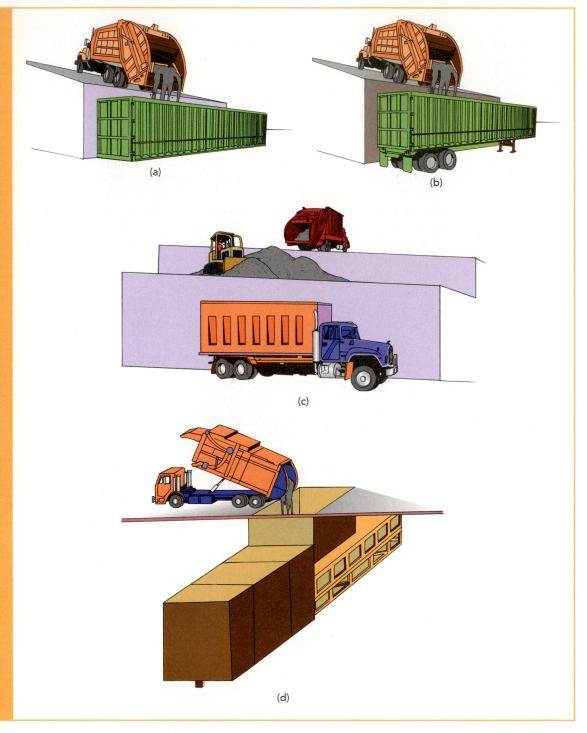

Figure 3-17 Several typical transfer stations. (a) Dump to container. (b) Dump to trailer. (c) Store and dump to truck trailer. (d) Dump to compactor.

rail to a remote landfill in Imperial County, California. On a much smaller scale, residents in rural areas such as the unincorporated area of San Diego County bring their waste to transfer stations with 40-yd^3 (30-m^3) roll-off bins, which are then hauled to the disposal site.

The decision to build or not to build a transfer station is often an economic or regulatory decision. If the one-way haul distance from the point of the full-collection vehicle to the discharge point is short, then it is likely that no transfer station is needed. On the other hand, if the discharge point is far away and the collection vehicle will have to be away from its primary role of collecting refuse for too long, then a transfer station might be warranted. The relationship is illustrated in Figure 3-18. Where the two curves cross is the breakeven point. Longer distance will warrant the construction of a transfer station, while shorter hauls will make it uneconomical.

In deciding if a transfer station is economical, the cost of direct haul must be compared to the cost of a transfer station. Direct haul is a variable cost based on the number of miles that have to be traveled. A transfer station also has a variable cost based on the transfer vehicle mileage but also has a fixed cost based on the capital and operating cost of the transfer station. The variable cost for direct haul and transfer vehicle is typically expressed as a cost per mile. The fixed cost for a transfer station is based on cost per ton.

As an example, if it is 40 miles round trip to a landfill and it costs $1.50 per mile to drive a garbage truck carrying 5 tons of garbage, the cost would be $12 per ton. If a transfer vehicle carrying 20 tons of garbage cost $2 per mile and the fixed cost at the transfer station is $5 per ton, the cost would be $9 per ton. Thus, a transfer station would be more economical.

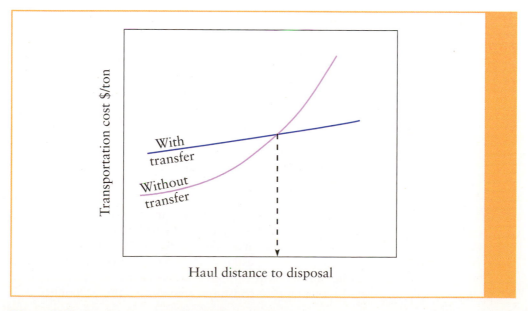

Figure 3-18 Breakeven point of transfer stations.

Sometimes a transfer station is required regardless of distance to a landfill. To minimize the traffic and air pollution impacts at a landfill, a permit may limit the landfill to only receiving waste from transfer stations. This significantly reduces the number of vehicles travelling to a landfill.

In situations where the sophistication of linear programming models is not warranted, a brute-force technique using a simple grid system can be of value. In this case, the region is divided into equal blocks on an $X-Y$ grid, and the solid waste generation is then estimated based on population. The sites for transfer stations and final disposal facilities are initially screened to eliminate obviously inadequate areas (e.g., urban areas for landfills). Trial-and-error siting of facilities is then used to obtain the most reasonable combination of solid waste disposal facilities.

3-4   COLLECTION OF RECYCLABLE MATERIALS

Recycling entered the mainstream of solid waste management in the 1990s. No longer was recycling conducted by underfunded, idealistic individuals, but rather, multinational garbage companies were now involved. Recycling, regardless of the price paid for recycled material, could be profitable. While some long for the days of the idealistic recycler, no one can dispute that more material is being recycled today. In 2008, approximately one-third of the MSW being generated is recycled.

Two factors have caused recycling to succeed. First, government has provided leadership in the area of waste reduction. Over 40 states have adopted waste reduction goals, and some states have set mandatory goals, with noncompliance resulting in fines of up to $10,000 per day. Second, the public has accepted the concept that recycling is not free. Just as residents pay for garbage service, they also now pay for recycling. Over 8600 curbside recycling programs have been implemented, resulting in 86.2 million tons—or 34.5% of all solid waste—being recycled. Table 3-1 highlights the progress of recycling in the United States.

Some groups, such as the Grassroots Recycling Network, see this as only the beginning and are calling for a zero waste goal.[11]

> To waste, to destroy our natural resources, to skin and exhaust the land instead of using it so as to increase its usefulness, will result in undermining in the days of our children the very prosperity which we ought by right to hand down to them amplified and developed.
>
> —Theodore Roosevelt, Seventh Annual Message, December 3, 1907

Progressive communities, such as San Francisco, have adopted a zero waste goal. The California Integrated Waste Management Board has embraced the zero waste goal and uses the slogan, "Zero Waste—You Make It Happen."

A major barrier to the increasing expansion of recycling is finding markets for the material. Recycling cannot occur without markets for the recycled material, and markets are created by using post-consumer material in the manufacture of a product.

Table 3-1 Collection of Recyclables, 2012

Material	Recovered	
	(millions of tons)	(percent of generation)
Paper and paperboard	44.4	64.6
Glass	3.0	27.7
Ferrous metals	5.6	33.0
Aluminum	0.7	19.8
Other nonferrous metals	1.4	68.0
Plastics	2.8	8.8
Rubber and leather	1.4	17.9
Textiles	2.3	15.7
Wood	2.4	15.2
Other materials	1.3	28.3
Food waste	1.7	4.8
Yard trimmings	19.6	57.7
Total recycled MSW	86.6	34.5

Source: [25]

Producers are being urged to take responsibility for the products they produce to ensure that the material is either recyclable or that it uses post-consumer recycled material. For example, the soft drink industry sells its products in aluminum cans that easily can be recycled, and these cans already contain a high percentage of recycled aluminum. On the other hand, the industry also sells products in PETE plastic containers that are much more difficult to recycle and are rarely used in the manufacture of new bottles (unlike in Europe and Australia, where the soft drink industry uses plastic containers with recycled plastic).

Because the recycling industry is still young, there has not been time to settle into a standard system that has a high probability of succeeding in every application. Each community has almost an infinite number of options in how to design its recycling program.

Many communities have switched to having their citizens place all the recyclable material into one container. A separate truck takes the recyclable to a materials recovery facility (*clean MRF*) for processing. Because the recyclables are not separated by commodity but instead placed commingled into one container, this is called *commingled* collection of recyclable materials (see Figure 3-19). The MRF for processing commingled recyclable materials (see Figure 3-20) is much smaller than a MRF that must process both recyclable and non-recyclable waste (*dirty MRF*). However, the tradeoff is that another truck must travel the same route to pick up the garbage. In addition, the residents must separate the recyclable materials and place them in a container for this option to work.

Some jurisdictions have used one truck to pick up both solid waste and commingled recyclables in a dual-compartment packer truck (Figure 3-21). The truck has two lifters and is fitted with dual compartments: one side for recyclable materials and one side for solid waste. The contents of the two compartments

Figure 3-19 Commingled recyclables on the receiving floor. (Courtesy William A. Worrell)

Figure 3-20 Storage bays for the separated recyclables. (Courtesy William A. Worrell)

Figure 3-21 (a) Dual-compartment packer truck, one side for garbage and one side for commingled recyclables. (b) Two tippers, one for the garbage side and one for the recyclables side. (Courtesy William A. Worrell)

remain split during the dumping process, and the mixed refuse then can be compacted to achieve a full load. The advantage of this system is that only one truck and driver is needed to collect both the solid waste and recyclable materials. However, the amount of material collected must be balanced so that both compartments fill at the same rate. If only one compartment is full and the other is not, the system loses its advantage. Because the truck now has to discharge at two locations, the landfill and MRF also must be near each other to maintain the efficiency of this option.

Another option for collecting commingled recyclable material is the *blue bag system* in which the recyclables are placed in a special blue bag provided to or by each customer. The same truck that picks up the solid waste also picks up the blue bags in a separate compartment. The bags are removed manually at the transfer station and the contents processed separately from the solid waste.

One disadvantage of any system that collects commingled recyclables is the potential contamination of the paper products by residual liquids in the glass, aluminum, or plastic containers. In addition, the sorting costs for commingled recyclables can be high. In the 1990s communities initiated collection programs where the residents separated the materials and placed them into two or three containers. For example, bottles and cans go in one container, newspaper in a second, and mixed paper in the third. This eliminates the need to sort newspaper, which is the most common recyclable (by weight) and shifts the cost of sorting to the waste generator. This system requires trucks with multiple compartments (Figure 3-22). While this method has a low initial cost, the participation rate may be low. In the past 10 years, as more MRFs were built, many communities switched to the single commingled waste wheeler for recyclables. When Los Angeles switched from this method to a commingled container, the recycling rate increased by 40%.[12]

Yard waste also typically is placed in a separate container. These yard waste programs are being expanded to include food waste and in some cases paper and compostable bags and utensils. A third truck is needed to collect the yard waste and take this material to a mulching or composting operation.

In addition to these items, some communities have implemented curbside used motor-oil collection. Residents are allowed to place their used motor oil at the curb, where it is collected to be recycled into new oil. Other communities have tried curbside collection of batteries, paint, and other products.

Both experience and mathematical modeling suggest that collection of commingled recyclables on a per-ton basis is more expensive than the collection of mixed refuse. As would be expected, the cost on a per-ton basis decreases with higher participation rate and conversion to a single stream automatic collection system. If the income from the sale of recycled materials is taken into account, however, the cost of collecting recycled items is not significantly different from the cost of collecting mixed, unsorted refuse.

A curbside recycling program can divert between 15% and 30% of the residential waste and costs between $2 and $5 per month per customer, depending on various options. An organics program can divert between 15% and 30% of the residential waste and also costs between $2 and $5 per month. One study concluded that the cost of a typical suburban recycling program is between $114 and $120 per ton of material collected, based on a 50% set-out rate.[13]

Everett et al. have developed a model that estimates the cost of collecting partially separated (commingled) recyclables at curbside.[14] The model estimates the time to collect such waste and to sort it at the truck into various individual components. Three variables in the model are travel time, sorting time, and waiting time (at stop lights, for example).

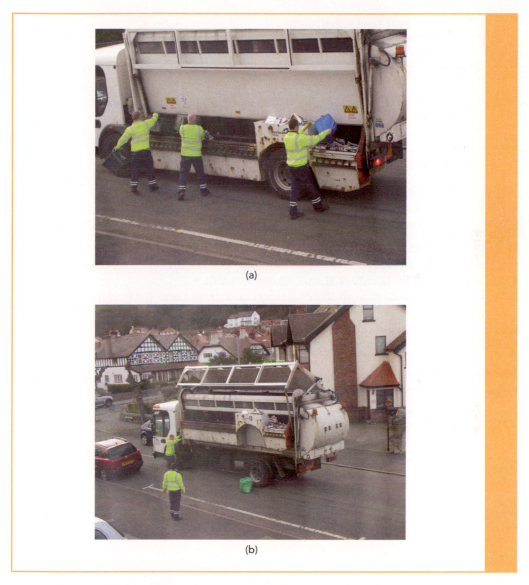

Figure 3-22 (a) Workers place different recyclables in separate bins (b) Bins are dumped into separate compartments in the truck. (Courtesy William A. Worrell)

Travel time is estimated as

$$T = \frac{D}{L(1 - e^{-kD})C}$$

where

T = travel time between two consecutive stops, seconds

D = travel distance between two consecutive stops, m (ft)

Table 3-2 Travel Time Coefficients

Crew Size	L (km m^{-1})	k (m^{-1})
One person	18.8	2.2×10^{-2}
Two persons	24.3	1.7×10^{-2}

Source: Based on Resources, Conservation and Recycling 22, Everett, S. Maratha, R. Dorairaj, and P. Riley, "Curbside Collection of Recyclables I: Route Time Estimation Model," Pages No. 177–192.

L = calculated coefficient, km h^{-1} (mile h^{-1})
k = calculated coefficient, m^{-1} (ft^{-1})
C = conversion factor, 1.467 for American standard units, 0.278 for SI units

The factors L and k vary with crew size, the maximum speed attained by the vehicle, the acceleration, driver ability, and road conditions. The values of L and k, as estimated by Everett et al., are shown in Table 3-2.

The time to walk to the containers and bring them back to the truck can be estimated as

$$W = 0.86A$$

where

W = walk time, seconds
A = average walk distance at a single stop, m

The sorting time is estimated as

$$S = 21.3 + 2.4B$$

for a one-person crew, and

$$S = 23.2 + 1.8B$$

for a two-person crew, where

S = average sorting time of a single set-out, seconds
B = average set-out amount, kg

On average, the set-out amount for a typical curbside program varies between 0.7 and 12 kg.

Finally, the truck has to navigate the streets, and if the traffic is heavy and/ or if there are many stop signs and traffic lights, there could be considerable wait time. This can be estimated as

$$E = M_S N_S + M_L N_L$$

where

E = average wait time at traffic lights and stop signs, seconds
M_S = mean time spent at stop signs, seconds
N_S = number of stop signs
M_L = mean time spent at traffic lights, seconds
N_L = number of traffic lights

The model (with these variables) has been found to predict the time for collection within 10% of the actual value.[14]

Whenever a curbside program is initiated in a community, the leaders want to know how successful it has been. Success can be measured in many ways, including these:

- Fraction of refuse components diverted from the landfill, tons/week
- Fraction of households participating in the program; participation can be defined in many ways, such as having set out recyclables at least once a month
- Fraction of households participating on any given week
- Profit from the sale of recyclables, $/week

All of these methods are useful, but they have to be well defined and used consistently.

In the world of MSW, the holy grail is to be able to weigh how much refuse each household or business generates. To determine how much refuse is generated, the weight of each container must be measured every time the container is emptied. If a household has three containers, one for garbage, one for recyclables, and one for green waste, then each container must have a unique identification number (radio-frequency identification (RFID) chip) and the associated empty weight. This data base has to be developed for all the containers in the city. When the truck empties the container, the truck must be able to record the container number and the actual weight of the container. With the unique identification number and the full weight, it is now possible to determine how much refuse was collected.

The difficulty is that trucks must be retrofitted with a scale, identification number (RFID chip), reader, and recording device. After much difficulty one community retrofitted identification numbers on containers, developed a data base, and added the scale and other equipment on the truck. What was measured was that the green-waste container weighted on average 49.8 pounds, the garbage container 33.8 pounds, and the recycling container 18.4 pounds. However, the set-out frequency for the containers was the opposite of the weight, with recycling containers being set out most frequently, followed by garbage and green waste the least frequently. Finally, the average generation by weight for the households was 43% garbage, 30% recyclables, and 27% green waste. While this was one of the first attempts to quantify the amount of refuse being generated on the household level, one must be cautious in extrapolating from the results because of the small sample size.

3-5 LITTER AND STREET CLEANLINESS

Litter is a special type of MSW. It is distinct from other types of MSW in that it is a solid waste that is not deposited into proper receptacles. As one travels around the world, it is fascinating to see all the different receptacles developed for garbage, recyclables, or organic material (see Figure 3-23). Unfortunately, if the public does not use these receptacles, litter is generated.

We usually think of litter as existing in public places, but litter could be on private premises as well. Although litter is usually considered to be a visual affront only, it also may be a health hazard. Broken glass and food for rats are but two

Figure 3-23 A collage of public receptacles. (Courtesy William A. Worrell)

examples. Litter is also a drain on our economic resources, because the public must pay to have it collected and removed when it is on public property.

The collection of litter is of secondary importance to a community, because it does not represent a critical public service as do police and fire protection, water treatment, and collection of refuse from residences and commercial establishments. Litter removal is expensive, costing municipalities in the United States and Europe billions annually.

The composition of roadside litter can vary considerably from place to place, as can the method of data collection. One major problem with any litter data analysis is that the reports fail to specify the guidelines used in the collection and identification of litter and seldom specify the way in which the percentages of the various components were calculated. For example, a broken bottle can be counted as either one item or many items, depending on the guidelines. Similarly, the results can be calculated as a percent of the total of each of the following items:

- Items by actual count
- Total weight of litter
- Volume of litter
- Visible items by actual count

Because of the lack of a standard counting technique, the following guidelines for conducting litter studies are suggested:[15]

1. Count as one item all pieces larger than 2.5 cm (1 in.). This count includes removable tabs from beverage cans.
2. Do not count rocks, dirt, or animal droppings.
3. Count as one item all pieces of any item clearly belonging together, such as a broken bottle. Otherwise, count each piece of glass, newspaper, and so on singly.
4. Do not count small, readily decomposable material, such as apple cores.
5. For roadside litter surveys, measure all items within the officially designated right of way.
6. Empty liquids out of all bottles and cans before collecting them.

The litter survey, if conducted along a road, should be started by driving along the road at a slow speed and having a passenger record the visible items into a tape recorder for future transcription. Next, the litter is identified, recorded, and manually collected. The items should be separated during collection into as many components as feasible. The collected items are then weighed and the volume measured. The relationship between visible items and total items along a roadside is shown in Figure 3-24. It is interesting that along fairly clean roads, the visible fraction is only about 6% of the total litter count! These data also confirm that a large fraction of our visible litter is bottles and cans.

For community litter surveys, the photometric technique developed for Keep America Beautiful, Inc. (KAB) may be used.[16] The blockfaces of a community are first numbered, and a preliminary sample size is established. About 5% of the blockfaces are usually adequate. Using the random number technique, the blockfaces and the locations on those blockfaces to be measured are selected. As shown in Figure 3-25, a marker is located in the front center of the survey area, and a chalkboard is used to identify the location and date. To facilitate the counting of litter from the developed photographs, a picture is taken of a clean pavement laid out with white marking tape in a 1 ft grid, 6 ft wide × 16 ft long, which is parallel to a street curb. A transparency of the grid is prepared, and the resulting 96-square grid is placed on top of each litter photograph. The litter is then counted and classified using a magnifying glass. The first photographs are used for establishing the baseline litter conditions. The litter rating (L) is calculated for each picture (location) as the squares containing some litter compared to the total number of

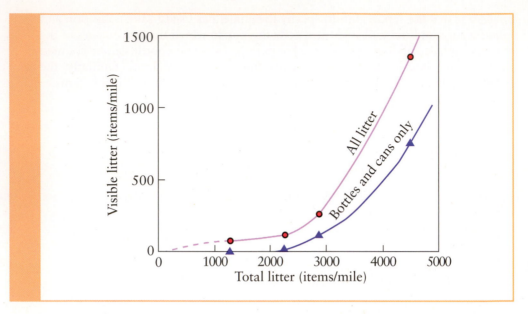

Figure 3-24 Results of a litter survey on a rural road. Source: Vesilind, P. A. 1976. *Measurement of Roadside Litter*. Durham, N.C.: Duke Environmental Center, Duke University.

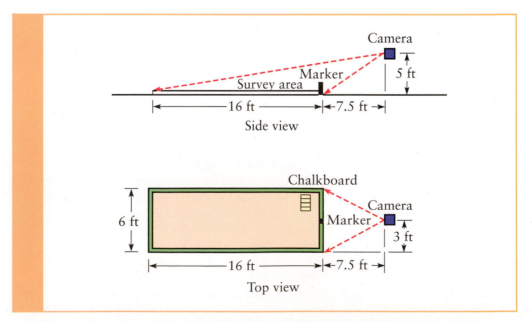

Figure 3-25 Keep America Beautiful litter measurement technique. Source: The Photometric Index. n.d. Stamford, Conn.: Keep America Beautiful.

squares (%). After initial baseline photographs are analyzed and the L is calculated, the number of sampling sites necessary can be calculated as

$$N = (22.8)S^2$$

where

N = sample size needed to make a 0.5-point difference between two average litter ratings (L) in an area, significant at the 90% confidence level

S^2 = variance of the litter ratings (L) of the initial photographs

The variance is calculated as

$$S^2 = \frac{\sum_{i=1}^{n} F(L_i)^2}{n-1} - \frac{\left(\sum_{i=1}^{n} FL_i\right)^2}{n(n-1)}$$

where

L_i = litter rating of the ith photograph
n = total number of photographs
F = frequency, or the number of photographs with any one L

EXAMPLE 3-6

Suppose that a town has 600 blockfaces and that a 5% sample, or 30 blockfaces, is photographed as explained previously. The litter ratings (L) are as shown:

	Number of Photographs		
L	F	FL	FL^2
1	6	6	6
2	3	6	12
3	1	3	9
4	4	16	64
5	1	5	25
		36	116

How many blockfaces need to be photographed for a litter survey? The S^2 is calculated as

$$S^2 = \frac{116}{30-1} - \frac{(36)^2}{30(29)} = 2.5$$

SOLUTION

If $S^2 = 2.5$, the number of sampling sites necessary is

$$N = 22.8 \,(2.5) = 57$$

In other words, $57 - 30 = 27$ more sites are needed in order to have a statistically satisfactory baseline.

Litter theoretically can be controlled by cognitive, social, and technological means. A cognitive solution would be convincing people not to litter; a social solution would be depriving the public of items that might become litter or fining people who litter heavily if they are caught; and a technical solution would be simply cleaning up after littering has occurred.

The first option demands an explanation of why people litter, a question requiring studies on the psychology of litterers. In one study,[17] the actions of 272 persons were observed when they bought a hot dog wrapped in paper. Of interest was the final deposition of the wrapper. Ninety-one people chose to dispose of the wrapper improperly (they littered). The probability of any one person littering, based on this sample, could be calculated as

$$E = 0.019 + 0.414(A) + 0.1654(C) + 0.1532(D)$$

where

E is the probability that a person would litter
A, C, and $D = 0$

except that

$A = 1$ if the person is 18 years old or younger
$C = 1$ if there are no trash cans conveniently located
$D = 1$ if the area is already dirty with litter

From the study, it is clear that age is quite important—younger people being much more likely to litter than are older persons. There was no statistical difference between 19- to 26-year-olds and persons older than 26 years. Gender was found to be statistically insignificant. Because the study was conducted in 1973, its validity to today's urban populations may be questionable. Intuitively, however, the role of younger persons as the major contributors to urban litter remains valid.

EXAMPLE 3-7

Calculate the probability of a 40-year-old person littering a dirty street that has no convenient trash cans.

SOLUTION

$$E = 0.019 + 0 + 0.1654 + 0.1532 = 0.33$$

That is, of 100 people answering that description, 33 would probably litter the street.

Such studies yield clues as to how persons might be induced not to litter (e.g., put out more trash cans and clean up the street) and who the target population is (e.g., young people).

Other psychology studies about litter have been directed at finding out what motivates people not to litter. In one study,[18] movie theater patrons during Saturday matinees were asked by several means not to litter the theater. The total quantity of litter was then measured and used as an indicator of the success of that control approach. The results showed that measures such as personal exhortation for cooperation and anti-litter cartoons had no effect on litter, but that payment of money for pieces of litter at the end of the showing resulted in about a 95% reduction in litter. The clear indication is that self-interest, such as placing a substantial deposit on beverage containers, is an effective force in convincing people not to litter.

The second method of litter control is to prevent items that might become litter from ever reaching the consumer. In the earlier example of the hot dog wrapper, it would seem reasonable to suggest that 100% litter-free results could be obtained by not giving customers a paper wrapper around their hot dog. The banning of single use plastic bags is a practical means of controlling this type of litter.

The third method of litter control is to clean up the mess once it has occurred. This system is commonly used in sports stadiums and other public areas where no effort is made to ask people to properly dispose of their waste. For roadside litter, it seems that the most economical litter control alternative is actually frequent cleanup.

An effective means of litter control is to enlist the help of the community by having organizations "adopt" a section of roadway. Organizations as varied as church groups, Rotary Clubs, sports teams, and even private businesses have agreed to keep sections of roadways clean by conducting periodic litter pickups. This method of litter control not only keeps roadways and streets cleaner, but it also brings the litter problem down to a personal level. Anyone who has contributed a Saturday to the hot and dirty job of collecting litter along the roadway will not throw trash out the car window and will be critical of those who do.

Attempts also have been made to design mechanical litter collection machines. One towed device has proven both inexpensive and effective. It works by having a series of rotating plastic teeth that fling the litter into a collection basket (much like a leaf collector connected to a lawn mower).[19] A more sophisticated and ambitious unit uses a vacuum arm on a truck to suck up the roadside litter.

Finally, street cleanliness can be negatively affected by the very people who collect the household refuse. In a study at the University of Florida, litter was measured using the KAB method on typical residential streets before and after garbage collection.[21] An almost 300% increase in litter was noted. The collection supervisor believed the reason was twofold. First, he believed that his automated collection vehicles were poorly designed and that they were not always able to get everything into the truck. The second reason was that the city had initiated a volume-based refuse collection charge. He suggested that the residents had been practicing the "Seattle Stomp"—a tricky two-step that originated in Seattle when that city switched to volume-based collection. Residents had figured they could get more refuse into a can if they stomped on the refuse and packed it into the can. This led to the garbage being stuck in the can and to the potential for spillage during the transfer to the truck.

3-6   FINAL THOUGHTS

In "the good old days," the garbage man came twice a week, went to your back-yard, and collected all your garbage. Today, some communities have three different-sized garbage cans, three recycling containers, a yard waste container, and a used oil container. Collection may involve up to 3 fully automated vehicles, driven by a man or women who collects these materials using a joystick instead of lifting a 70 pound can. The instructions on how to put your garbage out are almost as complicated as those for operating your cell phone. Yet, with all this complexity, it is working. Why? Because many people have an environmental ethic and recognize the need for integrated waste management. In 2012, only 40% of the United Stated population voted, while at the same time 58% routinely recycled.

The limits to recycling are still unknown. Learned people are debating whether it is possible to even think about recycling all of the materials we consider waste. Most likely, fundamental principles of physics[22] and thermodynamics[23] and mass transfer[24] make the goal of 100% collection of materials impossible. It is quite likely that, when the environmental effects of recycling are compared to the impact of disposal options, there will be some point at which more recycling will actually have increasingly detrimental effects on energy use and materials conservation. At the present time, however, one of the major objectives in solid waste engineering is to effectively and efficiently collect all of the materials people no longer want. Because the collection of municipal solid waste accounts for between 50% and 75% of the total cost of refuse management, it is important that solid waste engineers properly design collection systems.

Finally, it is important to recognize that garbage collection is still a labor-intensive job. Any system developed is going to be only as good as the men and women who operate it. As one solid waste director of a major city said, "There is nothing more beautiful than watching my fleet of one hundred garbage trucks leaving the yard silhouetted against the rising sun."

3-7   APPENDIX: DESIGN OF COLLECTION SYSTEMS

Systems analyses can be used to design collection systems so as to minimize the cost of such systems. Consider the simple system pictured in Figure 3-26. The waste generated at four sources (denoted by centroids of the collection area, which is a poor assumption, especially if the disposal sites are close to the collection routes) is to be allocated to two disposal sites. The objective is to achieve this in a minimum-cost manner.

At the same time, several requirements must be met (constraints in an optimization model).

1. The capacity of each disposal site (e.g., a landfill) is limited.
2. The amount of refuse disposed of must equal the amount generated.
3. The collection route centroids cannot act as disposal sites, or the total amount of refuse hauled from each collection area must be greater than or equal to zero.
4. Total cost equals hauling cost plus disposal cost.

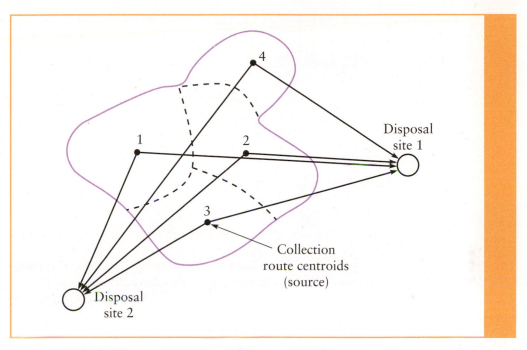

Figure 3-26 Elements of a simple solid waste management system.

The following notation is adopted:

x_{ik} = quantity of waste hauled from source i to disposal site k, per unit time
c_{ik} = cost per quantity of hauling the waste from source i to disposal site k
F_k = disposal cost per waste quantity at disposal site k (capital plus operating)
B_k = capacity of disposal site k, in waste quantity per unit time
W_i = total quantity of waste generated at source i, per unit time
N = number of sources i
K = number of disposal sites k

The problem then boils down to minimizing the following objective function:

$$\sum_{i=1}^{N}\sum_{k=1}^{N} x_{ik}c_{ik} + \sum_{k=1}^{K}\left(F_k\sum_{i=1}^{N} x_{ik}\right)$$

subject to the following constraints.

Constraint 1. The sum of all the solid waste hauled out of each section of the community must be equal to or less than the capability of the disposal sites to receive that waste, or

$$\sum_{i=1}^{N} x_{ik} \le B_k \text{ for all } k$$

Constraint 2. The sum of all the waste hauled from a section of the community (to any disposal site) has to equal the amount generated in that section, or

$$\sum_{k=1}^{K} x_{ik} = W_i \text{ for all } i$$

Constraint 3. The waste hauled out has to be positive (not negative), or

$x_{ik} \geq 0$ for all i, k

The first term in the objective function is transportation costs, and the second term is disposal costs. For the case shown in Figure 3-22, the objective function is

$$\text{Minimize } [x_{11}c_{11} + x_{21}c_{21} + x_{31}c_{31} + x_{41}c_{41} + x_{12}c_{12} + x_{22}c_{22} + x_{32}c_{32}$$
$$+ x_{42}c_{42} + F_1(x_{11} + x_{21} + x_{31} + x_{41}) + F_2(x_{12} + x_{22}$$
$$+ x_{32} + x_{42})]$$

subject to the following constraints:

$x_{11} + x_{21} + x_{31} + x_{41} \leq B_1$
$x_{21} + x_{22} + x_{32} + x_{42} \leq B_2$
$x_{11} + x_{21} + x_{31} + x_{41} = W1$
$x_{12} + x_{22} + x_{32} + x_{42} \leq W2$
$x_{11} \geq 0, x_{12} \geq 0, \ldots, x_{23} \geq 0, x_{24} \geq 0$

This problem can be solved using any linear programming algorithm. The *transportation algorithm* is particularly useful for such applications.

EXAMPLE 3-8

Assume the solid waste generation and disposal figure for the system pictured in Figure 3-26 is as follows.

Source i	Generation, W_i (tonnes/week)	Cost of Transport, c_{ik}	
		To Site 1 ($/tonne)	To Site 2 ($/tonne)
1	100	5	12
2	130	7	5
3	125	4	8
4	85	13	6

Disposal Site, k	Capacity, B_k	Cost, F_k($/tonne)
1	450	4
2	200	6

SOLUTION

We can hand calculate the cost for any option. Suppose all of the waste was to be sent to disposal in site 1 ($k = 1$). The cost for delivering the refuse from the first source ($i = 1$) is

100 tonnes/wk × $5/tonne = $500/wk

Similarly for the other sections, the cost would be $910, $500, and $1105 per week or a total cost of $3015/wk.

The disposal cost is

440 tonnes/wk × $4/ton = $1760/wk

The total cost is therefore $4775/wk.

This is only one solution, and it might not be the least-cost solution. To find the least-cost solution (minimize cost), we can use the transportation algorithm and find the following:

From Source i	To Disposal Site k	Waste Hauled (tonnes/week)	Transport Cost ($/week)	Disposal Cost ($/week)
1	1	100	500	400
2	1	25	175	100
2	2	105	525	630
3	1	125	500	500
4	2	85	510	510

Therefore, the total system minimum cost is $4350/week, substantially less than if all the refuse were shipped to disposal site 1. Note that considerable capacity remains unused in landfill 2.

Systems in which transfer stations are used also can be optimized by systems analysis, using the scheme introduced previously. Figure 3-27 shows the same community with four sources of waste and two disposal sites, but now a transfer station is placed in the town. The trucks now have K disposal points and J intermediate facilities. As before, these facilities have processing costs and annualized capital, plus operating costs. F_j is the annual cost for the transfer stations, and F_k is the annual capital and operating costs for the disposal sites. The other variables are

c_{ij} = cost per quantity of hauling the waste from source i to intermediate facility j

c_{jk} = cost per quantity of hauling the waste from intermediate facility j to final disposal facility k

x_{ij} = quantity of waste hauled from source i to intermediate facility j, per unit time

x_k = quantity of waste hauled from intermediate facility j to final disposal facility k, per unit time

B_j = capacity of intermediate facility j, in waste quantity, per unit time

P_j = proportion of waste at intermediate facility j that, after processing, remains for disposal (P_j = 1.0 if the facility is a transfer station, but P_j = 0.2 if it is an incinerator)

J = number of intermediate facilities j

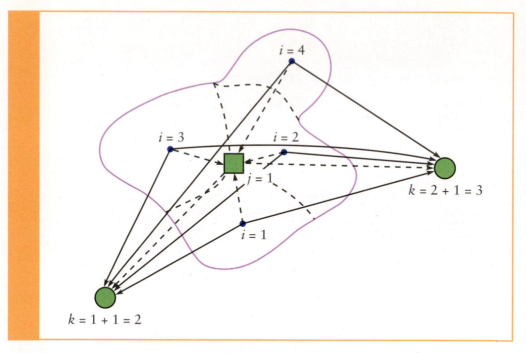

Figure 3-27 Adding transfer stations to the collection system plan.

The problem now is to minimize the following objective function:

$$\sum_{i=1}^{N}\sum_{j=1}^{J} c_{ij}x_{ij} + \sum_{i=1}^{N}\sum_{k=1}^{K} c_{ik}x_{ik} + \sum_{j=1}^{J}\sum_{k=1}^{K} c_{ik}x_{ik}$$

$$+ \sum_{j=1}^{J} F_j \sum_{i=1}^{N} x_{ij} + \sum_{k=1}^{K} F_k \sum_{i=1}^{N} x_{ik}$$

This objective function is subject to the following constraints:

Constraint 1. The quantity of waste generated at source i, W, must equal the sum of all the waste hauled from that source to the J intermediate sites and K disposal points.

$$\sum_{j=1}^{J} x_{ik} + \sum_{k=1}^{K} x_{ik} = W_i \text{ for all } i$$

Constraint 2. The capacity of the jth intermediate site, B_j, must be more than or equal to the total waste brought to it. If this constraint is omitted, the model can be used to determine the required capacity.

$$\sum_{i=1}^{N} x_{ij} \leq B_j \text{ for all } j$$

Constaint 3. B_k, the capacity of the final disposal site k (which might be influenced by the number of trucks a tipping floor or landfill site can handle, the compaction capacity at a landfill, and so on), should not be exceeded by the waste brought in directly from the collection sites or from the intermediate facilities.

$$\sum_{i=1}^{N} x_{ik} + \sum_{i=1}^{J} x_{jk} \leq B_k \text{ for all } k$$

Constaint 4. Whatever waste is shipped to an intermediate site must be shipped out to a disposal site. The proportion of waste that remains for disposal after any processing is denoted by P_j.

$$p_j \sum_{i=1}^{N} x_{ij} - \sum_{k=1}^{K} x_{jk} = 0 \text{ for all } j$$

Constaint 5. The nonnegativity constraints are

$$x_{ij} \geq 0; \ x_{ik} \geq 0; \ x_{jk} \geq 0 \text{ for all } i, j, k$$

3-8   APPENDIX: POTENTIAL SOLUTIONS TO THE PROBLEM OF LITTERING

Several Swiss towns have littering problems. Because tourism is an important economic sector in Switzerland and because the country is known for its cleanness, the littering problem has caught the attention of the authorities. The Swiss Federal Office of the Environment has noted that the greatest litter problem occurs in city centers.

This littering problem requires targeted actions at a local level, which makes it difficult to carry out national measures. In Switzerland various combinations of measures adapted to the local circumstances are being pursued, such as:

1. **Information and Awareness-Raising Campaign**
   In the daily struggle against littering, it is essential to keep the public informed and to raise awareness about the problems of littering and about the correct way to deal with potential items of litter such as cigarettes, take-out food packaging, drinks packaging, free newspapers, and chewing gum. This can be put into practice using advertising posters, newspaper advertisements, anti-litter ambassadors, or jointly organized paper rubbish campaigns in the community and parishes. Anti-litter posters can be ordered free of charge in all three national languages from the Interest Group for a Clean Environment (Interessengemeinschaft für eine saubere Umwelt—IGSU) or from the Summit Foundation. Communities and towns may also book, without cost and for a limited time, anti-litter ambassadors from the IGSU for litter-intensive events or cleanup of public places.

2. **Information and Education**
   One important measure against littering is education about the environment in schools, because the correct way to deal with waste cannot be learned early enough.

3. **Code of Behavior for the Retail Sector and Specimen Contract for Free Newspapers**
   With the Code of Behavior the retail outlets selling take-out food make a voluntary commitment to supporting the communities in the fight against litter. The Code regulates responsibilities and defines the cooperation between retail outlets and event promoters on the one hand (without regard to the size of the retail outlet or the scope of the event) and the local authorities on the other hand. On this basis anti-litter measures can be implemented in public places on an amicable basis. With regard to free newspapers dropped in public areas, there are several initiatives between the cities (and communities) and the free newspapers. A model contract, which includes a list of specifications for the free newspaper publishers and distributors, is offered by the Swiss Federal Office for the Environment as a guide to help communities achieve transparent, matching agreements in the fight against free newspaper litter.

4. **Financial Incentives for Consumers**
   For events that take place in a clearly defined area, a deposit on recyclable packaging such as plastic bottles or reusable cups increases the rate of return and reduces litter. This knowledge has been gathered and documented at various public events in recent years.

5. **Sanctions**
   Awareness-raising measures can make sense in relation to certain population groups, but with some groups they have no effect. In these cases, sanctions such as fines may be considered. Some cantons and cities, of which the Cantons of Thurgau, Solothurn, and St. Gallen and the City of Berne were the first, passed anti-litter laws that included fines for those caught littering. Other cities, however, deliberately avoided laws that included fines for littering because it can be difficult to enforce the law. The lack of a uniform approach has been controversial.

References

1. *Unit-Based Pricing in the United States: A Tally of Communities*. 1999. Washington, D.C.: EPA.
2. U.S. Department of Labor, Occupational Safety and Health Statistics Department. 1998. Washington, D.C. As reported by *Waste Age* (July): p. 20.
3. *Getting More for Less; Municipal Solid Waste and Recyclables Collection Workbook*. 1996. Washington, D.C.: Solid Waste Association of North America.
4. DeWeese, A. 2000. "Mandates Motivate to Automate." *Waste Age* (February).
5. Greenberg, M. R., et al. 1976. *Solid Waste Planning in Metropolitan Regions*. New Brunswick, N.J.: The Center for Urban Policy Research, Rutgers University.
6. Liebman, J. C., J. W. Male, and M. Wathne. 1975. "Minimum Cost in Residential Refuse Vehicle Routes." *Journal of the Environmental Engineering Division*, ASCE, v. 101, n. EE 3:339–412.
7. Kwan, K. 1962. "Graphic Programming Using Odd or Even Points." *Chinese Math*. 1:207–218.
8. Shuster, K. A. and D. A. Schur. 1974. *Heuristic Routing for Solid Waste Collection Vehicles*. EPA OSWMP SW-113. Washington, D.C.

9. Goldstein, N. 1997. "The State of Garbage in America." *BioCycle* 38, n. 4:60–67.

10. Franklin Associates. 1998. *Characterization of Municipal Solid Waste in the United States.* EPA 530-R-980–007. Washington, D.C.

11. *Zero Waste.* The Grassroots Recycling Network. www.grrm.org/zw5.html.

12. "Single Stream Recycling: The Future Is Now." 2000. *Recycling Today* 38, n 1:79.

13. Miller, C. 1993. *The Cost of Recycling at the Curb.* Washington, D.C.: National Solid Waste Management Association.

14. Everett, J. W., S. Maratha, R. Dorairaj, and P. Riley. 1998. "Curbside Collection of Recyclables I: Route Time Estimation Model." *Resources, Conservation and Recycling* 22:177–192.

15. Vesilind, P. A. 1976. *Measurement of Roadside Litter.* Durham, N.C.: Duke Environmental Center, Duke University.

16. *The Photometric Index.* n.d. Stamford, Conn.: Keep America Beautiful.

17. Finnie, W. C. 1973. "Field Experiments in Litter Control." *Environment and Behavior* 5 n. 2.

18. Burgess, R. L., et al. 1971. "An Experimental Analysis of Anti-litter Procedures." *Journal of Applied Behavioral Analysis* 4 no. 2.

19. Hart, F. D., et al. 1973. *Design and Development of a Machine to Remove Litter from the Roadside.* Raleigh, N.C.: School of Engineering, North Carolina State University.

20. National Center for Resource Recovery. 1973. *Municipal Solid Waste Collection.* Lexington, Mass.: Lexington Books.

21. Schert, J. 2000. The Florida Litter Study, Measuring and Managing Litter: Illegal Dumping, City Costs, KAB Litter Index Review. Center for Solid and Hazardous Waste Management, University of Florida, Gainesville, Florida.

22. Georgescu-Roegen, N. 1976. *Energy and Economic Myths.* New York: Pergamon.

23. Biaciardi, C., E. Tiezzi, and S. Ulgiati. 1993. "Complete Recycling of Matter in the Frameworks of Physics, Biology and Ecological Economics." *Ecological Economics* 8:1–5.

24. Converse, A. O. 1996. "Letter to the Editor on Complete Recycling." *Ecological Economics* 19:193–194.

25. United States Environmental Protection Agency. "Municipal Solid Waste Generation, Recycling, and Disposal in the United States: Facts and Figures for 2012."

26. Results based on an unpublished study conducted by William Worrell.

Abbreviations Used in This Chapter

EPA = Environmental Protection Agency
KAB = Keep America Beautiful, Inc.
MRF = materials recovery facility

MSW = municipal solid waste
OSHA = Occupational Safety and Health Administration

Problems

3-1. Determine the time required to complete a filling and emptying cycle for a refuse collection vehicle serving a residential area if the following conditions pertain:

Truck volume = 20 yd³

Each location has on average two containers of 80 gallons each at 75% full. Refuse is picked up twice per week. Truck has a compaction ratio of 1:2.

Pickup time is 1.58 min per service. Truck spends 20 min at the disposal site. It takes 35 min to drive to the disposal site.

Assume a density of 250 lb/yd³ yard for uncompacted refuse and 500 lb/yd³ yard for compacted refuse.

3-2. Design an innovative system for transporting refuse from the kitchen to the curb. Cost should not be a major consideration. Imaginative ideas that just might work are required.

3-3. On a map of your campus (or any other convenient map), develop an efficient route for refuse collection, assuming that each blockface must be collected.

3-4. Visit City Hall and obtain the accident records for city employees. Report on the relative accident rate of solid waste workers.

3-5. The haul distance is 15 miles as the crow flies, and the anticipated average truck speed is 35 mph. Estimate the one-way haul time.

3-6. Using a study hall or social lounge as a laboratory, study the prevalence of litter by counting the items in the receptacles versus the items improperly disposed of. Each day vary the conditions as follows:
Day 1: Normal (baseline).
Day 2: Remove all receptacles except one.
Day 3: Add additional receptacles to have more than normal.
Plot the percent of properly disposed items versus the number of receptacles. Discuss the implications of your results.

3-7. Estimate the time to collect recyclables in green roll-out containers from a 1000-foot-long street if the homes are 50 feet apart, on both sides of the road, and a two-person crew is doing the collecting. Estimate that it takes 30 seconds to empty the contents of each green container into the truck; use a set-out amount of 12 kg.

3-8. Using the principles of heuristic routing, develop a collection route for the streets shown in Figure 3-28. Each blockface must be collected

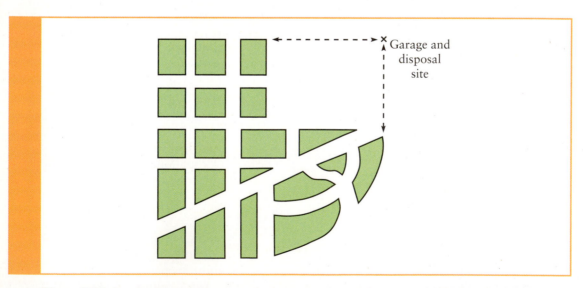

Figure 3-28 See Problem 3-8.

(i.e., one side of street collection). Eliminate all blind blockfaces (it is possible!) and minimize left-hand turns.

3-9. Design a refuse collection route for the suburban development shown in Figure 3-29. Note the large and busy four-lane highway and other features. What criteria should you use to judge the suitability of your route? Justify your route on the basis of these criteria. Assume the following:

Collection is required on all streets. The trucks come from town and return to town.

There is no median strip on the four-lane highway (left-hand turns are possible).

All streets other than the highway are small residential streets. The small bridge has a weight limit less than the weight of the empty collection vehicle.

3-10. How would you discourage littering in front of your university engineering building?

3-11. Suppose you are in charge of evaluating an anti-litter campaign for your campus. How would you obtain

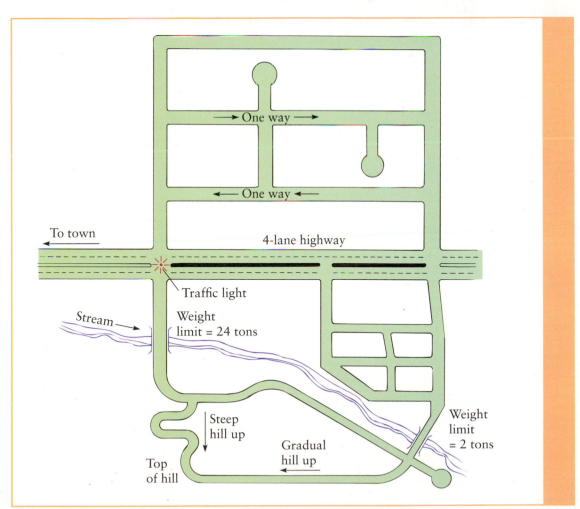

Figure 3-29 See Problems 3-9 and 3-14.

quantitative information on the success or failure of the campaign?

3-12. Two small communities, Alpha and Beta, each have a small landfill. They are considering building a joint landfill to serve both communities. (See map, Figure 3-30.)

a. Write an objective function that can be used to minimize the total cost of disposal where the towns would have the option of taking the refuse to their own landfill or to the common facility. They will not be able, however, to take the refuse to each other's landfill. Use the following variables, and state all the constraints:

F_i = fixed-cost landfill i ($100,000 per year, regardless of size)

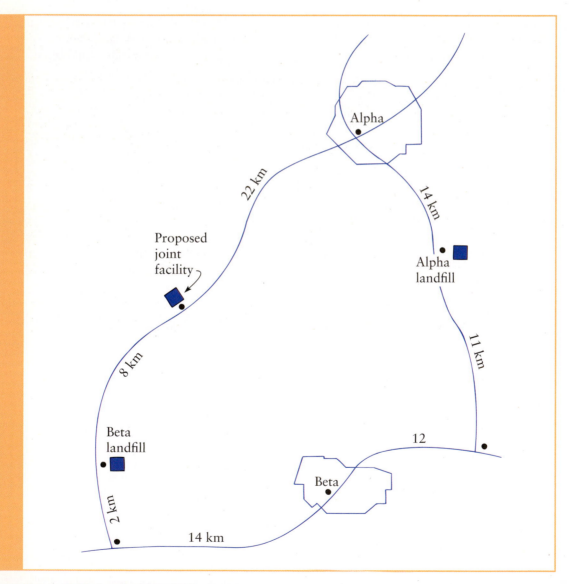

Figure 3-30 See Problem 3-12.

C_{ki} = transportation cost from town k to landfill i ($2.00 per km per ton)

x_k = refuse generated in town k (2000 metric tons per year for Alpha, and 3500 metric tons per year for Beta)

y_i = 1 if the landfill is operated, y = 0 if it is not

c_i = operating cost of landfill i ($10.00 per metric ton)

b.   Solve the objective function for two conditions: (1) when the towns use their own landfills exclusively and (2) when only the central shared facility is used by both towns.

c.   Use an optimization software package to find the optimum (least-cost) solution.

3-13. Your client is a small community of 5000 people. The community wants to begin a municipally operated refuse collection and disposal program and asks your advice on the purchase of a collection vehicle. The community wants to collect refuse once a week. What would you recommend?

3-14. Given the street configuration shown in Figure 3-29, and if all blockfaces are to be collected, is a unicoursal (Euler's) tour possible? Show your work.

3-15. A community has two landfills. The cost per ton of disposing of refuse in the landfills is exactly the same. There are three collection routes in town. Write an objective function and constraint equations for minimizing the transportation cost (only) from the three collection routes to the two disposal sites.

3-16. Write an objective function and constraint equations for minimizing the transportation cost (only) for the community shown in Figure 3-31. Assume that the cost of disposal is the same at both disposal sites, and therefore, the disposal cost can be ignored.

3-17. On the street map shown in Figure 3-32, determine first if an Euler's tour is possible. Show why or why not. Then show how you would route a truck so as to achieve the least number of deadheads. You are to collect on both sides of the street at once.

3-18. A community of 25,000 people wants to initiate public garbage collection.

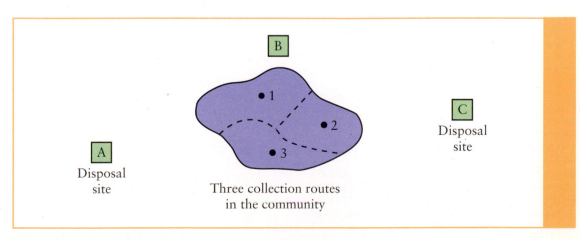

Figure 3-31 See Problem 3-16.

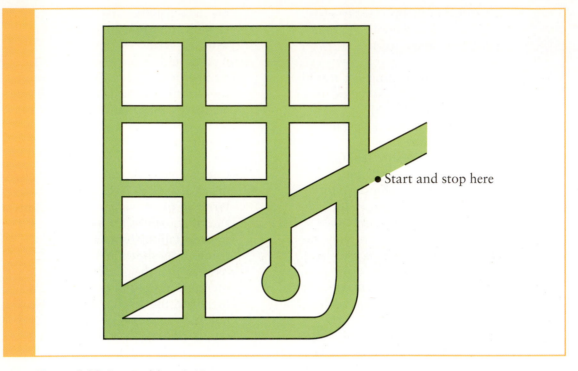

Start and stop here

Figure 3-32 See Problem 3-17.

If the collection is to be once a week, what size truck would the community need? *(Note:* Available truck sizes are 10, 14, 16, and 20 cubic yards.)

3-19. The mayor of Recycleville has asked you to develop an integrated waste management program. Residents currently pay $10 per month and can put out up to five 32-gallon cans per week. You offer the mayor an automated system with two price options:

Container Size (gallons)	Option 1	Option 2
32	$9.00	$12.00
64	$18.00	$15.00
96	$27.00	$18.00

In addition, you will provide the residents a 64-gallon commingled recycling container and a 96-gallon green-waste container.

a. What are the advantages of using the program you developed?

b. Are residents opting for the 96-gallon container paying more for this service? *(Hint:* Think of the cost per gallon of material.)

c. Which price option would likely have the higher recycling rate, and why? Look at the EPA web site for some case studies to support your reasoning.

d. What impact would collecting recyclables one week and collecting green waste the next week have on the recycling rate?

3-20. Estimate the number of commercial locations that can be serviced in one day by the collecting vehicle described below:

truck volume = 20 yd³ and density of 250 lb/yd³ for uncompacted refuse

compaction ratio = 2

container volume = 6 yd³

on average, containers are 80% full

drive between locations = 8 minutes

at site time (loading dumpster into truck) = 15 min

haul time (to transfer or disposal) = 20 min (one way)

to and from garage (at beginning and end of day) = 15 min

workday = 8 h

off-route factor = 0.25 (breaks, lunch, other necessities)

Assume all vehicles making the trip to transfer or disposal are full.

3-21. Assume it costs 30 cents per ton per mile to operate a garbage truck and 10 cents per ton per mile to operate a transfer truck. A transfer station also has a fixed cost of $10 per ton. If it is 30 miles to the landfill, what would be your least expensive transportation alternative (show your calculations)?

3-22. Rank the three factors given from highest impact to lowest impact on the probability of littering.

— Someone who is 18 years old or younger

— A location that does not have conveniently located trash cans

— Being in an area that is already littered

CHAPTER 4

Mechanical Processes

In almost all engineering disciplines, the designs of the processes used to produce a desired end product are based on the reasonable assumption that the nature of the raw material is known and can be defined accurately and precisely. This condition lends a feeling of confidence to the analysis of unit operations and breeds sophistication in the design of the process. Unfortunately, the MSW processing profession is not so blessed. Solid waste processing and materials recovery facilities must be able to accept almost all manner of solid waste. Some types of solid waste are easy to process, but occasionally materials that are difficult and/or dangerous to handle are also found in refuse. As a result, solid waste processing operations have large factors of safety and must be designed for extraordinary contingencies. This requirement often results in overdesign and underutilization in order to process all of the feed material.

In this chapter, solid waste handling prior to further materials separation or processing is discussed. None of the unit operations discussed here actually accomplishes materials separation but, rather, prepares the refuse for the separation operations that follow.

4-1   REFUSE PHYSICAL CHARACTERISTICS

Municipal refuse, either in its original or shredded state, has some material properties that make its processing and conveying hazardous and difficult. Although the art and science of solids conveying and storage are well studied in the mining and chemical engineering fields, the application of most of that knowledge is inappropriate for a heterogeneous, unpredictable, and time-variable material such as refuse. Hickman[1] lists the materials characteristics that must be considered in the design of any storage, conveying, or processing equipment.

Particle Size

Because of the nature of MSW, particle size is difficult to define. Sieves define size by only two dimensions; thus, a piece of wire could pass a sieve and still prove troublesome in conveying. The problems of defining particle diameter are discussed in Chapter 2, and the process of size separation by sieving is discussed further in Chapter 5.

Bulk Density

Shredded MSW, when stored in a storage pit, can achieve densities as high as 25 lb/ft³ (400 kg/m³). The variation in density has been found to be significant, as shown in Figure 4-1. This variation in density is an important factor in designing storage facilities.

Angle of Repose

The angle of repose is the angle to the horizontal to which the material will stack without sliding. Sand, for example, has an angle of repose of about 35°, depending

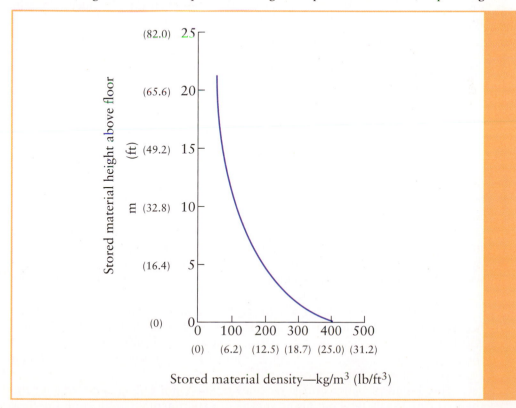

Figure 4-1 Variation of bulk density in MSW. Source: [1]

on the moisture content. Because of variable density, moisture, and particle size, the angle of repose of shredded refuse can vary from 45° to greater than 90°.

Material Abrasiveness

Refuse consists of many types of abrasive particles, including sand, glass, metals, and rocks. Removal of this abrasive material is often necessary before some operations (such as pneumatic conveying) can become practical.

Moisture Content

All of the foregoing properties are influenced by moisture content. The extent of this effect depends on the material. When the moisture level exceeds 50%, the high organic fraction can undergo spontaneous combustion if the material is allowed to stand undisturbed.

4-2 STORING MSW

The storage of MSW has long been a serious problem, especially at large combustor facilities. Most waste-to-energy combustors must be continuously fired and require sufficient storage for at least two days to allow for the unavailability of refuse over weekends. Similarly, if materials recovery systems operate through the weekend, they must store material in sufficient quantities to even out the fluctuations in supply.

Two major considerations in the design of MSW storage facilities are public health and fire. Rats and other rodents can inhabit storage areas unless special precautions are taken. The odor of slowly decomposing garbage can be overwhelming and can cause public relations problems downwind.

Spontaneous combustion is possible with the storage of MSW. The rule of thumb is that two days of storage is the safe maximum, with a week being dangerous. A fire in a storage pit is not only difficult to extinguish, but the resulting wet refuse after the fire is extinguished presents new disposal problems.

All storage facilities should be constructed as *first-in/first-out* systems. Unfortunately, this is not a simple task, and many of the existing storage systems tend to result in long-term storage of some fraction of the refuse.

A common storage system consists of a pit with an overhead bridge crane. Garbage and transfer trucks back up to the pit and discharge solid waste directly into it. An overhead crane, with an operator either directly on the crane or centered over the pit, is used to both spread the load in the pit and to retrieve the solid waste. The crane can drop the solid waste into a feed chute, onto a conveyor belt, or directly into a transfer vehicle.

Another common storage system relies on a large tipping floor. Solid waste is deposited onto the floor and is then stacked as high as 20 ft (7 m) by a front-end loader. In some facilities, a concrete or steel push wall is incorporated into the design. The front-end loader can also meter the solid waste onto a metal pan conveyor or directly into a transfer vehicle.

The design of better storage facilities requires not only a knowledge of the theory of materials flow but also a means of experimentally evaluating the flow rate of solid material in a storage chamber. The use of velocity probes, especially for solid waste, is clearly unacceptable. A number of potentially effective techniques are stereophotogrammetry, radio pills (transmitters that move with the solids in the bin), radiological tagging (e.g., with cesium 137), and X-ray methods.[2] With nonhomogeneous materials such as refuse, the radio pill or photogrammetry seems to be most applicable.

4-3   CONVEYING

Six basic types of conveyors are used for refuse:

1. Rubber-belted conveyors
2. Live bottom feeders
3. Pneumatic conveyors
4. Vibratory feeders
5. Screw feeders
6. Drag chains

The first three types are used primarily to move refuse; the last three are used to feed or meter refuse to a load-sensitive device such as a combustor.

Rubber-belted conveyors have been used to move unshredded raw material and are especially acceptable for less abrasive and less rugged loads, such as source-separated recyclables. Rigid metal interlocking belts (commonly referred to as *metal pan conveyors*) have been successfully applied for raw refuse conveying. Skirts (sides) are sometimes used for bulky material. A typical conveyor with skirts is shown in Figure 4-2. A trough-type rubber belted conveyor without skirts is shown in Figure 4-3.

The material on a conveyor with inclines greater than 20° will usually *tumble back*. This can be considered advantageous if the conveyor is feeding a shredder, since the movement of the refuse can even out the feed to the shredder.

One shredder manufacturer recommends angles from 38° to 40° to achieve a more even feed rate to the shredder.[3] In other applications, a very steep incline is required, and rubber belt conveyors with flights can be used.

The power requirements of belt conveyors can be estimated by a number of empirical equations, such as[4]

$$\text{horsepower} = \frac{LSF}{1000} + \frac{LTC}{990} + \frac{TH}{990} + P$$

where
L = length of conveyor belt, ft
S = speed of belt, ft/min
F = speed factor, dimensionless
T = capacity, tons/h
C = idle resistance factor, dimensionless
H = lift, ft
P = pulley friction, horsepower

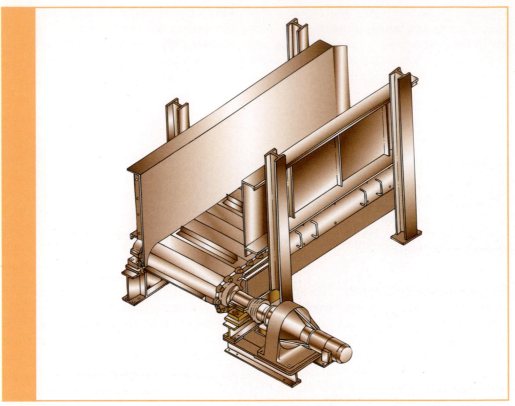

Figure 4-2 Typical feed conveyor. Source: Based on Planning and Specifying a Refuse Shredding System. Appleton, Wis.: Allis-Chalmers.

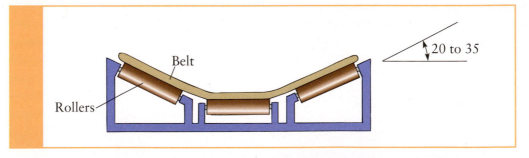

Figure 4-3 Conveyor commonly used for MSW. Capacities for this conveyor are shown in Table 4-1.

The first two terms on the right-hand side of the equation represent the power necessary to move the load horizontally, while the third term represents vertical movement. The last term represents power loss due to friction. Although the units in this equation are obviously spurious, the constants in the denominators are used to convert everything to power.

For specific materials, empirical evidence suggests some maximum loadings on conveyor belts without skirts, as shown in Table 4-1. These capacities are for a

Table 4-1 Rubber Conveyor Belt Capacities for Selected Materials at a Belt Speed of 100 Feet/Minute

Material	Belt Width (inches)	
	36	60
	Capacity, tons/hour	
Glass bottles	28.3	83.4
Plastic bottles	3.7	10.8
Aluminum cans	3.7	10.8
Newsprint	18.1	53.1
MSW from truck	40.0	117
Bulk material (100 lb/ft³)	127	375

Source: Based on Mitchell, J. R. 1971. "Designing for Batch and Continuous Weighers." Chemical Engineering (Feb. 28): 177; Belt Capacity Tables, B. W. Sinclair, Inc., http://www.bwsinclair.com /Downloads/Belt%20Conveyor%20Capacity%20Tables.pdf.

belt with a 20° trough (angle of the edge of the belt with the horizontal), as shown in Figure 4-3, and a belt speed of 100 ft/min (30.5 m/min).

Live bottom hoppers are related to vibrating feeders and are used to move MSW out of holding bins or transfer trailers. As the name implies, the bottom of the hopper has sliding interlocking beams that move at set speeds. By moving them slowly forward and then rapidly backward, the motion slowly moves the burden forward. Live bottom feeders have been installed in facilities where refuse has to be moved a short distance such as in transfer stations. Figure 4-4 is an illustration of a typical live bottom feeder.

Pneumatic conveyors have been used mainly for collecting raw bagged MSW in hospitals and other large buildings and in feeding shredded organic fractions to boilers as supplemental fuel. The required air velocities in pneumatic tubes can be estimated as

$$v_m = v_a - v_f$$

where

v_m = material velocity, ft/min
v_a = velocity of the air stream, ft/min
v_f = the *floating velocity*, or terminal velocity when falling in still air, ft/min

The floating velocity can be calculated by an empirical equation such as

$$v_f = 3250\sqrt{(SG)d}$$

where *vf* is the floating velocity in ft/min, *d* is the aerodynamic diameter of a representative particle in inches, and *SG* is the specific gravity of the material (relative to water).

The material velocity must be sufficient to be able to dislodge stuck particles and to even out the flow and can be estimated as

$$v_m = 585\sqrt{W}$$

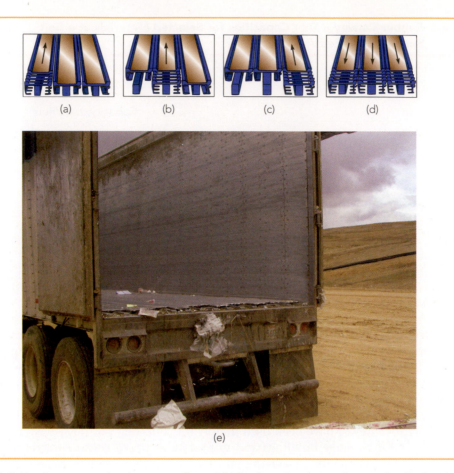

Figure 4-4 Truck equipped with moving floor. (a) The first slat group slips under the load to the rear of the load. (b) The second slat group slips under the load. (c) The third slat group slips under the load. (d) All slat groups and the load advance forward together. (e) A truck equipped with a walking floor has just emptied its contents. (Courtesy KEITH Mfg. Co.)

where W is the bulk density of the material in lb/ft³. Because of the problems of measuring specific gravities in heterogeneous mixtures, the specific gravity can be estimated as

$$SG = 0.1(W)^{2/3}$$

The total recommended air velocity is thus

$$v_a = 3200\sqrt{(SG)d} + 585\sqrt{W}$$

where v_a is the recommended air velocity in ft/min. The problem of accurately estimating the representative diameter of a particle makes the practical use of this equation difficult. Experience has shown, however, that maintaining air velocities of about 4500 ft/min (1400 m/min) is sufficient for maintaining the materials flow in vertical tubes. Table 4-2 lists some recommended air velocities for common

Table 4-2 Recommended Air Velocities for the Pneumatic Conveying of Some Representative Materials

Material	Minimum Air Velocity (ft/min)[a]
Coal, powdered	4000
Cotton	4500
Iron oxide	6500
Shavings	3500
Vegetable pulp	4500
Paper	5000
Rags	4500

[a] To obtain m/sec, multiply ft/min by 0.00508.
Source: [7]

materials. Operational experience has shown that a materials-to-air concentration of 0.1 (e.g., 0.1 kg of paper/1.0 kg of air) is reasonable.[6] The friction loss within a duct due to materials flow is less than 10%, which is well within a factor of safety commonly used in fan design.

Pneumatic conveyors suffer from wear problems, especially if glass is being conveyed. Elbows in pneumatic lines can be expected to abrade quickly, and sacrificial pieces must be used and frequently replaced.

Vibrating feeders are advantageous because they also even out materials flow. These devices are used to move small quantities of rigid material. For example, a vibrating conveyor can be used to feed glass to a hand-sorting table. A 2 cm stroke at a frequency of 900 strokes/min is common.

Screw conveyors are used to meter shredded refuse into a furnace, because the screw serves as an air lock, and the feed rate of fuel can be adjusted easily by changing the rotational speed of the screw. The volume of material moved by screw conveyors can be estimated by recognizing that the capacity of the conveyor in the *flooded* condition (i.e., all the space between the blades is full, as might occur when a screw conveyor is used in the bottom of a hopper) is

$$Q = CNRV$$

where

Q = delivery of refuse, m^3/min
C = the efficiency factor
N = number of conveyor leads
R = rotational speed of screw, rpm
V = volume of refuse between each pitch, m^3

The number of leads means the number of blades that are wrapped around the conveyor shaft. A common wood screw, for example, has one. Theoretically, if the distance between adjacent conveyor blades is constant, the forward motion of the material being conveyed is directly proportional to the number of leads. The units for this equation can be any volume and time, such as cubic meter and seconds. These terms are defined further in Figure 4-5.

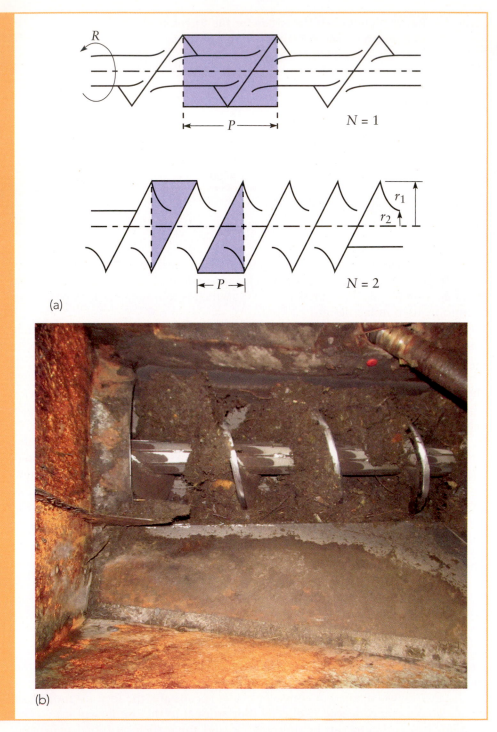

Figure 4-5 (a) Screw conveyor (*Note:* Top screw has $N = 1$, and bottom screw has $N = 2$.)
(b) Screw conveyor used to transport organic material.

The volume within each pitch can be calculated approximately as

$$V = P\pi\left(r_1^2 - r_2^2\right)$$

where

P = pitch (distance between adjacent conveyors' blades if the number of leads, N = 1), m
r_1 = radius to the conveyor tip, m
r_2 = radius to conveyor shaft, m

The dimensionless efficiency factor, C, is a function of the amount of slippage that occurs. Ideally, a screw conveyor operates by allowing the material to slide freely on the blade and thus prevent radial rotation. Any rotation by the material (sticking to the blade) lowers efficiency. In live bottom bins, where the conveyor screws are not in individual troughs, considerable slippage is expected.

EXAMPLE 4-1

A live bottom bin has eight screw conveyors, each with r_1 = 15 cm, r_2 = 6 cm, N = 1, P = 50 cm, and R = 10 rpm. Assuming that C = 0.5, calculate the total material flow.

SOLUTION

$$V = P\pi(r_1^2 - r_2^2) = 50(3.14)(225 - 36) = 29{,}673 \text{ cm}^3$$

For each conveyor,

$$Q = CNRV = (0.5)(10)(29{,}673)(10^{-6}) = 0.15 \text{ m}^3/\text{min}$$

or total flow is 8 × 0.15 = 1.2 m³/min.

If the screw conveyor is not flooded, its capacity cannot be determined theoretically, because the rate is influenced by a large number of variables.[8]

Drag chain conveyors are used to move solid waste in applications such as waste-to-energy plants that are burning refuse-derived fuel. A drag chain conveyor consists of an open- or closed-top metal rectangular pan. A chain runs the length of the pan along each side. Across the chain at 5- to 10-foot intervals are metal or wood flights. The chain drags the flights, which move the refuse. On the bottom of the metal pan can be slide door openings to chutes. Opening the door allows the refuse to fall into the chutes. At the bottom of the chutes are screw feeders to introduce the fuel into the furnance. Refuse that does not fall into the chutes can be returned to the beginning of the drag conveyor.

As a general rule, refuse should be conveyed and transferred as little as possible. It is thus good engineering design to eliminate, or at least minimize, points of transfer and conveying within a facility.

4-4   COMPACTING

One problem in the disposal of MSW is the low density of the material, which requires large volumes for its collection, handling, and final disposal. Compacting MSW can lead to significant cost savings. The structure of refuse can be pictured as an assemblage of particles interspaced with open air spaces called *voids*. Because these voids are large, and since many of the particles are absorbent, any moisture is absorbed in the material and is not in the voids. The total volume of material is made up of the solids plus the voids, given as

$$V_m = V_s + V_v$$

where

V_m = volume of material
V_s = volume of solids (including the moisture)
V_v = volume of voids

The *void ratio* is defined as

$$e = \frac{V_v}{V_s}$$

and the *porosity* is

$$n = \frac{V_v}{V_m}$$

By weight, the total material is made up of the solids plus moisture, given as

$$W_m = W_s + W_w$$

where

W_m = weight of material, including moisture
W_s = weight of solids
W_w = weight of moisture

The *bulk density* is defined as

$$\rho_b = \frac{W_m}{V_m}$$

Note that this is on a *wet basis*. The entire sample is weighed *as is*, and its volume is calculated. Most densities in compaction literature are expressed in terms of bulk density, mainly because it is easy to measure and can be readily used in comparative studies. Bulk densities also can be expressed on a *dry basis* if the sample is then dried and the weight of the moisture subtracted. The compaction of refuse in this text and in most literature is expressed as the increase in bulk densities (wet basis).

When MSW is compacted, the density is increased as a result of the crushing, deforming, and relocating of individual items in the refuse. Hollow containers, such as bottles and cans, begin to collapse at different pressures, depending on their orientation and strength. For example, cans collapse at pressures of 10 to 30 psi (0.1 to 0.3 N/m^2), and glass bottles crush at 5 to 35 psi (0.05 to 0.35 N/m^2).[9]

The compaction of some materials is irreversible, in that when the pressure is released, the material does not spring back to its original volume. MSW, however, contains many items that contribute to reversible compaction. At normal compaction pressures, 20% expansion can occur within a few seconds after the release of pressure, and this expansion can be as much as 50% after a few minutes.[10] The greater the pressure, the greater will be the bale integrity (its resistance to falling apart). A typical compression curve, using a small sample of refuse in a laboratory press, is shown in Figure 4-6.

4-5 SHREDDING

Strictly speaking, *shredding* is one form of *size reduction*—others being such processes as *cutting*, *shearing*, *grinding*, *crushing*, and other imaginative terms, many of which originated in mining engineering. In solid waste work, however, shredding is the generic term for size reduction even though, strictly speaking, shredding is only one method of size reduction. In this text, as in solid waste literature, shredding encompasses all of the processes used for making little particles out of big particles.

Many types of shredders are presently on the market, and almost all of them were developed originally for an application and feed material other than refuse. Most of our present refuse shredding technology comes from the mining industry, which has for many years used shredders for ore processing. The application of this technology to refuse, however, is not an easy matter, because these devices were developed for homogeneous feeds that have well-established breakage characteristics.

4-5-1 Use of Shredders in Solid Waste Processing

The first applications of shredders to MSW were to facilitate disposal with little consideration for materials recovery. The pioneering work on shredding for disposal was done by Robert Ham and his colleagues at the University of Wisconsin. They found that shredded MSW had a more uniform particle size, was fairly homogeneous, and compacted more readily than unshredded waste, mainly because the larger voids had been eliminated.[11] After shredding, MSW looks not unlike confetti and has a light, bulky nature. In fact, the overall density of the material is decreased by over 50%, from 350 to 400 lb/yd^3 (200 to 240 kg/m^3) to 125 to 150 lb/yd^3 (75 to 90 kg/m^3). Shredding reduces required landfill volume, since shredded refuse compacts better within the landfill. In addition, the landfill will have more uniform settlement, and this helps maintain the integrity of the top cap.

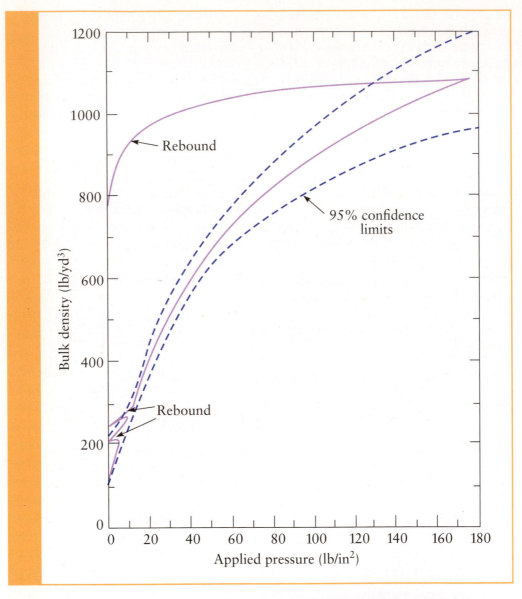

Figure 4-6 Compression curve for a sample of MSW in a laboratory. The rebound curves occur once the compressive pressure is released. Source: [9]

Experience with landfilling of shredded MSW indicates that shredded refuse does not require a daily earth cover. Extensive testing has shown that the conditions that make an earth cover necessary in a conventional landfill no longer exist with shredded refuse. Earth cover in a shredded refuse landfill is considered unnecessary because of the following reasons:[12]

• *Odor.* If the shredded refuse is well mixed, it can retain its aerobic character when spread in reasonably thin layers so that odors are minimized.

- *Rats*. There are no large food particles in shredded refuse that could support a rat population.

- *Insects*. The drier refuse, regularly covered with new layers, suppresses insect breeding, and maggots are killed during shredding.

- *Blowing Paper*. Small pieces are not caught by the wind and do not blow away, while large pieces of paper and plastics found in a conventional landfill would be readily transported by wind.

Because of the lack of cover, leachate from shredded refuse is produced, and this leachate is at a higher concentration of pollution than leachate from normal landfills. But because the shredded MSW is allowed to be open to the air, significant drying takes place, and the total leachate production is minor. Because of the superior compaction characteristics of shredded MSW, shredders also have been used before high-compression baling.[13]

Although the advantages of shredding prior to landfilling seem impressive, most regulatory agencies insist that daily cover still be applied even if the refuse is shredded. Such a requirement effectively eliminates the cost advantage of shredding refuse prior to landfilling.

A second use of shredding is in the production of *refuse-derived fuel* (RDF). The breaking apart of the various constituents within MSW results in a shredded waste that has a more uniform heating value and requires less excess air in a furnace compared to unshredded refuse, thus saving on air pollution control equipment and costs.

A third use of shredders is in the processing of yard waste as well as demolition debris, branches, and other organic material to produce a mulch that then can be composted or used as a ground cover. Shredders used for this purpose are (strictly speaking) *grinders*, which consist of a tub in which grinding gears rub the feed particles against the inside wall and break them up into smaller pieces. A typical grinder is shown in Figure 4-7.

Perhaps the most important use of shredders is in materials recovery. Most wastes, from municipal solid waste to electronic waste or a car, are composed mostly of materials that are physically attached, and it is not easy or simple to separate them. Consider as an example a typical kitchen appliance. The electric motor has perhaps 10 or more materials, both plastics and metals, and the housing is of perhaps two different types of plastic or metal. Disassembling this device to recover the various materials would be prohibitively expensive. The alternative is to shred it up into small pieces, and at least some of these small pieces would then be of a single material, which could be separated (the topic of the next chapter).

4-5-2 Types of Shredders Used for Solid Waste Processing

The list of size-reduction equipment used in both the mining and chemical industries is surprisingly long. One well-known mining and ore dressing handbook, for example, lists over 50 different devices that could be applied to solid waste shredding.[4] A review of size-reduction equipment widely used in chemical engineering applications lists 21 different devices.[14]

Figure 4-7 Tub grinder for yard waste. (Courtesy P. Aarne Vesilind)

The application of a specific device for MSW has not been a simple matter, however, since the material is significantly different from ore and other homogeneous feeds encountered in these industries. For example, coal can be counted on to shatter upon impact, and thus a shredder that would process coal would also shatter glass bottles rather well. On the other hand, the same shredder probably would not shred metal cans, which must be cut or torn apart within a shredder.

One of the first size-reduction devices used for solid waste processing was the *hammermill*, which consists of a central rotor on which are pinned radial hammers that are free to swing on the pins. The rotor, enclosed in a heavy-duty housing, is an integral part of the shredding operation. In the *horizontal hammermill*, the rotor is supported by bearings on either end, and the feed is by gravity (free drop) or conveyor (force fed) (Figure 4-8). A discharge grate placed below the rotor determines the size of the product, since a particle cannot pass through this grate until it is smaller than the grate opening in two dimensions. Some hammermills are symmetrical, so the direction of the rotor can be changed to alternate wear surfaces without necessitating hammer maintenance after each run.

The *vertical hammermill*, as the name implies, has a vertical shaft, and the material moves by gravity down the sides of the housing (Figure 4-9). These

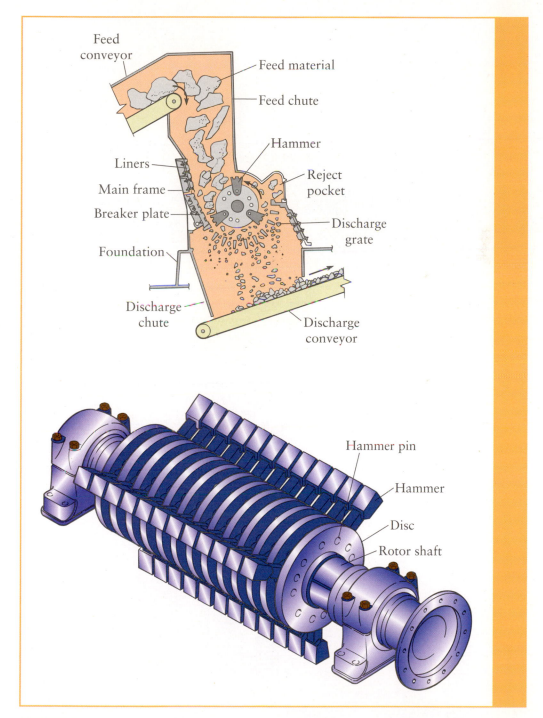

Figure 4-8 Horizontal hammermill shredder. Source: Based on Franconeri, P. 1976. "Selection Factors in Evaluating Large Solid Waste Shredders." Proceedings ASME National Waste Processing Conference. Boston.

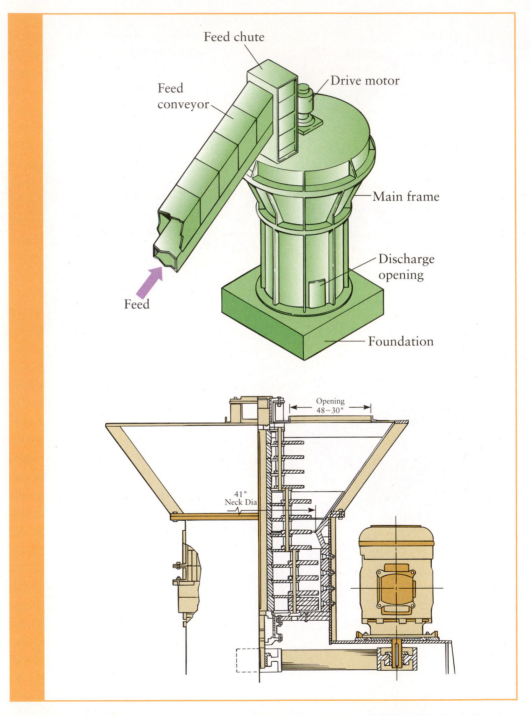

Figure 4-9 Vertical hammermill shredder. Source: Based on Franconeri, P. 1976. "Selection Factors in Evaluating Large Solid Waste Shredders." Proceedings ASME National Waste Processing Conference. Boston.

mills usually have a larger clearance between the housing at the top of the mill and progressively smaller clearances toward the bottom, thus reducing the size of the material in several steps as it moves through the machine. Since there is no discharge grate, the particle size of the product must be controlled by establishing a proper clearance between the lower hammers and the housing (Figure 4-10).

Another type of shredder used for MSW is known as a *hog*, which was originally used in pulp and paper manufacturing from wood chips. The hog is used to shred green waste. Care must be taken to ensure that no abrasive items such as glass or metals are involved.

Slow-speed *shear shredders* were originally used to slice whole tires prior to disposal. If a whole tire is placed in a landfill, it can create problems by eventually floating to the surface due to its low density (if the space within the tire is still filled with air). Some states, such as California and Florida, require tires to be cut into pieces prior to landfilling. Shear shredders have been used increasingly for the processing of mixed solid waste as well because of their lower energy requirements, lower rates of wear, and, most

Figure 4-10 Inside a vertical hammermill shredder. (Courtesy William A. Worrell)

importantly, the reduced chance of explosions. A typical shear shredder is shown in Figure 4-11.

Finally, a *flail mill* consists of arms with elbows that beat at the plastic bags to open them so that the contents can be processed further. Often glass bottles are broken in flails, but other than that, minimal size reduction occurs.

(a)

(b)

Figure 4-11 Shear shredder. (a) Close-up of the shredder blades. (b) In operation shredding tires. (Courtesy William A. Worrell)

4-5-3 Describing Shredder Performance by Changes in Particle-Size Distribution

One of the most important design and operational parameters to be considered in size reduction is the change in particle-size distribution of the feed and the final product. The effect from shredding can differ for various material components in solid waste. Figure 4-12 is a graphical description of 13 different categories and shows the size distribution after shredding in a hammermill.[9] The wide variation in size is obvious and is one of the primary attributes of shredding, which allows for subsequent separation of the various material components. The following discussion is focused on describing the composite curve, which is not a picture of a homogeneous material but is made up of various components within the different particle-size categories.

Laboratory measurement of particle size is in itself difficult, since the material is in odd shapes. A piece of wire, for example, presents a difficult problem in classification because it is clearly quite small in two dimensions (and thus can escape further reductions in size in the shredding operation), but the effect of its length on subsequent separation operations—such as air classification—can be troublesome.

The method of measuring particle size can also influence the results of any given study. The common procedure for measuring particle-size distribution is by sieving, yet the shape of both the particles and the sieve openings can affect the number of particles that can pass through an opening.[9] In addition to shape factors, problems with providing an adequate duration of sieving, wear and tear on the sieve and the material, variations in the sieve apertures, and errors in observation and sampling all suggest that there may be problems involved in comparing size-distribution data obtained at various laboratories.[16] Particle-size measurement is discussed more fully in Chapter 5.

The size distribution of particles generally cannot be expressed by any single-valued function and is instead expressed by an equation describing the distribution of various size fractions. The general nomenclature[17] used for these equations is

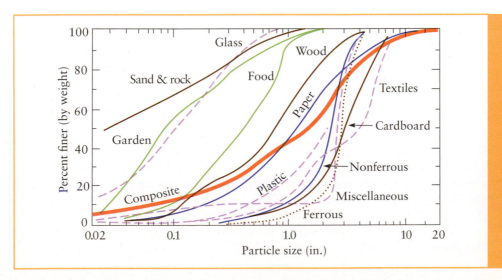

Figure 4-12 Size reduction of various MSW components after shredding. Source: [10]

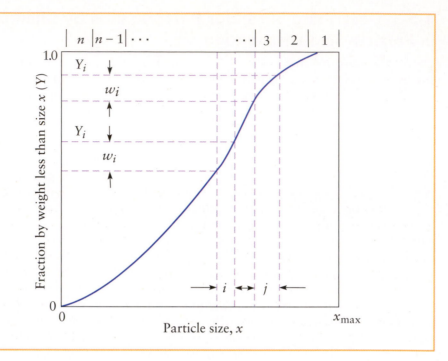

Figure 4-13 Cumulative particle-size distribution curve.

defined in Figure 4-13, which shows a plot of the cumulative weight Y less than size x, which is plotted versus the particle size x. The particle sizes are broken into an arbitrary number of intervals (usually according to sieve sizes)—with the top size (or grade) called number 1—on down to the nth grade. The ith interval has within it w, which is the weight fraction of the material.

In the breakage process, as the curve is shifted to the left (greater number of smaller particles), some portion of the materials in the ith fraction remain (are not broken), and some portion originates from some larger sizes, or some general interval j. Yj is the cumulative fraction by weight less than size j.

A number of equations have been proposed for describing the particle-distribution curve. Gaudin,[14] for example, suggested the following equation to describe the particle-size distribution for brittle materials:

$$Y = \left(\frac{x}{q}\right)^p$$

where Y is the cumulative fraction of material by weight less than size x, and q and p are constants specific to the material processed and the conditions under which the breakage occurs. In this case, q is the theoretical maximum size, and p defines the slope of the line on log–log coordinates.

The most widely used particle-size descriptor is the Rosin-Rammler model,[18] first proposed in 1933 and stated as

$$Y = 1 - \exp\left(\frac{-x}{x_0}\right)^n$$

where n is a constant and x_0 is the ***characteristic particle size*** (or just ***characteristic size***) defined as the size at which 63.2% $(1 - 1/e = 0.632)$ of the particles (by weight) are smaller. The Rosin-Rammler equation is a generalized expression for sigmoidal curves, such as those in Figure 4-14. Note that the constant n is the slope of the line $\ln(1/(1 - Y))$ versus x on log–log coordinates, since the linear form can be derived as[38]

$$Y = 1 - \exp\left(\frac{-x}{x_0}\right)^n$$

$$\ln\left(\frac{1}{1 - Y}\right) = \left(\frac{x}{x_0}\right)^n$$

$$\log\left[\ln\left(\frac{1}{1 - Y}\right)\right] = n \log\left(\frac{x}{x_0}\right)$$

$$\log\left[\ln\left(\frac{1}{1 - Y}\right)\right] = n[\log x - \log x_0]$$

The value of x_0 is also defined as that size wherein $(1/(1 - Y)) = 1.0$, or, equivalently, where $1/e$ of the particles are larger than x_0. The equation suggests that for a specific value of x_0, as the constant n increases due to changes in machine or feed characteristics, the value Y decreases, meaning that a coarser product is obtained. Conversely, for a given n, a larger x_0 also defines a coarser particle size of the product.

The Rosin-Rammler equation is plotted on log–log coordinates in Figure 4-14 to illustrate the definition of both n and x_0. Table 4-3 presents a compilation of some Rosin-Rammler coefficients for various refuse shredders.

The characteristic size can be calculated from a specification such as 90% passing a given size, which is common in ore comminution practice. Example 4-2 (after Trezek and Savage[20]) illustrates this procedure.

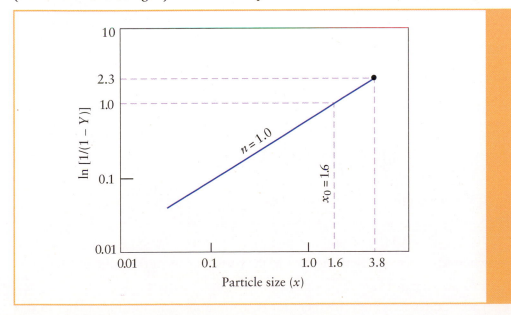

Figure 4-14 Rosin-Rammler particle-size distribution curve.

Table 4-3 Rosin-Rammler Exponents for Shredded Refuse

	n	X_0
Washington, DC	0.689	2.77
Wilmington, DE	0.629	4.56
Charleston, SC	0.823	4.03
San Antonio, TX	0.768	1.04
St. Louis, MO	0.995	1.61
Houston, TX	0.639	2.48
Vancouver, BC	0.881	2.20
Pompano Beach, FL	0.587	0.67
Milford, CT	0.923	1.88
St. Louis, MO	0.939	3.81

Source: Based on Stratton, F. E., and H. Alter. 1978. "Application of Bond Theory to Solid Waste Shredding." Journal of the Environmental Engineering Division, ASCE, 104, n. EE1.)

EXAMPLE 4-2

Suppose that a sample of refuse must be shredded to produce a product with 90% passing 3.8 cm. Assume that $n = 1$. Calculate the characteristic size.

SOLUTION

Since $Y = 0.90$, $\ln(1/(1 - Y)) = \ln(1/(1 - 0.90)) = \ln 10 = 2.3$. Plot $x = 3.8$ cm versus $\ln(1/(1 - Y)) = 2.3$ on log–log paper, as shown in Figure 4-14. For $n = 1$, the slope of the line is 45°, which can be drawn. Lines for any other slope (n) can be constructed similarly by measuring the slope with a ruler. The characteristic size x_0 is then found at $\ln(1/(1 - Y)) = 1.0$, as $x_0 = 1.6$ cm.

The conversion from x_0 to 90% passing is possible by recognizing that

$$x_0 = \frac{x}{\left[\ln\left(\frac{1}{1 - Y} \right) \right]^{1/n}}$$

If $Y = 90\%$, this expression reduces to

$$x_0 = \frac{x_{90}}{2.3^{1/n}}$$

where x_{90} = screen size where 90% of the particles pass. If the value of n is 1.0,

$$x_{90} = 2.3 \, x_0$$

These expressions are convenient for design purposes, as illustrated later in this chapter.

The characteristic size and the slope *n* also can be useful in measuring the effectiveness of a shredder in achieving breakage.[21] By sampling and sieving, a particle-size distribution of the shredder output can be obtained, and the characteristic size and slope *n* can be calculated. Such calculation is facilitated by the use of **Rosin-Rammler Paper**, first developed by the Bureau of Mines in 1946. Figure 4-15 shows a sample of such paper, and its use is illustrated by Example 4-3.

EXAMPLE 4-3

A product from a MSW shredder was sieved with the following results. Calculate the Rosin-Rammler characteristic size and the slope *n*.

Sieve size x (in.)	Percent Retained on Sieve	Percent Finer than Y
6	11	89
3	9	80
1	17	63
0.5	21	42
0.1	22	20
Fines	20	—

Note that if 11% of the material is retained on the first sieve of 6 inches, 89% of the material has to be finer (smaller) than this size. If 9% is retained on the second sieve, 11 + 9 = 20% of the material has been retained, and 80% has to be finer than 3 in.

SOLUTION

These values of *x*, particle size, and percent finer than *Y* (or percent passing) are plotted as shown in Figure 4-15. The characteristic size x_0 is read off the graph at *Y* = 63.21 passing as 1.6 in. The slope *n* is calculated by measuring with a ruler as 3.0/5.5 = 0.55.

The use of such calculations is shown in Figure 4-16, which represents actual data from a shredder operation. The concern that prompted the study was hammer wear and whether or not the operation of the shredder deteriorated over time. As can be seen from the figure, the characteristic size and the slope *n* did not change much over the nearly 15,000 tons of refuse processed.[20]

Another breakage model that assumes only a single fracture of materials is the Gaudin-Meloy model,[21] which is written as

$$Y = 1 - \left(1 - \frac{x}{x'_0}\right)^r$$

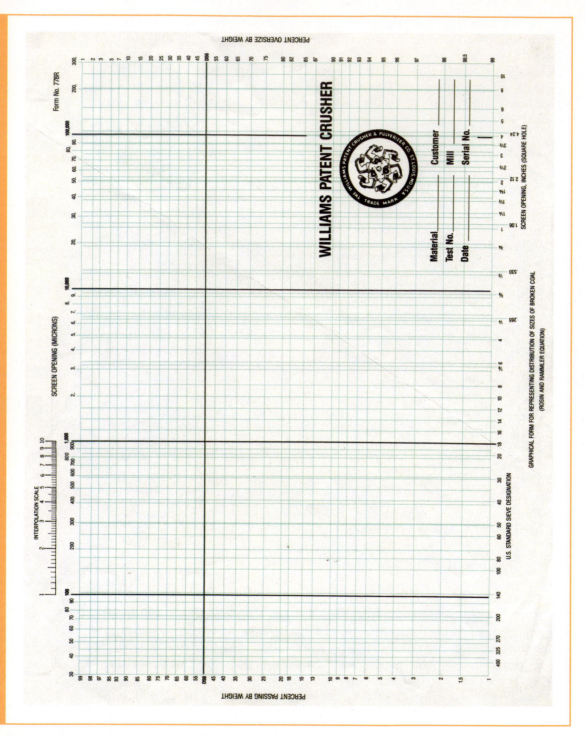

Figure 4-15 Rosin-Rammler paper for plotting particle-size distribution. Source: Williams Patent Crusher

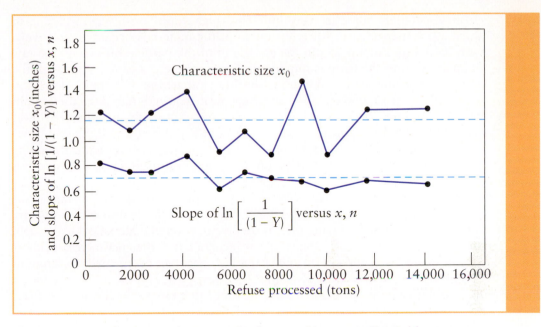

Figure 4-16 Data from an evaluation study of a vertical hammermill shredder.
Source: Vesilind, P. A., A. E. Rimer, and W. A. Worrell. 1980. "Performance Characteristics of a Vertical Hammermill Shredder." Proceedings 1980 National Waste Processing Conference. Washington, D.C. ASME. With permission from ASME.

where x'_0 is again called the *characteristic size* of the feed material. Note the distinction between x'_0 and the Rosin-Rammler x_0. The size ratio r is defined as the ratio of the size of the broken piece to the size of the original piece.

A number of other particle-size distribution functions have been suggested,[22,23,24] but none of these appears to improve the required precision over the models discussed here. They merely add complexity to a problem where the lack of precision in measurement of particle size makes further sophistication questionable.[25] The π breakage theory is an exception, since it provides a complete description of the product from a shredding operation. The explanation of this theorem is complex, and therefore it is included as an appendix to this chapter.

4-5-4 Power Requirements of Shredders

Only limited information is available about the power requirements of shredders. Various estimates indicate that the hammermill is grossly inefficient, with only about 0.1% to 2.0% of the energy supplied to the machine appearing as increased surface energy of the product solids.[26] Part of the explanation for such low efficiencies lies in the plastic deformation and viscoelastic flow that accompanies shredding; these deformation processes require many times more energy than the creation of new surface area.

The efficiency of a shredding operation depends on how the energy is applied and on how the material reacts to it. For example, a roll mill, which crushes materials but does very little shearing, would waste energy trying to shred a newspaper. Also, since fracture in brittle materials occurs progressively, with flaws building up within the particle until the particle breaks up, the rate of application of the force is important. Because there is a time lag between the application of the force and eventual fracture, a machine that applies load rapidly is inherently inefficient, since more energy is required than if the force could be applied more slowly. It follows that for some high-speed machines, a slower speed could result in less energy use, at least to the point at which the rotor inertia is too low and the energy requirement once again increases with decreasing speed. Higher speeds, nevertheless, produce the finest product particle size, but this requires more energy.

Energy requirements also increase with feed rate. Plotting the feed rate versus energy use in kilowatts results in straight lines for both secondary and primary shredding.[27] For low feed rates, there is simply not enough material going through, and the machine is loafing. As the feed rate increases, the efficiency increases (lower specific energy use), and eventually the machine becomes overloaded. The amount of moisture in the feed refuse also influences energy use, and the minimum energy requirement appears to be in the 35 to 40% moisture range.[20] Moisture also affects particle size, with drier feed resulting in smaller product particle size.

Accurate estimates of the energy requirements for size reduction have not been possible, even for homogeneous materials. Several empirical relationships have been suggested. Those that consider only the fracture process are all based on the assumption that

$$\frac{dE}{dL} = -CL^{-n}$$

which states that the energy dE required to achieve a small size change dL in a unit mass of material is inversely proportional to the size of the article L. The symbols n and C are constants. If $n = 1$, for example, the equation can be integrated to yield

$$E = C \log\left(\frac{L_1}{L_2}\right)$$

where L_1/L_2 is the size-reduction ratio and E is the work done to reduce the particles from size L_1 to size L_2. C, as before, is a constant. Physically, this expression states that the work done is proportional to the number of new smaller particles created from the larger ones. This is known widely as *Kick's law*.[26]

If $n = 2$, the integrated form of the equation yields

$$E = C\left(\frac{1}{L_2} - \frac{1}{L_1}\right)$$

which is known as *Rittinger's law* and physically represents the assumption that the energy required is proportional to the amount of new surface formed by the size-reduction process.[26] Rittinger's law seems to agree better with the results of rough grinding operations, whereas Kick's law is a better approximation of fine grinding.

If $n > 1$ but is otherwise undefined, the integrated form of the equation is

$$E = \frac{C}{n-1}\left(\frac{1}{L_2^{n-1}} - \frac{1}{L_1^{n-1}}\right)$$

The *Bond work index*[28] is based on the assumption that the work done in crushing and grinding is directly proportional to the total length of new cracks formed in the material being reduced in size.[29] In this case, it is suggested that $n = 1.5$, and the general equation becomes the *Bond law*:

$$E = E_i\frac{\sqrt{L_F} - \sqrt{L_P}}{\sqrt{L_F}}\sqrt{\frac{100}{L_P}}$$

where E = specific work (kWh/ton) required to reduce a unit weight of material with 80% finer than some diameter LF in micrometers to a product with 80% finer than some diameter L_P in micrometers. In the preceding equation, E_i = work index, a factor that is a function of the material processed. This value is also the theoretical work required to reduce a unit weight from infinite size to 80% finer than 100 μm; units are (kWh/ton).

This expression is more conveniently expressed as

$$E = 10E_i\left(\frac{1}{\sqrt{L_P}} - \frac{1}{\sqrt{L_F}}\right)$$

The work indices for some common industrial materials are tabulated in Table 4-4. The 10 in the equation is the square root of 100 μm, hence the units of E_i are (kWh/ton).

The Bond work index can also be estimated by the dimensionally incorrect empirical relationship[29] as

$$E_i = 2.59\frac{C_s}{SG}$$

where

E_i = work index (kWh/ton)
C_s = impact crushing resistance, ft-lb/in. of thickness required to break
SG = specific gravity

Table 4-4 Work Indices for Common Industrial Materials

Material	Work Index (kWh/ton)
Coal	11.4
Glass	3.1
Granite	14.4
Slag	15.7

Source: [29]

4-5-5 Health and Safety

One need only reflect on the types of materials people thoughtlessly throw away into refuse cans to realize the serious health and safety aspects of shredding MSW. Although the health and safety aspects of size reduction deserve considerable study, only a brief summary of the problem is presented here. Specifically, shredders can be hazardous because of noise, dust, and explosions.

The noise levels around a 3-ton/h hammermill range from 95 to 100 dBA, with much of the noise produced being a low-frequency rumble. (dBA is a standard method of noise measurement and stands for decibels on the A scale of the sound-level meter. This scale is an attempt to duplicate the hearing efficiency of the human ear.)[30] In addition to a high, constant noise level, materials recovery facilities processing MSW produce considerable impact noise, which is difficult to measure and the effect of which on human beings is poorly understood. The existing federal Occupational Safety and Health Act (OSHA) standard limits noise to 90 dBA over an eight-hour working day. The corresponding limit set by the EPA is 85 dBA, which is equal to the value used in Switzerland[40] and Germany.[41] It seems likely that shredder operators will need to wear ear protection, and noise reduction should be considered in the design of future resource recovery facilities.

Dust can cause several problems: It can be a vector for the transmission of pathogenic microorganisms, it can itself have a detrimental effect on health by affecting the respiratory system, and it can explode. This last problem is discussed in the next section.

OSHA standards presently limit dust inhalation to 15 mg/m³ of total dust over an eight-hour day. The value for the maximal allowed working place concentration (MAK value) in Switzerland is 3 mg/m³ for particles that can access the alveoli and 10 mg/m³ for particles that can be respirated.[42] Limited studies of dust production in resource recovery facilities have shown that dust levels are from 7 to 13 times higher than the OSHA standard. This finding, which would not surprise shredder operators, dictates the use of face masks while working.[31]

Plate counts at shredding operations have indicated that the total bacterial counts during the shredder operation are as much as 20 times greater than the ambient, which contains about 880 organisms/m³ of air. Studies have shown that in resource recovery facilities where shredders are used, coliform counts can jump from 0 to 69 per cubic foot, and fecal streptococci from 0 to over 500 per cubic foot.[32] These indications clearly show the potential danger of disease transmission by the air route during shredding of MSW.

The high temperature and metal-to-metal contact in shredders has caused numerous explosions at existing shredder installations. Actually, small explosions such as those due to the breakage of aerosol cans occur regularly, and shredders are designed to accept these without suffering damage. Larger explosions, however, can damage the shredder and have resulted in fatalities.

Two types of explosions are generally recognized: dust explosions and those caused by explosive materials (such as gunpowder and partially filled gasoline cans). Some operators argue that the dust will never become explosive by itself and that all explosions are caused by combustible materials. This contention remains to be investigated.

Dust can cause explosions when the concentration of a combustible dust is sufficiently high and there is adequate oxygen and a spark. Below a certain dust concentration, the heat of combustion is not sufficient to propagate combustion, and an explosion cannot occur. If the concentration of oxygen is reduced to below about 10%, explosions should not occur. Other than continuously flooding the chamber with an inert gas, however, it is difficult to keep oxygen out of the shredder. The only realistic means of preventing explosions is to maintain a high level of surveillance on what is being fed to the shredder. Accordingly, almost all shredder installations have people scanning the feed conveyor for such potentially dangerous items as cans of paint thinner, gasoline tanks from cars, lawn mower engines, and so on.

Even keen surveillance, however, is not foolproof. In one instance, a gas tank was removed from the conveyor belt, but some of the gasoline spilled on the refuse. This occurred just at the change of shifts, and the new crew was not informed of the spilled gas. When the shredder was again fed, the gasoline vaporized and ignited from space heaters above the workers, sending a fireball back into the shredder.

Two methods presently used to reduce the damage when explosions occur in refuse shredders are venting and flame suppression. All shredders are constructed with blowout doors so that the pressure within the shredder housing can escape. Some are equipped with flame suppressors, which are designed to be released as soon as the pressure builds up to a critical level. This system will work for most types of explosions, but it is not fast enough for dynamite, gunpowder, and other explosives.

4-5-6 Hammer Wear and Maintenance

Because of the relatively unsophisticated and brute-force nature of the shredding process, the wear and tear on shredders can be substantial. One of the major maintenance headaches (and expenses) at shredding facilities is the wear of the hammers. The pattern of wear on hammers is illustrated in Figure 4-17. The wear is almost exclusively on the bottom edge of the hammer's crushing face, since this is the area of impact when the material is crushed against the grate. The wear seems to be due mostly to abrasion, although severe impact with very hard objects can also contribute to wear. Hammer wear can be reduced by both hard-facing the hammers with abrasion-resistant alloys and by slowing the speed of shredding.

As the hammers wear down, the shredding performance decays. The percent of materials passing the sieve (Y) can be related to the tons of refuse processed (T) by the equation

$$Y = b_0 + b_1 \exp(-b_2 T)$$

where b_0, b_1, and b_2 are all constants.[33] For example, for a 0.185-in. particle size, the relationship is

$$Y = 8 + 35e^{-0.0081T}$$

Hammer wear for horizontal hammermills can be expected to be in the range 0.05 to 0.10 lb/ton when shredding MSW. Experience has shown, however, that much higher rates of hammer wear are also possible.

Figure 4-18 shows hammer wear distribution for a vertical shredder used for raw refuse.[34] As labeled on the Figure 4-18, station 3 is at the top of the vertical shredder while station 9 is at the bottom. The actual hammers are shown on Figure 4-10. The hammers close to the top of the shredder are used for material breakage and do not suffer much wear, while the hammers close to the bottom show much higher rates of wear.[34]

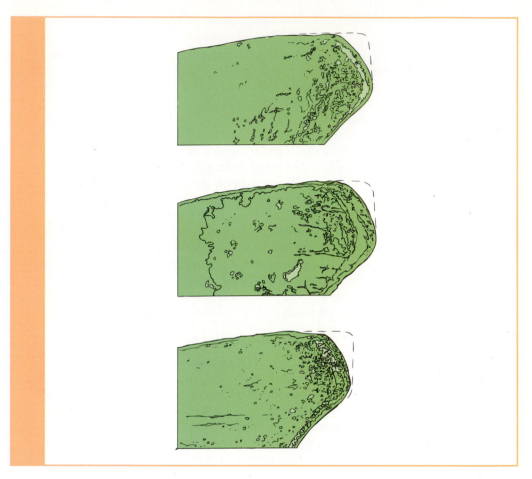

Figure 4-17 Hammer wear in hammermill shredders. (Courtesy of G. J. Trezek)

4-5-7 Shredder Design

Shredders are not, in the strict sense, designed by a consulting engineer or the purchaser. They are actually selected much as water pumps are selected for a specific application. Such specifications as the speed, motor horsepower requirements, and the rotor inertia must be specified.

The rotor inertia is usually expressed as WR^2, where W is the mass of the rotor assembly and R is the radius to the hammermill tips. This is not an accurate measure of rotor inertia but is simply a convenient parameter for comparative purposes. A wide range of WR^2 is offered by manufacturers, from about 50,000 to 150,000 lb-ft^2 (2000 to 6000 kg-m^2).[15]

The motor horsepower is designed on the basis of starting horsepower. If the motor inertia WR^2 is expressed as lb-ft^2, the rotor speed N in rpm, and starting time t in seconds, then the torque T is

$$T = \frac{WR^2 N}{9.6gt}$$

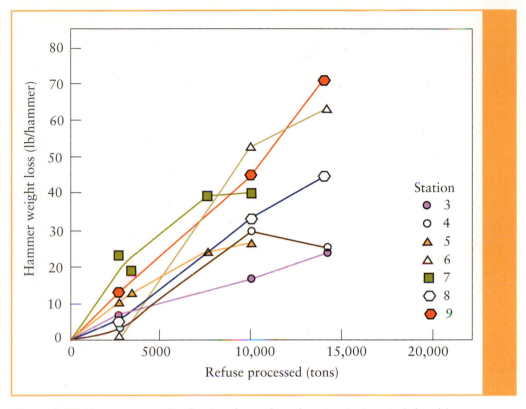

Figure 4-18 Hammer wear distribution depends on location in a vertical shredder. Source: Vesilind, P. A., A. E. Rimer, and W. A. Worrell. 1980. "Performance Characteristics of a Vertical Hammermill Shredder." Proceedings 1980 National Waste Processing Conference. Washington, D.C. ASME.

where g is the gravitational constant at 32.2 ft/sec^2. The horsepower is then calculated as

$$\text{horsepower} = \frac{2\pi TN}{33,000}$$

where T is torque in ft-lb.

Typically, a shredder used for MSW has four rows of hammers with a width/diameter ratio greater than 1.0, a hammer weight of 150 lb (70 kg), a rotor inertia of 35,000 lb-ft^2 (1500 kg-m^2), a hammer tip speed of 14,000 ft/min (4260 m/min), and a starting time of 30 sec. As a rule of thumb, an MSW shredder must be designed for at least 15 kWh/ton. For horizontal hammermills, the grate openings can be changed to achieve different particle-size distribution of the product. Similarly, final clearance in vertical mills can be changed to produce various-sized products. The characteristic size x_0 (discussed in the previous section) and the specific energy can be related to the grate spacing or exit clearance, as shown in Figure 4-19.

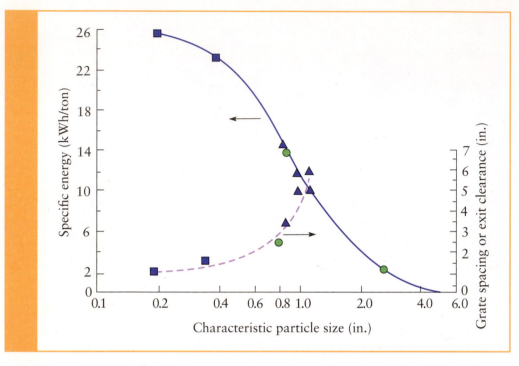

Figure 4-19 Relationship between grate spacing, energy use, and product size. Source: Trezek, G. I., and G. Savage. 1975. "Report on a Comprehensive Refuse Comminution Study." *Waste Age* (July): 49–55.

EXAMPLE 4-4

For x_0 = 0.64 in. (1.62 cm) and n = 1, find the grate spacing (clearance) and the motor power requirements for a feed of 5 tons/h.

SOLUTION

Using Figure 4-19, enter at x_0 = 0.64 and find the grate opening as 2.54 in. and specific energy requirement as 18 kWh/ton, which translates to 90 kW (120 hp).

The Bond work index may be used for shredder design by noting first that the Rosin-Rammler equation can be written as

$$Z = \exp\left(\frac{-x}{x_0}\right)^n$$

where

   Z = cumulative fraction greater than some stated size x
   x_0 = characteristic size
   n = constant

That is, $Z = 1 - Y$, where Y is previously defined as the cumulative fraction finer than some size x.

At $Z = 0.2$ (meaning that 20% of the feed is larger), then $x = L_p$, which is the screen size through which 80% of the product passes. Solving for L_p,

$$0.2 = \exp\left(\frac{-L_p}{x_0}\right)^n$$

$$L_p = x_0(1.61)^{1/n}$$

Substituting into the Bond work index equation,

$$E = \frac{10E_i}{[x_0(1.61)^{1/n}]^{1/2}} - \frac{10E_i}{L_F^{1/2}}$$

This equation now allows for an estimation of shredder power requirements (given a definition of the required product) and an estimation of the work index E_i.

Although no data on refuse shredder power consumption versus particle size are yet available, it is possible to back-calculate for the Bond work index. Table 4-5 is such a tabulation and shows that (on average) the E_i for refuse is about 430 (kWh/ton).

Table 4-5 Bond Work Index for Shredding of Refuse and Refuse Components

Shredder Location	Material Shredded	Bond Work Index, E_i (kWh/ton)
Washington, DC	Refuse	463
Wilmington, DE	Refuse	451
Charleston, SC	Refuse	400
San Antonio, TX	Refuse	431
St. Louis, MO	Refuse	434
Houston, TX	Refuse	481
Vancouver, BC	Refuse	427
Pompano Beach, FL	Refuse	405
Milford, CT	Refuse	448
St. Louis, MO	Refuse	387
Washington, DC	Glass	8
Washington, DC	Paper	194
Washington, DC	Steel cans	262
Washington, DC	Aluminum cans	654

Source: [19]

EXAMPLE 4-5

Assuming that E_i = 430 kWh/ton, x_0 = 1.62 cm, n = 1.0, and L_F = 25 cm (about 10 in., a realistic estimate as shown in Table 4-5), find the power requirement for a shredder processing 5 tons/h.

SOLUTION

$$E = \frac{10(430)}{\left[16,200(1.61)^{1/1}\right]^{1/2}} - \frac{10(430)}{(250,000)^{1/2}} = 18 \text{ kWh/ton}$$

or

$$18 \frac{\text{kWh}}{\text{ton}}\left(5 \frac{\text{tons}}{\text{h}}\right) = 90 \text{ kW}$$

Note that this is the same value as found in Example 4-4.

4-6   PULPING

Wet pulping, although a well-developed process in the pulp and paper industry, has been applied to solid waste processing on a very limited basis. Pictured in Figure 4-20 is a tub 12 ft (3.6 m) in diameter with a high-speed cutting blade on

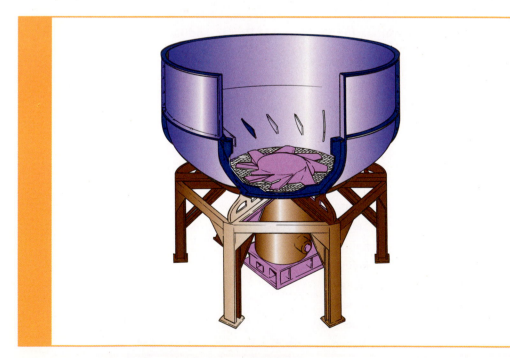

Figure 4-20 Pulper used for processing MSW.

the bottom driven by a 300-hp motor. The raw refuse is pulped, and all pulpable and friable materials are reduced in size to fit through the holes immediately below the cutting blade. The resulting slurry has a solids content of about 4%. Pieces of metal and other nonbreakable materials are ejected from the pulper through an opening on the side of the tub, washed, and put through a ferrous recovery system. The slurry can be centrifuged to remove the organics. No such facilities are operating in the United States; the last one, in Dade County, Florida, was closed in the late 1980s. However, pulping is used to prepare source-separated organic waste for feed to a low solids anaerobic digester, such as the digestor in Toronto, Canada.

4-7   ROLL CRUSHING

Roll crushers are used in resource recovery operations for the purpose of crushing brittle materials (such as glass) while merely flattening ductile materials (such as metal cans), hence allowing for subsequent separation by screening. Roll crushers were first employed in materials recovery facilities for the reclamation of metals from incinerator residue and have found use in processing partially source-separated refuse composed of glass containers and aluminum and steel cans.

A variation of the roll crusher is the roll crusher/perforator. Curbside recycling programs typically collect PET, also known as PETE (soda bottles), and HDPE (milk containers) from residents. Even though instructed not to, some residents screw the lids back onto the empty containers. When these containers are baled, they then do not compress and can cause problems with the finished bales. To solve this problem, some processors have placed roll crushers/perforators in front of the baler. Puncturing the container prior to baling eliminates the problem.

Roll crushers work by capturing and forcing the feed through two rollers operating in opposite directions. The first objective of roll crushing is to capture the pieces that are to be crushed. This capture depends on the size and characteristics of the particles and the size, gap, and characteristics of the rollers. To illustrate the importance of this capture, imagine attempting to force a basketball through a clothes wringer. A small rubber ball, on the other hand, could be captured readily and flattened.

The variables involved in the analysis of roll crushing are shown in Figure 4-21. The diameters of the two rolls are D, while the diameter of the particle to be crushed is d. The normal force between the particle and the rollers is N, and the tangential force is T. If the resultant force R is pointed downward, the particle will be captured and crushed. If it points upward, the particle will ride on the rollers.

The vertical component of N is

$$N_v = N \sin \frac{n}{2}$$

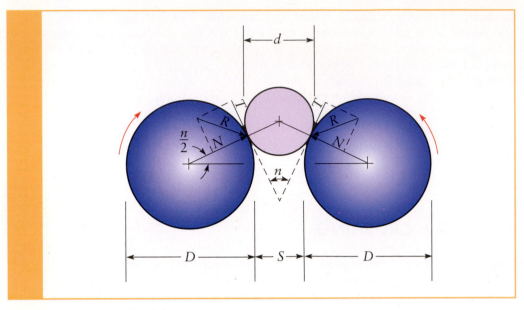

Figure 4-21 Roll crusher.

where n is the angle between the two tangential forces and $n/2$ is the angle between the horizontal and the line connecting the centers of the feed particle and the roller. Similarly, the vertical component of the tangential force is

$$T_v = T \cos \frac{n}{2}$$

At the point where crushing is possible, $N_v = T_v$

$$N \sin \frac{n}{2} = T \cos \frac{n}{2}$$

or

$$\tan \frac{n}{2} = \frac{T}{N}$$

In this instance, the angle n becomes known as the ***angle of nip***. Since T/N = the coefficient of friction (φ), the necessary condition for crushing to take place is

$$\tan \frac{n}{2} \leq \varphi$$

From Figure 4-21,

$$\frac{S}{2} + \frac{D}{2} = \frac{D}{2} + \frac{d}{2} \cos \frac{n}{2}$$

where S is the separation between the rollers. It follows that

$$\cos \frac{n}{2} = \frac{D + S}{D + d}$$

EXAMPLE 4-6

It is intended to crush pieces of a glass of a nominal diameter of 2 cm (feed) to a maximum diameter of 0.5 cm (product). The coefficient of friction between steel and glass is 0.4. Find the diameter of the rollers that will capture and crush this material.

SOLUTION

$$\tan \frac{n}{2} = 0.4$$

$$\frac{n}{2} = 21.8$$

$$\cos \frac{n}{2} = 0.928 = \frac{D + S}{D + d} = \frac{D + 0.5}{D + 2}$$

and $D = 18.8$ cm.

The capacity of a roll crusher can be estimated as the maximum volume squeezed through the rollers (Figure 4-22):

$$C = kMDWS\rho$$

where

C = capacity, tonnes/h
k = dimensional constant = 60 if all dimensional units are as below
M = speed of rollers, rpm
D = diameter of rollers, m
W = length of rollers, m
S = separation, m
ρ = density of material, g/cm³

EXAMPLE 4-7

For the rollers in Example 4-6, find the capacity if $M = 60$ rpm, $W = 0.5$ m, and $\rho = 2.5$ g/cm³.

SOLUTION

$$C = kMDWS\rho = (60)(60)(0.189)(0.5)(0.005)(2.5) = 4.2 \text{ tonnes/h}$$

A related use of roll crushers is as *can flatteners*. Run at slow speed, can flatteners are placed in processing facilities after the magnetic separation of aluminum from steel cans and are economical if the cans are to be shipped long distances.

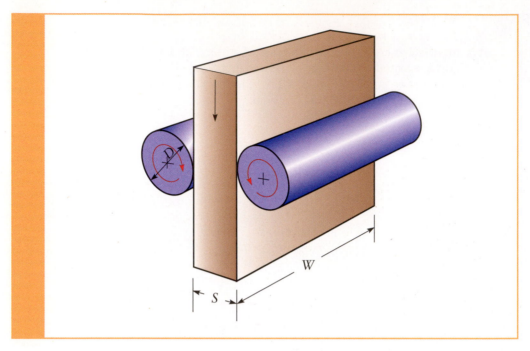

Figure 4-22 Capacity of a roll crusher.

4-8 GRANULATING

For some materials, such as plastic bottles, the high energy and cost of hammermills are not warranted. Size reduction can be achieved far better using *granulators*, which are slow-speed shears that cut instead of shatter. Granulators can be economically effective if the plastic is to be shipped long distances, since the granulated plastic has a far higher density than compacted bottles.

4-9 FINAL THOUGHTS

People who have had extensive experience in the storage and handling of MSW are confirmed cynics. A new piece of equipment, no matter how highly touted by the manufacturer, is received with a wry, knowing smile. The question always is: "Yes, but will it work with MSW?" Too many times, they have seen the unsuccessful transfer of technology and hardware from another application. Always the problem is the same: Solid waste is a heterogeneous and unpredictable material, and equipment designed for a simpler feed cannot handle MSW. With an increasing need for MSW processing, we will see more and more equipment specifically designed for refuse and proven in the field using the real stuff. Only then will these cynics be convinced.

4-10 APPENDIX: THE PI BREAKAGE THEOREM

Breakage of particles can be described by the breakage function, $B(x, y)$, which is defined as the fraction by weight of products that have a size less than x when particles of original size y are broken once. For example, $B(20, 500) = 0.4$ means that 40% by weight of an original particle of size 500 falls below size 20 after a single breakage. The breakage function also can be normalized, so that $B(x/y)$ describes the particle-size distribution regardless of the size of the original particle. For example, $B(0.04) = 0.26$ means that 26% by weight of the product falls in sizes below 0.04 of the original size.

Much of the original development in breakage theory is credited to Epstein,[35] with the matrix application discussed next by Broadbent and Callcott.[36] Trezek at the University of California at Berkeley was the first to apply this technique to MSW,[30] but the development shown here was originally developed by Vesilind, Pas, and Simpson.[37]

The basic assumption in this development is that during the breakage process, a fraction of the particles entering as feed are not broken and exist as part of the product. For any given particle size, therefore, the product consists of some particles that were originally that size and were not broken and some that are the result of larger pieces breaking into smaller ones—the latter being referred to as a *complement*.

A second assumption is that the particle-size *distribution* of the product due to the breakage of any single particle can be described by a continuous function. In other words, a large particle will break into many smaller pieces, which—when analyzed by sieving—will yield a smooth particle-size distribution. In the matrix analysis technique developed by Broadbent and Callcott, the distribution of particles following such breakage is described by the function

$$B(x, y) = \frac{1 - \exp(-x/y)}{1 - \exp(-1)} \qquad \text{(Equation 4-1)}$$

where y is the particle size being broken, and $B(x, y)$ is the cumulative fraction of the product equal to or smaller than x. Although this equation was chosen in the Broadbent-Callcott analysis, other equations may describe the particle-size distribution as well or better and may be found to be superior when this analysis is applied to MSW.

Consider now a series of sieves, where the sieve sizes are related by a constant factor a, so the range of sizes passing the largest sieve is 1 to a, the next sieve is a to a^2, the next sieve a^2 to a^3, and so on to a^{n-1} to a^n. (Such geometric gradation is not necessary for this analysis, but it happens to be convenient for illustrative purposes.) The feed particle-size distribution then can be described by the fractions f falling into these size ranges or grades so that fraction f_1 of the feed is composed of particles between size 1 and size a, f_2 is between a and a^2, and so on. Obviously,

$$\sum_{i=1}^{n} f_i = 1.0$$

Furthermore, assume that the particle-size distribution within each grade can be described by the geometric mean or

$$a^{1/2}, a^{3/2}, a^{5/2}, \ldots, a^{(2n-1)/2}$$

Assuming now, as discussed previously, that the fraction of the feed within each grade that actually breaks is π, the amount of the various fractions that are broken is

$$\pi f_1, \pi f_2, \pi f_3, \ldots \pi f_n$$

The fraction π of the topmost grade (f_1) breaks into smaller pieces according to

$$B(x, y) = \frac{1 - \exp(x/y)}{1 - \exp(-1)}$$

so that of the π fraction breaking, the cumulative fraction of those particles equal to or finer than those in the top grade after breakage is

$$B_{11} = \frac{\left[1 - \exp\left(\dfrac{-a^{1/2}}{a^{1/2}}\right)\right]}{[1 - \exp(-1)]}$$

Note that $B_{11} = 1.0$, since $y = a^{1/2}$ (the particle size being broken), and $x = a^{1/2}$ (the size of the product).

Some fraction of the particles in the top grade break into a smaller size $a^{3/2}$. The cumulative fraction of particles equal to or smaller than this size originating in the top grade of particle size $a^{1/2}$ is

$$B_{21} = \frac{\left[1 - \exp\left(\dfrac{-a^{3/2}}{a^{1/2}}\right)\right]}{[1 - \exp(-1)]}$$

and so on. The breakage of the top grade is summarized in Table 4-6.

Of the topmost feed grade, πf_1 was designated for breakage where $(1 - \pi f_1)$ did not enter the breakage process. The cumulative mass of the original feed that is equal in size to (or smaller than) $a^{1/2}$ after breakage is $B_{1,1}(\pi f_1)$. If $b_{11} = 1.0$, as we have assumed, all of the particles of the product are equal to or finer than $a^{1/2}$.

Similarly, the products of the breakage of the topmost grade are now distributed throughout the smaller grades, so $B_{2,1}$ represents the cumulative fraction of particles equal to or finer than $a^{3/2}$, $B_{3,1}$ is the cumulative fraction equal to or finer than $a^{5/2}$, and so on. The fraction of the product particles in the top grade after breakage is thus $b_{11} = B_{1,1} - B_{2,1}$. This also can be interpreted as the fraction of particles which (although broken) remained in the top grade. Similarly, the fraction of particles in the next smallest grade is $b_{21} = B_{2,1} = B_{3,1}$ and so on.

The example given previously is for the breakage of particles that were originally in the topmost grade only with a geometric size $a^{1/2}$. But breakage occurs within all grades, and this relationship must apply equally well for any other grade of the

Table 4-6

Grade	Cumulative Fraction of Original (πf_1) Particles Equal to or Finer than the Grade After Breakage
1 to a	$B_{11} = \dfrac{1 - \exp(-a^{-1/2}/a^{1/2})}{1 - \exp(-1)}$
a to a^2	$B_{21} = \dfrac{1 - \exp(-a^{-3/2}/a^{1/2})}{1 - \exp(-1)}$
a^2 to a^3	$B_{31} = \dfrac{1 - \exp(-a^{-5/2}/a^{1/2})}{1 - \exp(-1)}$
$\cdot$	$\cdot$
$\cdot$	$\cdot$
$\cdot$	$\cdot$
a^{n-1} to a^n	$B_{n1} = \dfrac{1 - \exp(-a^{-(2n-1)/2}/a^{1/2})}{1 - \exp(-1)}$

feed. Summing up the particles in the product that originated from breakage of larger particles, we get the equations in Table 4-7, where the first b subscript refers to grade of product, and the second subscript defines the origin of that fraction.

For example, the product in the second grade (a to a^2) is now made up of feed particles that originated as the top grade (1 to a) and were broken (the quantity $\pi b_{21} f_1$). The particles within the second grade (although broken) remained there as $\pi b_{22} f_2$. The product of the third grade is composed of particles that originated on the first grade ($\pi b_{31} f_1$), the second grade ($\pi b_{32} f_2$), and the third grade ($\pi b_{33} f_3$). To state it another way, the parameter b_{ij} is defined as the fraction of material in size interval j that falls into the size interval i after breakage. Thus the products of size interval 1 are distributed so that b_{21} falls into size interval 2, b_{31} falls into interval 3, and so on. The sum of the values of by over all the values of i is 1.

Table 4-7

Grade	Product Resulting from Breakage
1 to a	$P_1 = \pi b_{11} f_1$
a to a^2	$P_2 = \pi b_{21} f_1 + \pi b_{22} f_2$
a^2 to a^3	$P_3 = \pi b_{31} f_1 + \pi b_{32} f_2 + \pi b_{33} f_3$
$\cdot$	$\cdot$
$\cdot$	$\cdot$
$\cdot$	$\cdot$
a^{n-1} to a^n	$P_n = b_{n1} f_1 + b_{n2} f_2 + b_{nn} f_n$

Table 4-8

Grade	Total Product
1 to a	$P_1 = \pi b_{11} f_1 + (1 - \pi) f_1$
a to a^2	$P_2 = \pi (b_{21} f_1 + b_{22} f_2) + (1 - \pi) f_2$
a^2 to a^3	$P_3 = \pi (b_{31} f_1 + b_{32} f_2 + b_{33} f_3) + (1 - \pi) f_3$
.	.
.	.
.	.
a^{n-1} to a^n	$P_n = \pi (b_{n1} f_1 + b_{n2} f_2 + \cdots b_{nn} f_n) + (1 - \pi) f_n$

These equations can be summarized in a single equation using matrix notation. If $\mathbf{P}'$ is the product vector, $\mathbf{f}$ is the feed vector, and $\mathbf{B}$ is defined as the breakage matrix derived from Equation 4-1, then

$$\mathbf{P}' = \mathbf{B}(\pi)\mathbf{f}$$

Recall that not all of the particles in any one size fraction were broken in the process. Hence, the product contains a fraction that originally existed in the feed and was not broken: $(1 - \pi)f_n$. The final product within each grade consists of these original particles plus those that resulted from the breakage process. Hence, the equations resulting are given in Table 4-8.

These equations can similarly be expressed in matrix form as

$$\mathbf{P} = \pi\mathbf{Bf} + (1 + \pi)\mathbf{f}$$

In the above equation, the feed vector $\mathbf{f}$ and the product vector $\mathbf{P}$ can be obtained from sieve analysis, and the breakage matrix $\mathbf{B}$ is defined by the quantities b_1, b_2, b_3, . . . and can be calculated. The only unknown quantity is the scalar value p, which could well be a valuable index for describing the shredding operation. The validity of the above equation depends on the validity of the original assumptions. If p is indeed constant for a given shredding operation, any one size particle is as likely to be broken as any other. This is not altogether true, since particles tend to become more resistant to fracture as they become smaller.

The question of the breakage function is still open to argument. Trezek et al. investigated the applicability of several models and decided that either the Gaudin-Meloy function or a modified Broadbent-Callcott function gave excellent results.[20] The latter was written as

$$B(x) = \frac{1 - \exp\left[-(x/x_0)^n\right]}{1 - \exp(-1)}$$

where n is a positive index requiring back-calculations and varying from 0.845 to unity. The term x_0 is the *characteristic size* as previously defined. Figures 4-23 and 4-24 show the results of some experiments using these two functions. The first figure is an estimate of primary shredding, and the second shows the results from secondary shredding. In both cases, either breakage function seems to predict the particle-size distribution adequately.

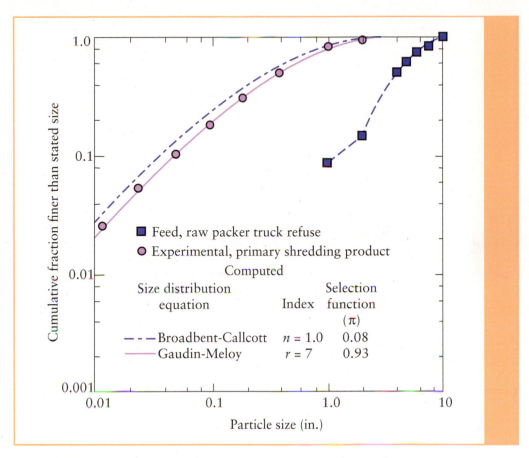

Figure 4-23 Experimental and calculated particle-size distribution for primary shredding. Source: Trezek, G. I., and G. Savage. 1975. "Report on a Comprehensive Refuse Comminution Study." *Waste Age* (July): 49–55.

Another refinement to the model is to eliminate the original assumption that the fraction of any particle size that enters the breaking process is constant for all sizes—in other words, that p is constant. To do this, we can argue that p can be replaced by a matrix S such that it characterizes the size-reduction process and is a diagonal matrix with entries $S_1, S_2, S_3, \ldots S_n$ along the diagonal (zeros elsewhere). The final product equation can be expressed as

$$P = BSf + (I - S)f$$

where I is a unit matrix.

A radioactive tracer has been used to measure the breakage process and to evaluate some of the assumptions made in the foregoing analysis.[39] In this study, the breakage function $B(x, y)$ did not vary with the time of grinding given a certain feed size. It was also found reasonable to normalize $B(x, y)$ to $B(x/y)$. Finally, this study found by back-calculating the values for S and $B(x/y)$ that these values were within reasonable agreement with experimental data.

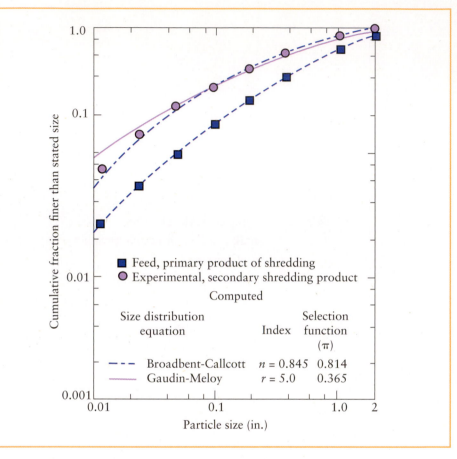

Figure 4-24 Experimental and calculated particle-size distribution for secondary shredding. Source: Trezek, G. I., and G. Savage. 1975. "Report on a Comprehensive Refuse Comminution Study." *Waste Age* (July): 49–55.

EXAMPLE 4-8

For the following feed, calculate the fraction of particles in each size range if it is assumed that $\pi = 0.5$ and the breakage function is $B(x/y) = (1 - \exp(-x/y))/(1 - \exp(-1))$.

Average Size in Sieve (mm)	Fraction of Feed
100	0.80
75	0.20
50	0
25	0

SOLUTION

The basic equation is

$$P = \pi Bf + (1 - \pi)f$$

where P is the product vector, B is the breakage matrix, and f is the feed vector. Written in longhand:

$$\begin{bmatrix} P_1 \\ P_2 \\ P_3 \\ P_4 \end{bmatrix} = (\pi) \begin{bmatrix} b_{11} & 0 & 0 & 0 \\ b_{21} & b_{22} & 0 & 0 \\ b_{31} & b_{32} & b_{33} & 0 \\ b_{41} & b_{42} & b_{43} & b_{44} \end{bmatrix} \begin{bmatrix} f_1 \\ f_2 \\ f_3 \\ f_4 \end{bmatrix} + (1 - \pi) \begin{bmatrix} f_1 \\ f_2 \\ f_3 \\ f_4 \end{bmatrix}$$

$$B(x, y) = \frac{1 - \exp(-x/y)}{1 - \exp(-1)}$$

$$B(100, 100) = \frac{1 - \exp(-100/100)}{1 - \exp(-1)} = 1.0$$

$$B(75, 100) = \frac{1 - \exp(-75/100)}{1 - \exp(-1)} = 0.83$$

This means that of the particles in size interval 1 (100 mm) that enter the breakage process, 83% break down to size 75 mm or smaller. Hence, the fraction of the particles that remain in the size interval 1 (even after breakages) is $b_{11} = 0.17$. Similarly,

$$B_{(50, 100)} = \frac{1 - \exp(-50/100)}{1 - \exp(-1)} = 0.62$$

or 62% of the material that broke becomes 50 mm or smaller. Hence $0.83 - 0.62 = 0.21$, or 21% of the original size material is now in size interval 2, or $b_{21} = 0.21$. Similarly, $B_{25,100} = 0.35$ and $b_{31} = 0.62 - 0.35 = 0.27$, and the last size interval must have a contribution of 0.35. Note that these fractions (0.17, 0.21, 0.27, and 0.35) sum to 1.0.

The material that was at 75 mm also breaks, so that

$$B(50, 75) = \frac{1 - \exp(-50/75)}{1 - \exp(-1)} = 0.77$$

or $b_{22} = 0.23$. Also,

$$B(25, 75) = \frac{1 - \exp(-25/75)}{1 - \exp(-1)} = 0.45$$

and $b_{32} = 0.77 - 0.45 - 0.32$, and $b_{43} = 0.45$. There was no feed material at the 50- and 25-mm sizes, and therefore, $b_{33} = b_{43} = b_{44} = 0$. In summary,

$$\begin{bmatrix} P_1 \\ P_2 \\ P_3 \\ P_4 \end{bmatrix} = (\pi) \begin{bmatrix} 0.17 & 0 & 0 & 0 \\ 0.21 & 0.23 & 0 & 0 \\ 0.27 & 0.32 & 0 & 0 \\ 0.35 & 0.45 & 0 & 0 \end{bmatrix} \begin{bmatrix} 0.8 \\ 0.2 \\ 0 \\ 0 \end{bmatrix} + (1 - \pi) \begin{bmatrix} 0.8 \\ 0.2 \\ 0 \\ 0 \end{bmatrix}$$

The solution to the simultaneous equations is

$P_1 = 0.47$

$P_2 = 0.20$

$P_3 = 0.15$

$P_4 = 0.18$

Note that the sum of the product fractions equals 1.0, as it should.

References

1. Hickman, W. B. 1976. "Storage and Retrieval of Prepared Refuse." *Proceedings ASME National Waste Process Conference,* Boston.

2. Resnick, W. 1976. "Flow Visualization Inside Storage Equipment," *Proceedings International Conference on Bulk Solids—Storage, Handling and Flow.* London: Powder Advisory Centre.

3. *Planning and Specifying a Refuse Shredding System.* Appleton, Wisconsin: Allis-Chalmers.

4. Taggart, A. F. 1945. *Handbook of Mineral Dressing.* New York: John Wiley & Sons, Inc.

5. Mitchell, J. R. 1971. "Designing for Batch and Continuous Weighers." *Chemical Engineering* (Feb. 28):177.

6. Tanzer, E. K. 1968. "Pneumatic Conveying for Incineration of Paper Trim." *Proceedings ASME National Incineration Conference,* New York.

7. Madison, R. D. (Ed.). *Fan Engineering.* Buffalo, New York: Buffalo Forge Company.

8. Bates, L. 1976. "Performance Features of Helical Screw Equipment." *Proceedings International Conference on Bulk Solids—Storage, Handling, and Flow.* London: Powder Advisory Centre.

9. Ruf, J. A. 1974. "Particle Size Spectrum and Compressibility of Raw and Shredded Municipal Solid Waste." Ph.D. thesis, University of Florida.

10. American Public Works Association. 1972. *High Pressure Compaction and Baling of Solid Waste.* EPA-OSWMP SW-32d. Washington, D.C.

11. Ham, R. K. 1975. "The Role of Shredded Refuse in Landfilling." *Waste Age* 6, no. 12:22.

12. Reinhard, J. J. and R. K. Ham. 1974. *Solid Waste Milling and Disposal on Land Without Cover.* EPA (NTIS PB-234 930 and PB-234 931). Washington, D.C.

13. *Baling Solid Waste to Conserve Sanitary Landfill Space.* 1974. EPA-OSWMP. Washington, D.C.

14. Gaudin, A. M. 1926. "An Investigation of Crushing Phenomena." *Transactions,* AIME 73:253.

15. Franconeri, P. 1976. "Selection Factors in Evaluating Large Solid Waste Shredders." *Proceedings ASME National Waste Processing Conference.* Boston.

16. Trezek, G. J. 1974. *Significance of Size Reduction in Solid Waste Management.* EPA-600/2–77–131, Cincinnati.

17. Austin, L. G. 1971–72. "Introduction to the Mathematical Description of Grinding as a Rate Process." *Power Technology* 5, no. 1.

18. Rosin, P. and E. Rammler. 1933. "Laws Covering the Fineness of Powdered Coal." *Journal,* Institute of Fuel 7:29–36.

19. Stratton, F. E. and H. Alter. 1978. "Application of Bond Theory to Solid Waste Shredding." *Journal of the Environmental Engineering Division,* ASCE, 104, no. EE1.

20. Trezek, G. I. and G. Savage. 1975. "Report on a Comprehensive Refuse Comminution Study." *Waste Age* (July): 49–55.

21. Gaudin, A. M. and T. P. Meloy. 1962. "Model and a Comminution Distribution Equation for Single Fracture." *Transactions,* AIME, 223:40–43.

22. Evans, I. and C. D. Pomeroy. 1966. *The Strength, Fracture, and Workability of Coal.* Elmsford, New York: Pergamon Press, Inc.

23. Harris, C. C. 1969. "The Application of Size Distribution Equations to Multi-event Process." *Transactions,* AIME, 244:187–190.

24. Bennet, J. G. 1936. "Broken Coal." *Journal,* Institute of Fuel 10, no. 22.

25. Gawalpanchi, R. R., P. M. Berthouex, and R. K. Ham. 1973. "Particle Size Distribution of Milled Refuse." *Waste Age,* pp. 34–45.

26. Coulson, J. M. and J. F. Richardson. 1955. *Chemical Engineering.* Elmsford, New York: Pergamon Press, Inc.

27. Diaz, L. F. 1975. "Three Key Factors in Refuse Size Reduction." *Resource Recovery and Conservation* 1:111–113.

28. Bond, F. C. 1952. "The Third Theory of Comminution." *Transactions,* AIME, 193:484.

29. Perry, J. H. (Ed.). 1963. *Chemical Engineering Handbook.* New York: McGraw-Hill Book Company.

30. Trezek, G. J., D. M. Obeng, and G. Savage. 1972–73. *Size Reduction in Solid Waste Processing.* EPA, Grant No. R801218, Second Year Progress Report.

31. Drobny, N. L., H. E. Hull, and R. F. Testin. 1971. *Recovery and Utilization of Municipal Solid Waste.* EPA-OSWMP SW-Ioc. Washington, D.C.

32. Diaz, L. F., et al. 1976. "Health Considerations Associated with Resource Recovery." *Compost Science* (Summer): 18–24.

33. *Final Report on a Demonstration Project at Madison, Wisconsin, 1966–1972.* 1973. EPA-OSWMP. Washington, D.C.

34. Vesilind, P. A., A. E. Rimer, and W. A. Worrell. 1980. "Performance Characteristics of a Vertical Hammermill Shredder." *Proceedings 1980 National Waste Processing Conference.* Washington, D.C. ASME.

35. Epstein, B. 1948. "Logarithmic-Normal Distribution in Breakage of Solids." *Industrial and Engineering Chemistry.* 40:2289–2291.

36. Broadbent, S. R., and T. C. Calcott. 1956, 1957. "Coal Breakage Processes." *Journal,* Institute of Fuel 29:524 and 30:13.

37. Vesilind, P. A., E. I. Pas, and B. Simpson. 1986. "Evaluation of the Pi Breakage Theory for Refuse Components." *Journal of the Environmental Engineering Division,* ASCE 112, no. 6.

38. Vesilind, P. A. 1980. "The Rosin-Rammler Particle Size Distribution." *Resource Recovery and Conservation* 5:275–278.
39. Gardner, R. P. and L. G. Austin. 1962. "The Use of Radioactive Tracer Technique and a Computer in the Study of the Batch Grinding of Coal." *Journal,* Institute of Fuel 35:174.
40. Suva, 2015. http://www.suva.ch, visited April 27.
41. BMJV, 2015. Federal Ministry of Justice and Consumer Production, Germany (Bundesministerium der Justiz und für Verbraucherschutz) http://www.gesetze-im-internet.de, visited April 27.
42. Suva, 2015. Grenzwerte am Arbeitsplatz 2015. Suva, Bereich Arbeitsmedizin, Überarbeitete Ausgabe: January 2015, Order Number:1903.d.

Abbreviations Used in This Chapter

dBA = decibels on the A scale of the sound level meter
EPA = Environmental Protection Agency
HDPE = high-density polyethylene
MAK = Maximale Arbeitsplatz-Konzentration, i.e., maximal allowed working place concentration

MSW = municipal solid waste
PETE = polyethylene terephthalate (often also named PET)
RDF = refuse-derived fuel
OSHA = Occupational Safety and Health Act
SG = specific gravity

Problems

4-1. A pneumatic conveyor is to move wood chips with a maximum diameter of 1 in. Estimate the air velocity required.

4-2. Why are *first-in/first-out* storage methods necessary in the processing of MSW?

4-3. Two shredders are used to process nearly identical municipal solid waste. The products are sieved, and the data are presented as Rosin-Rammler plots. From these plots, it is concluded that the characteristic size (x_0) and *n* value are as follows:

Shredder	Characteristic Size (in.)	*n* value
"Trash Mauler"	0.34	1.0
"Gobbler"	0.46	1.2

    a. Which shredder did a better job of reducing the size of the solid waste? How do you know?

    b. Which shredder yielded a more uniform product? How do you know?

    c. Using a Rosin-Rammler plot, show the curves for each shredder. Assume a linear curve on the Rosin-Rammler plot.

4-4. Calculate the shredder energy requirements for shredding the feed to the product as shown in Figure 4-25. Use the Bond work index method. Make any assumptions required.

Sieve Size (inches)	Fraction of Feed Retained on Sieve
10	0.05
8	0.20
4	0.32
2	0.18
1	0.20

4-5. Shredded solid waste was run through a set of sieves with the following results: What is the Rosin-Rammler

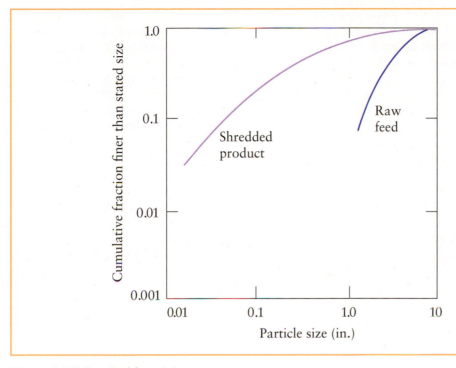

Figure 4-25 See Problem 4-4.

characteristic size (x_0) of this product? Use a graph to obtain your answer.

4-6. Aluminum beer cans are to be processed in a roller mill to a maximum thickness of 0.5 cm. Calculate the size of rollers required.

4-7. Estimate the theoretical maximum capacity of a single lead screw conveyor. The dimensions are measured as pitch = 20 in., radius to conveyor tip = 18 in., and radius to conveyor hub = 6 in. The speed is measured as 30 rpm. How much will the final answer be sensitive to a 10% error in measurement for each of the variables?

4-8. A product from a shredder follows the Rosin-Rammler particle size distribution with $n = 1.0$ and a characteristic size (x_0) of 2 cm. What is the 90% passing size?

4-9. Show, using a mathematical derivation, that the characteristic size (x_0) in the Rosin-Rammler equation is defined as the point where 63% is finer than that particle size.

4-10. A shredded refuse is classified by particle size as follows:

Sieve Size (cm)	Percent by Weight Passing
10	85
5	65
2	48
1	40

What is the characteristic size (x_0) as defined by a Rosin-Rammler plot?

4-11. A shredder has a feed and product as shown:

	Percent by Weight Finer Than	
Size (in.)	Feed	Product
4	80	95
2	15	65
1	5	25
0.5	0	10

a. What is the characteristic size (x_0) of the feed? Of the product?
b. Do both distributions fit the Rosin-Rammler particle-size distribution function?
c. Draw and label typical product particle-size distribution curves for each of the following.
    i. The feed becomes wetter (higher moisture content).
    ii. The shredder is run at a higher speed.
d. What is the effective power requirement (kWh/ton) if the Bond work index is 400? You are reviewing an engineering report recommending the construction of a materials recovery facility. A shredder for processing raw MSW (right off the truck) is recommended as a part of this facility.

4-12. You turn to the cost estimate and discover that the only two costs associated with the shredder are annualized—capital cost and the power cost. What other costs would you recommend be included in this calculation?

4-13. A horizontal hammermill shredder is to process 8 tons/hour of MSW to a characteristic size (x_0) of 0.7 in. What grate spacing and horsepower would be required to accomplish this?

4-14. A horizontal hammermill shredder is to process 8 tons/hour of MSW to a characteristic size (x_0) of 0.7 in. Estimate the required horsepower using the Bond work index method. Assume that the feed characteristic size is 4 in. and the particle-size distribution follows the Rosin-Rammler model, with $n = 1$. (*Note:* You will need log–log graph paper.)

4-15. The city of Durham, North Carolina, has a population of 100,000, and the daily per capita production of MSW is 4.0 pounds. A horizontal hammermill shredder is to be installed to process the raw waste to a product that is 90% finer than 6 cm. The shredder is to operate 8 hours per day. Assume the n in the Rosin-Rammler equation is 1.0, the Bond work index is 400 kWh/ton, and the raw refuse is 80% passing 20 cm. Estimate the size of the electric motor needed to power this shredder.

4-16. A shredder is to reduce the particle size of 80% passing 10 cm to 80% passing 2 cm at a solids flow rate of 10 tonnes (10,000 kg) per hour. Size the motor for this shredder.

4-17. A hammermill shredder is to process 8 tons/hour of mixed municipal solid waste to a characteristic size of 0.7 in. What grate spacing and horsepower would be required to accomplish this? Use the Bond work index method and assume that the feed characteristic size is 4 in. and that the feed particle-size distribution follows the Rosin-Rammler model with $n = 1.0$.

4-18. Describe the approaches used to control the destructive force of explosions in hammermills.

4-19. A horizontal shredder processing municipal solid waste was found to produce shredded product as shown:

Particle Size (in.)	Percent Finer Than (%)
10	95
5	85
2	50
1	34

What is the characteristic size for shredded product?

4-20. What is the Bond law used for? Why is it useful?

4-21. Using Figure 4-18, what action could you take to make the weight loss more uniform between hammers?

4-22. Why is it so difficult to process MSW?

CHAPTER 5

Separation Processes

This chapter is devoted to various means of separating selected components from mixed municipal refuse and/or previously separated recyclables. Materials recovery facilities (MRFs, pronounced "murfs") that process mixed waste are called *dirty MRFs*, while those that process partially separated material (the recyclables) are called *clean MRFs*.

All of the separation devices included in this chapter are based on a principle of *coding* and *switching*. Some property of the material is used as a recognition code (such as magnetic/nonmagnetic or large/small), and switches (such as magnets or screens) are then used to achieve separation. All materials separation devices (including human beings) operate on the same principle: First, there must be a recognizable code to differentiate the materials in question, and then this code must be used in a switching device that physically separates the materials.

5-1 GENERAL EXPRESSIONS FOR MATERIALS SEPARATION

In separating various pure materials from a mixture, the separation can be either *binary* (two output streams) or *polynary* (more than two output streams). For example, a magnet capturing ferrous material is a binary device, whereas a screen with a series of different sized holes, producing several products, is a polynary separation device.

5-1-1 Binary Separators

A schematic of a binary separator is shown in Figure 5-1. The input stream is composed of a mixture of x and y, and these are to be separated. The mass per time (e.g., tons/hour) of x and y fed to the separator is x_0 and y_0, respectively. The mass per time of x and y exiting in the first output stream is x_1 and y_1, and the second output stream is x_2 and y_2.

Assume that the device is intended to separate the x into the first output stream and y into the second. If the separator is totally effective, then all of x goes to the first output and all of y to the second. In practice, this is seldom achievable, and the first stream is contaminated with some y and the second with some x. The effectiveness of the separation then can be expressed in terms of recovery. The recovery of component x in the first output stream is R_{x1}, defined as

$$R_{x1} = \left(\frac{x_1}{x_0}\right)100$$

where recovery is expressed as a percentage and x_1 and x_0 are in terms of mass/time. Similarly, the recovery of y in the second output stream is expressed as

$$R_{y2} = \left(\frac{y_2}{y_0}\right)100$$

Since the mass balance holds,

$$x_0 = x_1 + x_2$$

then

$$R_{x1} = \left(\frac{x_0 - x_2}{x_1 + x_2}\right)100$$

The effectiveness of a separator cannot be judged only on the basis of recovery. Consider for a moment what would happen if the binary separator were run so as to achieve $x_2 = y_2 = 0$. In other words, all of the feed is exited as output number one. In that case, the recovery of x is 100%, but the device is not performing its desired function, since no separation occurs. A second operational parameter is therefore required, and this is usually an expression of purity stated as

$$P_{x1} = \left(\frac{x_1}{x_1 + y_2}\right)100$$

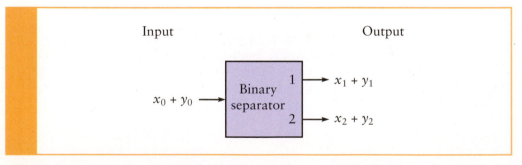

Figure 5-1 Binary separator.

where Px_1 is the purity of the first output stream in terms of x, which is expressed as a percentage. Similarly, the purity of the second output stream in terms of y is

$$P_{y2} = \left(\frac{y_2}{x_2 + y_2}\right)100$$

Usually, both purity and recovery are needed for a complete and accurate description of binary separation performance.

At times, the input and output streams are more conveniently expressed in terms of concentrations instead of mass/time. The equations for calculating the effectiveness of the separation can be shown in this case to be

$$R_{x1} = \frac{[x_1]([x_0] - [x_2])\,100}{[x_0]([x_1] - [x_2])}$$

where x_1 is the concentration of x in output stream 1, x_0 is the concentration of x in the input stream, and x_2 is the concentration of x in output stream 2 with all concentrations expressed as percentages. A similar expression can be written for component y (with the 1 and 2 subscripts reversed, of course). The purity of x would be

$$P_{x1} = \frac{[x_1]\rho_x}{[x_1]\rho_x + [y_1]\rho_y}100$$

where ρ_x and ρ_y are the densities of x and y, respectively.

Often a binary separator is designed to extract one type of material from a waste stream. For example, a magnet draws off ferrous materials as the desired output. Such an output is often called the *product* or *extract*, and the second output is the *reject*. Literally, the magnet extracts the ferrous materials and rejects the rest. In subsequent discussions of materials separation in this text, one output is referred to as the extract and the other as the reject.

5-1-2 Polynary Separators

Two types of polynary systems are possible, as shown in Figure 5-2. In the first case, x_0 and y_0 are the two components in the feed, and the separator has more than two output streams with x and y appearing in all of them (but in different amounts). In such a system, the recovery of x in the first output stream is

$$R_{x1} = \left(\frac{x_2}{x_0}\right)100$$

as before, where x is in mass/time units. Similarly, the purity of x in the first output stream is

$$P_{x1} = \left(\frac{x_1}{x_1 + y_1}\right)100$$

The recovery of x in the mth output stream is

$$R_{xm} = \left(\frac{x_m}{x_0}\right)100$$

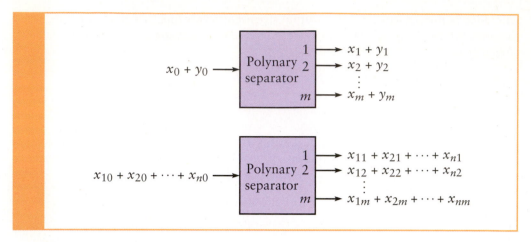

Figure 5-2 Polynary separators.

The second type of polynary separator is the most general case where the feed contains n components $(x_{10}, x_{20}, x_{30}, \ldots x_{n0})$, and these are to be separated into m outputs. The notation is shown in Figure 5-2; x_{11} is the x_1 that ended up in the first output, x_{21} is the x_2 that ended up in the first output, and so on. The recovery of x in the first output is thus

$$R_{x11} = \left(\frac{x_{11}}{x_{10}}\right)100$$

where x_{11} is the x_1 that ended up in the first output and x_{10} is the x_1 in the feed. The purity of this stream in terms of x_1 is

$$P_{x1} = \left(\frac{x_{11}}{x_{11} + x_{21} + \cdots + x_{n1}}\right)100$$

with all x terms having units of mass/time.

5-1-3 Effectiveness of Separation

Because of the inconvenience of using two measures of separation effectiveness (recovery and purity) to define the operation of a materials separator, a single-value parameter would be useful. One such parameter sometimes used in literature, but not recommended, is overall recovery, defined as

$$OR_{x,y} = \left(\frac{x_1 + y_1}{x_0 + y_0}\right)100$$

This parameter is useful only for process design, such as sizing conveyor belts. Since this term is not a measure of separation effectiveness (i.e., 100% overall recovery can be achieved by simply bypassing or turning off the separator), it should not be used in describing the operation of materials separation.

Rietema[1] reviewed these efforts and suggested a measure of separation effectiveness. Stated for a binary separation with input of x_0 and y_0, Rietema defined effectiveness as

$$E_{x,y} = 100 \left| \frac{x_1}{x_0} - \frac{y_1}{y_0} \right|$$

Another means of obtaining a single value of binary separator performance, developed by Worrell[2] and modified by Stessel,[3] is to multiply the fraction of x in the first output stream by the fraction of y in the second output stream and take the square root of the product, or

$$E_{x,y} = \left(\frac{x_1}{x_0} \frac{y_2}{y_0} \right)^{1/2} 100$$

Both Rietema's and the Worrell-Stessel definitions of effectiveness are rational in that if perfect separation occurs (all of x_0 goes to the first output, so that $x_1 = x_0$ and $y_2 = y_0$), the effectiveness is 100%. That is, the device is 100% effective in performing its intended operation. Likewise, if no separation occurs ($x_0 = x_1$ and $y_0 = y_1$), then both measures of effectiveness are zero.

EXAMPLE 5-1

A binary separator has a feed rate of 1 tonne/h. It is operated so that during any 1 hour, 600 kg reports as output 1 and 400 kg as output 2. Of the 600 kg, the x constituent is 550 kg, while 70 kg of x ends up in output 2. Calculate the recoveries and the effectiveness of the separation using the methods discussed previously.

SOLUTION

The recovery of x in the first output is

$$R_{x1} = \left(\frac{x_1}{x_0} \right) 100 = \frac{(550)100}{550 + 70} = 88\%$$

The purity of this output stream is

$$P_{x1} = \left(\frac{x_1}{x_1 + y_1} \right) 100 = \frac{(550)100}{600} = 92\%$$

Using Rietema's definition of effectiveness,

$$E_{x,y} = 100 \left| \frac{x_1}{x_0} - \frac{y_1}{y_0} \right| = 100 \left| \frac{550}{620} - \frac{50}{380} \right| = 76\%$$

and according to the Worrell-Stessel effectiveness equation,

$$E_{x,y} = 100 \sqrt{\frac{550}{620} \frac{330}{380}} = 88\%$$

5-2 PICKING (HAND SORTING)

The most primitive method for the separation of materials from waste (and historically the first) is *hand sorting* or *picking*. Ever since civilization began, scavengers have been an integral part of society. Selectively accepting other people's waste, collecting and processing it, and selling it at a profit is a time-honored profession and, in recent times, quite a profitable one. The first hand-sorting facility in the United States was built by Colonel Waring for New York City in 1898.[4] The refuse from 116,000 people was sorted, and in over $2\frac{1}{2}$ years of operation, about 37% of the refuse was recovered, a major part of which was rags. The recovered material yielded an income of about $1 per ton. The income from this and other plants was not sufficient to maintain them, however, and the job of scavengering was given back to private entrepreneurs, who paid the city about $1 per ton for the privilege.

In many countries scavenging at landfills is discouraged because of health considerations and the potential for accidents. However, throughout the world many people survive by scavenging at landfills. The plight of these workers has been become more known through films such as *Waste Land* and *Landfill Harmonic*. *Waste Land*, a 2010 documentary that was nominated for an Academy Award, follows the *catadores* (waste pickers) in Brazil at the world's largest garbage dump, Jardim Gramacho.

Landfill Harmonic is the story of [should read "story of children"] children who live at Cateura, the trash dump for Asuncion, Paraguay; these children wanted to be in an orchestra. Since there was no money for real instruments, they made instruments from trash—violins and cellos from oil drums, flutes from water pipes and spoons, and guitars from packing crates.

In modern MRFs, pickers (or more properly, *hand sorters*) have two major functions. First, they recover (pick out) items of value from the material being processed. Commonly, corrugated cardboard, bundles of newspaper, and large pieces of metal (reinforcing bars, etc.) are recovered by the pickers. This is known as *positive sorting* because the valuable resources have been removed from the other material. Their second function is to remove all those items that that have no value or could cause damage to the rest of the processing system, such as explosives, as discussed in the previous chapter. This type of sorting is called *negative sorting*.

The functions of hand pickers (salvage and protection) and positive and negative sorting can be combined. For example, at a MRF there could be hand pickers to remove cardboard. Cardboard may be less than 10% of the material at the MRF, and it is very easy to identify. Thus, hand sorters could remove the cardboard using positive sorting. On the other hand, at the end of the MRF there could be a lot of mixed paper with maybe 5% contamination. It would be easier to remove the contamination using a negative sort because the amount of material being removed would be much less. In addition, the hard sorters are always looking for items that could damage the MRF. One large processing facility recovered a piece of titanium 60 cm in diameter and 10 cm thick off the conveyor belt leading to the shredder. Not only is this a valuable piece of metal, but it would have completely destroyed the shredder if it had been allowed to go in.

The *coding* and *switching* functions in hand sorting are simple to define. The material is recognized visually (coding) by such properties as color, reflectivity, and opacity; verified by sensing its density; and removed (separated) by hand picking. Hand sorting is usually done on the conveyor belt after the bags have been mechanically opened in a trommel or a bag-opening flail mill. At a clean MRF, the material may arrive in the loose form or in paper bags, and no opening is needed.

At some facilities, no such preprocessing is used, and the sorting operation is hence highly inefficient. Typically, the conveyor belt is loaded, and the material is leveled out by a skimmer. The pickers stand on either side of the conveyor belt and remove the selected materials. Experience has shown that pickers can salvage up to about 1000 lb/h/person. However, the quantity sorted is highly dependent on the density of the material. For example, the picker removing cardboard removes far more material by weight than the picker removing film plastic.

A picking belt should be no more than 24 in. (60 cm) wide for one-sided picking, or 36 to 48 in. (90 to 120 cm) wide for pickers on both sides, and should not move faster than 30 to 40 ft/min (about 9 m/min), depending on the number of pickers.[5] If at all possible, the picking operation should be done in daylight. Artificial light, especially fluorescent bulbs, gives off a narrow band of light, and this makes identification (coding) of the various components difficult. Large sky-lights should be installed if outside operation is impractical.[6] Picking material from MSW is a dirty and dangerous profession—not recommended for the squeamish. A typical picking belt is pictured in Figure 5-3.

Figure 5-3 Pickers in a clean MRF. (Courtesy William A. Worrell)

5-3 SCREENS

Screening is a process of separation by size. A series of uniform-sized apertures allows smaller particles to pass while rejecting the larger fraction. A particle can pass a screen if it is smaller than the opening in at least two dimensions. Material that passes through the holes is called the *extract*, and material that does not pass through the holes is called the *reject*.

Screening in material recovery operations has been used commonly toward the end of a series of unit operations and is intended primarily for glass removal, since glass would by then have been crushed to fine particles. Screens also have been used for reclaiming a high-organic (garbage) fraction from shredded waste[7] and as rough sorters at the beginning of materials recovery facilities. The breakage and removal of much of the glass in primary screening has been found to be highly beneficial in reducing wear on downstream shredders.

Screens, like other separation devices, cannot be expected to attain 100% recovery. In other words, some undersized material (smaller than the screen apertures) will report as reject and not be removed as extract. (In screening, remember that the *reject* does not pass through the holes, while the *extract* does pass through the holes.) The equation expressing the recovery of undersize materials from a screen is based on the recovery equation for a binary materials separation operation as

$$R_{x1} = \left(\frac{x_1}{x_0}\right)100$$

where

R_{x1} = screen recovery, %
x_1 = amount of material recovery as extract, that is, the material falling through the holes, mass/time
x_0 = amount of undersized material that *could have* fallen through the holes, mass/time

All of the oversize material entering as feed reports as reject—not falling through the apertures—which is only logical, since oversize is defined by screening. However, much of the material in refuse is flexible, and a large plastic bag may fall through a hole size of 4 in. By our definition, the plastic bag is then part of the extract (or the undersize fraction). In binary separation, if we define the extract (exit stream 1 in Figure 5-1) as x_1, then $y_1 = 0$, and the above definition of recovery is adequate to describe screen operation. In other words, the purity of the extract is always 100%. Theoretically, it is possible to achieve very high recoveries with screening, but only at the cost of limiting the throughput. To maintain adequate throughput, most screens are operated between 85% and 95% recovery.[8]

A misleading and yet often used expression of screen effectiveness is

$$R_{x1} = \frac{x_1}{x_0 + y_0}$$

Recovery expressed in this way is more correctly termed a *split* (or that fraction of the feed that exited from output stream number 1). This expression is not a fair indication of how well the screen performed in separating the small particles that it was theoretically capable of separating.

5-3-1 Trommel Screens

A popular screen for processing municipal refuse and compost is the revolving screen or the trommel, which is an inclined cylinder mounted on rollers with holes in the side, as shown in Figure 5-4. The drums roll at slow speeds of 10 to 15 rpm, thus using very little power. The main advantage of the trommel screen is its resistance to clogging. Some of the material within the screen might tend to hang on but will eventually drop off. Trommel screens also can be equipped with spikes to break open plastic bags. This has been used in mixed-waste material recovery facilities at the beginning of the process line.

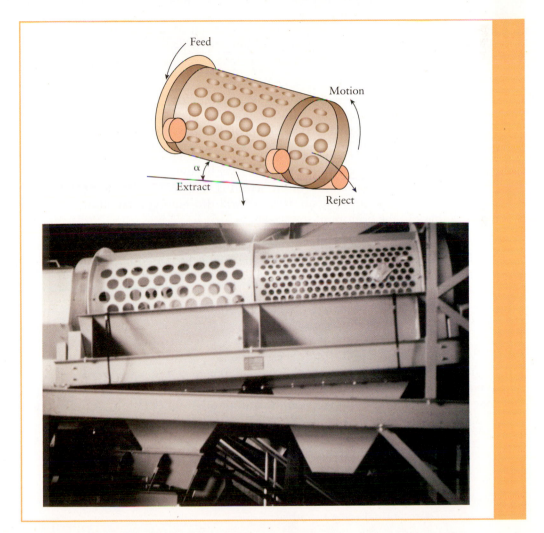

Figure 5-4 Trommel screen. (Courtesy P. Aarne Vesilind)

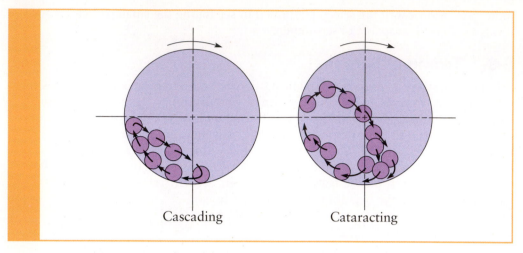

Cascading   Cataracting

Figure 5-5 Two types of particle paths in a trommel screen.

The trommel screen works by allowing the refuse in the screen to tumble around until the smaller pieces find themselves next to the apertures and fall through. The tumbling motion may be of two kinds, as shown in Figure 5-5:

1. *Cascading:* The charge is lifted up by the circular motion of the screen and then tumbles down on top of the layer heading upward.
2. *Cataracting:* The speed of the screen is sufficiently great to actually fling the material into the air, where it will drop along a parabolic trajectory back to the bottom of the screen.

Cataracting produces the greatest turbulence, and the trommel should achieve the greatest efficiency. As the drum speed is increased further, a third type of motion is eventually attained—*centrifuging*. In this case the material adheres to the drum and never drops off, resulting in low recovery.

With reference to Figure 5-6, consider a particle p in contact with the inside of the screen. The centrifugal force acting to press it against the inside wall is c, and the w_1, a component of the gravitational force w, acts to pull it away. The angle between the vertical and the line Op is α_1, and $w_1 - w \cos \alpha_1$. If $c > w_1$, the particle remains in contact with the screen. However, if $w_1 > c$, the particle will fall off. If c remains greater than w_1 as α_1 decreases to zero (particle at the top), the particle never does drop off but remains on the wall through the rotation. At the point of separations, $c = w \cos a$, where a is the angle at which separation occurs.

The centrifugal force is

$$c = \frac{w}{g}(rw^2)$$

where

w = rotational velocity, rad/sec
r = radius, cm
g = acceleration due to gravity, cm/sec^2

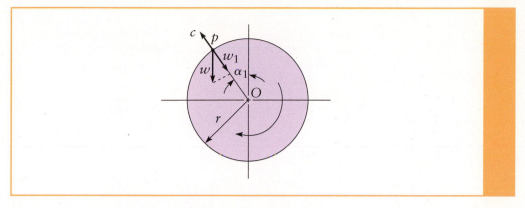

Figure 5-6 Definition of terms for trammel screen analysis.

Combining the two equations,

$$w \cos \alpha = \frac{w}{g}(rw^2)$$

$$\cos \alpha = \frac{rw^2}{g}$$

and since $v = 2\pi n$, where n is the speed of the screen in revolutions per second,

$$\cos \alpha = \frac{4\pi^2 rn^2}{g}$$

From this relationship, it is clear that the angle α at which the particle leaves the wall of the screen and begins its free flight varies with both r and n. The critical point is at $\alpha = 0$, or $\cos \alpha = 1$, so that the *critical speed* is

$$n_c = \sqrt{\frac{g}{4\pi^2 r}}$$

If $w_1 > c$ at α, the particle will lose contact with the wall and begin its flight in a parabolic path until it once again hits the screen (Figure 5-7). The equation of the parabola at the origin p_1 is

$$y = x \tan \alpha - \frac{gx^2}{2V_1^2 \cos^2 \alpha}$$

where

V_1 = initial velocity of the particle p_1 as it leaves the wall
x, y = coordinates

The equation of the circular path of the screen is

$$x^2 + y^2 = (2r \sin \alpha)\, x + (2r \cos \alpha)\, y = 0$$

The simultaneous solution of these equations gives the coordinates of the point d, where these two curves intersect. This point is at the coordinates

$$x = 4r \sin \alpha \cos^2 \alpha$$
$$y = -4r \sin^2 \alpha \cos \alpha$$

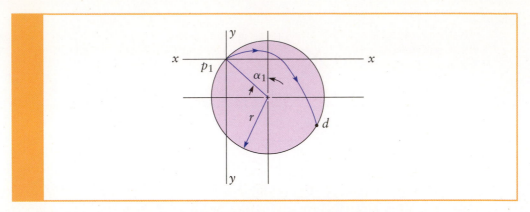

Figure 5-7 Flight path of a particle leaving the inside wall of a trommel screen.

The development here is for a single particle within a screen. This is, of course, an unrealistic assumption, and the motion of particles has to be analyzed, taking into account the action of other particles.

The critical speed and the fraction of the screen occupied by the refuse are related as shown in Figure 5-8. The plot is in terms of bulk volume, which is defined as

$$F = \frac{S}{V}$$

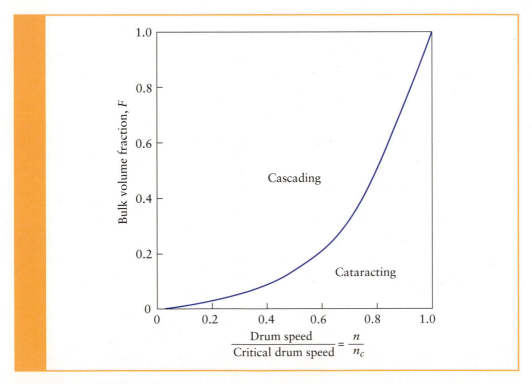

Figure 5-8 Effect of volume occupied by solids on trommel flow characteristics.
Source: [9]

where

F = bulk volume fraction
S = volume occupied by the solids and the air spaces between the solids
V = total volume inside the trommel screen

If the screen is totally full ($F = 1.0$), only cascading is possible for speeds less than critical (there is no space for the particles to fall back through the air). At lower volume fractions, cataracting becomes possible at speeds lower than critical. At the limit, with only one particle in the screen, cataracting occurs at even low speeds, since there is no particle-particle interference.

EXAMPLE 5-2

Assume a 2.7-m-diameter trommel. Calculate the critical speed.

SOLUTION

$$n_c = \sqrt{\frac{g}{4\pi^2 r}} = \left[\frac{980}{4(3.14)^2(270/2)} \right]^{1/2} = 0.43 \text{ rotation/sec}$$

or 26 rpm.

The recovery obviously varies with the speed. A 9-ft-diameter (2.7-m) trommel screen operates at its highest effectiveness at about 45% of the critical speed. The rule of thumb is that recovery is greatest at a speed where the load rides one-third of the distance to the top of the screen.

Another operating variable (not considered here) is the slope of the drum. Within limits, as the slope is increased, the solids retention time is decreased, and the percent of product recovered is decreased because the particles have less chance of finding a hole through which to drop. Where trommel screens have different-sized holes along the drum, the slope affects the amount of material within each size range captured. Figure 5-9 shows some laboratory results with shredded MSW. The trommel for these tests had a section of 1-in. holes.[10] The recovery was measured as the percent of particles less than 1 in. captured. These data for under-loaded conditions show that recovery drops off rapidly with the angle of incline. A small slope, however, would have a low throughput, and a proper balance between recovery and throughput must be achieved.

For unshredded MSW, optimum trommel performance can be obtained if the solids retention time is between 30 seconds and 1 minute,[11] and the material makes 5 to 6 revolutions within the drum.[12] A pilot plant obtained 95 to 100% recovery of 3-in. shredded refuse at a rate of 2 tons/h. At 2.5 tons/h, the recovery dropped to 91%. The 4-ft-diameter, 6-ft-long trommel rotated at 18 rpm.

Figure 5-10 shows how the power requirements for a trommel screen vary with loading. At a very low load, the extra power demand is zero and increases as

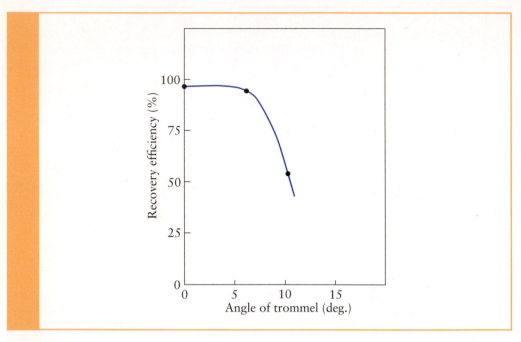

Figure 5-9 Trommel screening of MSW. Angle of the trommel affects efficiency of solids recovery. Source: [10]

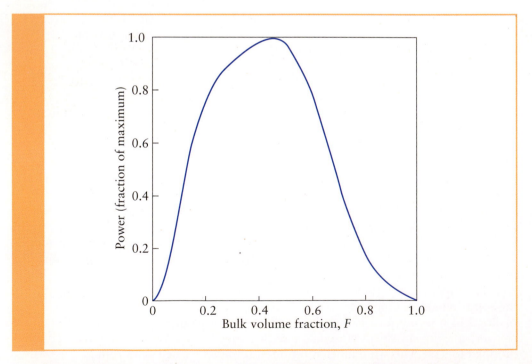

Figure 5-10 Power requirements of a trommel screen as a function of load. Source: [9]

the trommel lifts more and more of the charge. When the trommel fills up and no separation occurs, the extra power demand is again zero.

5-3-2 Reciprocating and Disc Screens

The second type of screen used is a disc screen, which consists of rotating discs that move solid waste across the screen. These screens are very rugged and can process large quantities of solid waste (Figure 5-11).

A third type of screen is an inclined or horizontal shaking screen. This screen, however, is readily plugged by rags, paper, and other objects and is limited in its application to cleaner feeds. One application of such a screen has been the removal of the small pieces of glass so as to produce uniformly sized pieces that might be color sorted. Figure 5-12 is an estimate of the capacity of a vibrating screen when separating glass from shredded refuse.

Disc and reciprocating screens in materials recovery facilities can be used for removing glass after crushing because the glass would be in small pieces, while the size (in two dimensions) of other materials—such as cans, plastics, and paper—would not be reduced. Screens also have been used for reducing the amount of inert (noncombustible) material in facilities that produce refuse-derived fuel (RDF), since again, the shredded paper would be much too big to fit through the holes.

Figure 5-11 Disc screen. (Courtesy William A. Worrell)

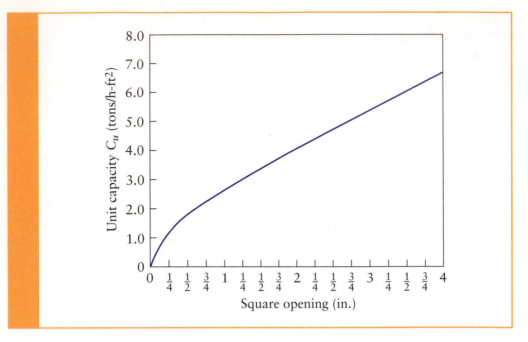

Figure 5-12 Unit capacity of a vibrating screen in separating glass from MSW.
Source: Based on Matthews, C. W., "Screening," *Chemical Engineering*, July 10, 1972.

5-4 FLOAT/SINK SEPARATORS

The *code* for all float/sink separators is the settling (or rising) velocity of a solid particle within a fluid. One of the earliest float/sink separators was the process of winnowing, in which the farmer threw the mixture of grain seed and chaff into the wind and the wind was able to carry off the chaff without suspending the grain (Figure 5-13). More accurately, the settling velocity of the grain was much higher than the settling velocity of the chaff, and the grain thus fell quickly to the earth.

In solid waste separation, a number of operations use the float/sink principle, including air classification (not unlike winnowing), heavy-media separation, jigging, flotation, and others. In all cases, the theory of operation is the same.

5-4-1 Theory of Operation

The motion of a solid particle suspended in a fluid, such as air or water, is governed by three forces:

F_E = some external force such as gravity or centrifugal force
F_B = buoyant force
F_D = drag force

The motion of the particle is described by Newton's law as acceleration = force/mass:

$$\frac{dv}{dt} = (F_E - F_D)\frac{1}{\rho_s V}$$

Figure 5-13 Winnowing, from an old wood cut. Source: Beardsley, J. B. 1937. *From Wheat to Flour*, pp. 28–29. Chicago: Wheat Flour Institute

where

$$V = \text{particle volume, m}^3$$
$$\rho_s = \text{particle density, kg/m}^3$$
$$dv/dt = \text{particle acceleration, m/sec}^2$$

A particle falling in a fluid under gravity has two phases in its motion: acceleration and terminal velocity. The latter is attained when the three forces—F_E, F_D, and F_B—are balanced and the acceleration $dv/dt = 0$. Thus, during a steady fall (terminal velocity),

$$F_B = F_D + F_E$$

These three forces can be expressed as

$$F_E = \rho_s V a$$
$$F_B = \rho V a$$
$$F_D = \frac{C_D v^2 \rho A}{2}$$

where

$$\rho s = \text{density of the solid particle, kg/m}^3$$
$$V = \text{volume of the particle, m}^3$$
$$a = \text{acceleration due to some external force, m/sec}^2$$
$$\rho = \text{density of the fluid, kg/m}^3$$
$$C_D = \text{drag coefficient}$$
$$v = \text{differential velocity between the particle and the fluid, m/sec}$$
$$A = \text{projected area of the particle, m}^2$$

Assume for the sake of convenience that the particle is a perfect sphere,

$$A = \frac{\pi d^2}{4} \quad \text{and} \quad V = \frac{\pi d^3}{6}$$

where d is the particle diameter. Further assuming that the external force causing the acceleration is gravitational with $a = g$, the equation reduces to

$$v = \left[\frac{4(\rho_s - \rho)gd}{3C_D\rho}\right]^{1/2}$$

which is the well-known *Newton's law*. Assuming laminar flow conditions, the drag coefficient is $C_D = 24/N_R$, where N_R = the Reynolds number,

$$N_R = \frac{v\rho d}{\mu}$$

where

v = velocity, m/sec
ρ = density, kg/m^3
d = diameter, m
μ = viscosity, kg/(sec $\times$ m)

Substituting yields the familiar *Stokes law*:

$$v = \frac{d^2g(\rho_s - \rho)}{18\mu}$$

Although this expression is not applicable for air classifiers—because the Reynolds number in air classification is about 10,000, placing the flow well into the turbulent flow regime—it yields some clues as to efficiency of float/sink classification. The objective is to have as large a difference in the settling velocities (v) as possible, and it is clear that the diameter (d) plays an important role, since v is a function of d^2. Often, with such uncontrolled, irregular, and unpredictable material as shredded refuse, the diameter is difficult to define. One solution to this problem is to define an *aerodynamic diameter* (or *hydrodynamic diameter* if the fluid is water), which can be back-calculated using known velocities. Using this technique, the aerodynamic diameters for shredded MSW light fraction have been found to be about 40% of actual diameters, as defined by screening.[13]

Alternatively, a modified drag coefficient might be used. Although C_D is approximately 1.0 for disc-shaped particles under ideal conditions, experimental evidence suggests that $C_D \approx 2.5$ for refuse particles falling in air. Example 5-3 illustrates this concept.

EXAMPLE 5-3

Assuming that the drag coefficient is 2.5, calculate the air velocity necessary to suspend 2 cm (screened) particles of shredded aluminum. Note that $\rho_s = 2.70$ g/cm^3 and $\rho = 0.0012$ g/cm^3.

SOLUTION

$$v = \left[\frac{4(2)(2.70 - 0.0012)980}{3(2.5)(0.0012)}\right]^{1/2} = 1{,}533\,\text{cm/sec}$$

In some cases, instead of adjusting the drag coefficient or back-calculating an aerodynamic diameter, it might be more convenient to define an *effective diameter*, which might be the average dimension of the particle as it is presented to the fluid stream. Since

$$A = \frac{\pi d^2}{4}$$

the effective diameter can be defined as

$$d_a = \left(\frac{4A}{\pi}\right)^{1/2}$$

This may be reasonable where flat objects, such as pieces of paper, are suspended. In other cases, the effective diameter equally well could be defined by the volume as

$$d_v = \left(\frac{6V}{\pi}\right)^{1/3}$$

A further discussion of particle-size analysis is included in the appendix to Chapter 2.

Another problem with applying Newton's law to float/sink classification is that the analysis assumes a single-particle settling in an infinite fluid (no boundary conditions). This is obviously not possible, and some accommodation must be made for the problems of interparticle actions and the effect of the walls.

In the turbulent regime, the effect of the wall can be accounted for by a correction factor[16] given as

$$m = 1 - \left(\frac{r}{R}\right)^{3/2}$$

where

r = radius of the sphere
R = radius of the tube
m = correction factor

This correction factor can be used to adjust the terminal velocity and thus account for the effect of walls, so that

$$v = m\left[\frac{4d(\rho_s - \rho)g}{3C_D\rho}\right]^{1/2}$$

When the particles are sufficiently concentrated, they act as a body with little interparticle movement. This can occur in a float/sink separator when a large slug of feed enters the throat section and is carried upward in a mass with the fluid stream. The velocity of the suspension, v_c, can be expressed relative to the velocity of a single particle, v, as defined by the Newton equation.[14] Under laminar flow conditions (Reynolds number less than about 500):

$$\frac{v_c}{v} = f_1\left(\varepsilon, \frac{d}{D}\right)$$

where D is the diameter of settling or flotation column and ε = void fraction. For turbulent conditions,

$$\frac{v_c}{v} = f_2\left(\frac{vd\rho}{\mu}, \varepsilon, \frac{d}{D}\right)$$

Experimental evidence[14] suggests that, for laminar flow conditions,

$$v_c = v\varepsilon^{[4.65 + 19.5d/D]}$$

For turbulent conditions, d/D is not significant, and

$$v_c = v\varepsilon^{4.65}$$

Finally, this discussion assumes that the particles are all made of rigid materials and that the fluid stream does not change their shape. Obviously, many of the materials in shredded refuse (such as paper and plastics) are flexible and porous, and the shape they present to the fluid in the separation is unpredictable and dynamic.

To summarize, the code in float/sink separators is the aerodynamic (or hydrodynamic) velocity of the particle in the fluid. But this is not the material property that should be used as the code for separation. The particles should be separated by their *density* only and not on the basis of diameter, shape, rigidity, and surface roughness. What is needed is a method of using the float/sink process such that the only basis for separation is density and not the irrelevant material characteristics.

Such an objective can be achieved by first considering the settling column shown in Figure 5-14. In such a simple classifier, if Stokes's law holds, the terminal

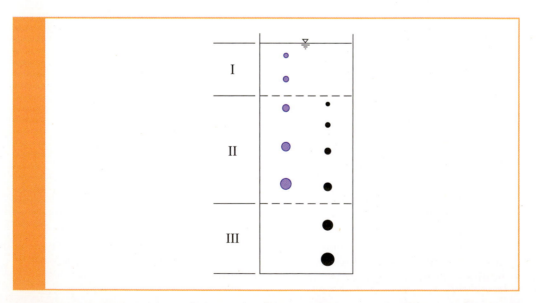

Figure 5-14 Vertical stratification of settling particles based on difference in settling velocities.

settling velocity of particles with densities greater than the fluid would be (based on the Stokes law):

$$v \propto (\rho_s - \rho)d$$

where

v = terminal settling velocity
d = particle diameter
ρ_s = particle density
ρ = fluid density

With reference to Figure 5-14, because the terminal velocity is governed by both size and density differences, the separation achieved is in three phases: the small lights, the large heavies, and a mixture of large lights and small heavies. Starting from rest, these particles accelerate until they attain terminal velocity. This velocity is reached by having the drag force increase from zero (at rest) until it becomes equal to the net gravitational force. At any instant during the acceleration phase, the motion of the particle can be described as

$$F_E - F_B - F_D = \frac{V\rho_s}{g}\frac{dv}{dt}$$

where

F_E = force due to gravity
F_B = buoyant force
F_D = drag force
V = volume of particle
ρ_s = density of particle
g = acceleration due to gravity
v = particle velocity
t = time

At time zero, with $F_D = 0$ and recalling that $F_E = V\rho_s$ and $F_B = V\rho$ (where ρ = fluid density), the initial acceleration is

$$a_0 = \frac{dv}{dt} = \left(1 - \frac{\rho}{\rho_s}\right)g$$

Thus, particles of the same density (or material), regardless of their size, have the same initial acceleration. Of course, the moment resistance enters the picture, this no longer holds, because particle size is a factor in drag forces. For any two particles of different density, the denser particle has a greater initial acceleration than does the less dense particle. Although this difference is small and short-lived, it is the basic principle on which the separation on the basis of density is achieved.

Consider now two particles of the same density but of different size: A = large and B = small. In a still fluid with both particles starting from rest, the large particle will accelerate faster, but since its terminal velocity is greater, it will still be accelerating when the small particle has attained its terminal velocity. This is shown schematically in Figure 5-15. The distance traveled by either of the particles in any given time t is the area under the curve or

$$\int_0^t v\,dt$$

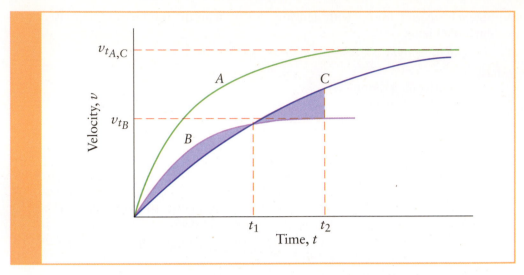

Figure 5-15 Acceleration of particles with different densities and sizes. Particle *A* is high density and small, while *C* is low density and large, and they have equal settling velocities. Particle *B* is also high density, but smaller than *A*.

A third particle, *C*, which is of a greater density but smaller size, has a terminal velocity equal to the large particle *A*. It would be difficult to separate these two particles in a device where both density and size are important. It is possible, however, to take advantage of the different accelerations and achieve separation by imposing a series of short accelerations starting from rest. The distance traveled during any one short spurt is small, but the difference of travel (when summed over many accelerations) is sufficient to have the more dense particle *A* travel farther than the less dense (but larger) particle *B*, thus achieving stratification and eventual separation.

Figure 5-15 illustrates that the denser but smaller particle *B* has—until time t_1—a higher settling acceleration than does the larger but lighter particle *C*. After time *t*, the lighter particle has a higher acceleration and catches up to the smaller particle at time t_2. (The difference between the areas under the curves to time t_1 equals the area from t_1 to t_2.) If the heavy but small particle *B* is to be separated from the lighter but larger particle *C*, the acceleration time has to be well below t_2. This time is a design variable that determines the size of particle to be separated as part of the more dense fraction.

For any float/sink process, if the separation is to be on the basis of density only and not on the basis of other particle characteristics such as size, the objective is to *never allow the particles to attain their terminal velocity*. The particles should be continually accelerated and decelerated—never allowing them to reach terminal velocity. Some of the types of float/sink separators discussed next take full advantage of this opportunity.

5-4-2 Jigs

A *jig* is a device that achieves the separation of less dense from more dense particles by using the differences in their abilities to penetrate a shaken bed. One very simple type of jig is the miner's pan used in gold mining days. This pan is filled with dirt and gravel from a stream bottom and shaken until the grains of gold penetrate the pan content and lodge on the bottom, appearing as the remainder of the pan contents is poured off. The separation of the sand and pebbles from the gold grains and nuggets is not by size but by density. The modern jig works in essentially the same way.

Figure 5-16 shows a diagram of jig. The feed may be thought of as comprising a less dense fraction, a more dense fraction, and a middling fraction (an intermediate mixture, which contains contaminants as well as some light and heavy fraction, depending on the efficiency of the operation). The mixture in the jig is subjected to pulsating forces produced by a plunger (diaphragm, air, or other mechanism) in the water medium so that the entire bed is lifted up and settles back. The particles (as they are settling) never have a chance to attain their terminal settling velocities, and the separation in a jig is on the basis of density only.

As the bed expands, the heavy particles of sufficiently large size and proper shape crash through the bed since the bed is in the *quick* condition and offers little resistance to such settlement. As the pulsations continue, the larger, more dense pieces end up at the bottom, and the less dense particles move to the top of the bed. A screen allows the dense particles to fall to the bottom hopper while the less dense particles (with or without the middling) are drawn off the top.

In addition to the common plunger jig (also called the *Harz jig*) shown in Figure 5-16, many other variations on the same theme are available commercially. Most of these vary the motion of the plunger so that the pressures

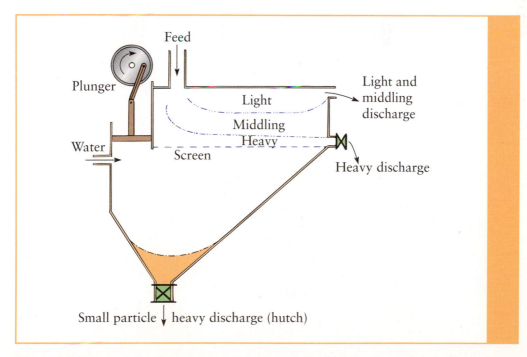

Figure 5-16 A common plunger jig.

imparted on the bed have different frequency diagrams, thus achieving specific bed movement.

5-4-3 Air Classifiers

The objective of air classification is to separate the less dense, mostly organic materials from the more dense, mostly inorganic fraction, using air as the fluid. The basic premise is that the less dense materials will be caught in an upward current of air and carried with the air, while the more dense fraction will drop down, unable to be supported by the air currents. The light fraction entrapped in the air stream must be separated from the air. Commonly, this is done with a *cyclone*, but it can be accomplished equally well with a large box or bag into which the particles drop while the exit air is filtered and escapes. The air can be either pushed or pulled, and the fan can be placed either before or after the cyclone. Except for smaller installations, placement of the fan in order to suck the material through the blades is not recommended because of the wear and tear suffered by the blades. The various arrangements for air classifiers are shown in Figure 5-17.

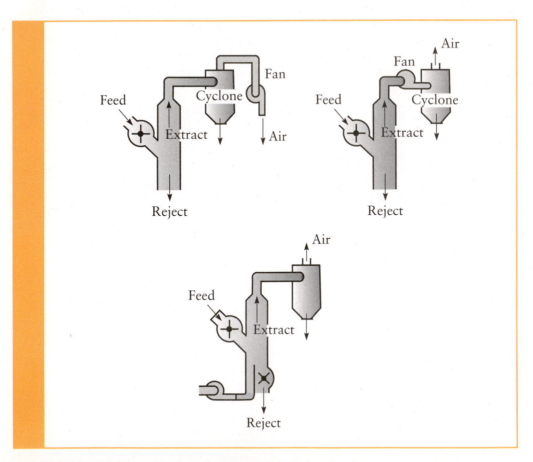

Figure 5-17 Three basic arrangements for air classification.

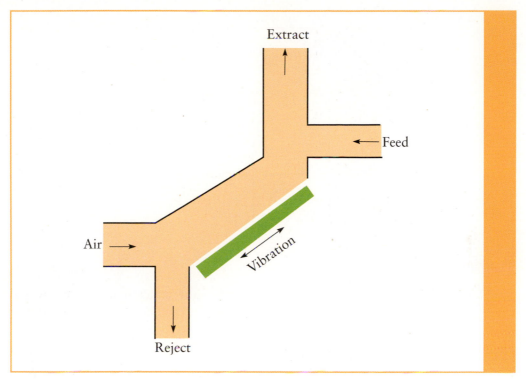

Figure 5-18 Vibrating air classifier.

Air classification is more successful if the particles in the air classifier do not adhere to each other. One means of breaking up the clumps of material is to use vibration. The *vibrating air classifier* is shown in Figure 5-18. Vibrating air classifiers combine the separation achieved due to the vibration with air entrainment. The feed vibrates along a sloping surface with the light material shaken to the top, where the air stream carries it around a U-shaped curve.

Similar in function (but slightly different in construction) is a series of classifiers more properly labeled *air knives*, where the air is blown horizontally through a vertically dropping feed. The aerodynamically light particles will be carried with the air stream while the heavy ones will have sufficient inertia to resist a change in direction and drop through the air stream. This technique also allows for separation into more than two categories, as shown in Figure 5-19. One other use of an air knife has been to help keep light contamination from carrying over during magnetic separation. Air is blown opposite the direction of travel of the metal under a magnet. This helps separate the lights from the metals and keeps the lights from being carried over onto the metals conveyor.

Considerable confusion exists in the terminology for the product (output) streams of air classifiers. Some authors call the material rising with the air stream the *light fraction* and the material falling the *heavy fraction*. These terms are misleading, since they imply ideal separation. Better terms for air classifiers are *overflow* and *underflow*, but these imply vertical geometry. In keeping with the

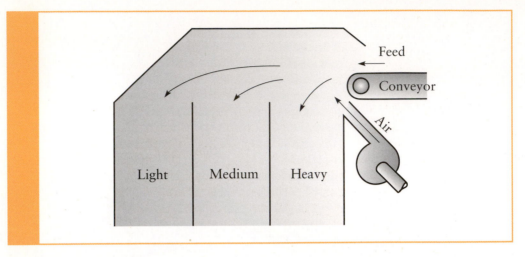

Figure 5-19 Horizontal air knife.

terminology introduced previously, in this text the material suspended by and removed by the air stream is called the *extract*, and the material not so removed is the *reject*.

Theoretical Best Performance of Air Classifiers

Because the *code* for air classifiers is aerodynamic velocity, it should be possible to estimate the effectiveness of air classification by measuring the aerodynamic velocities of the particles to be separated. One such method is the *drop test*, in which representative particles to be separated are dropped from a reasonable height in the absence of air turbulence, and the time needed to reach the ground is measured. Figure 5-20 shows the results of one such test for four different components: paper, plastic, aluminum, and steel.

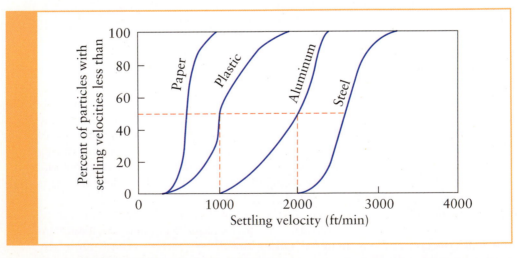

Figure 5-20 Typical results from a drop test.

The theoretical effectiveness of air classification (that is, if all the particles exited the air classifier just as they were expected to) can be estimated from such a test. For example, using the results shown in Figure 5-20, as the air velocity of a classifier is increased from zero, the first material to start to report to the extract (float up) is paper. At an air velocity of 1000 ft/min, all of the paper reports to the extract. But at this velocity, about 50% of the plastic also reports to the extract. If the air velocity is increased beyond 1000 ft/min, some of the aluminum would start to rise with the extract. At an air velocity of 2000 ft/min, all of the paper and plastic would be captured as the extract, but also 50% of the aluminum. The steel fraction would not report to the extract until the air velocity exceeded 2000 ft/min. At 3000 ft/min, almost all of the feed would report to the extract.

If the objective in the air classification of this mixture is to produce a refuse-derived fuel (RDF) consisting of only paper and plastic, the air velocity cannot be allowed to exceed 1000 ft/min, or the RDF will contain some aluminum. If the recovery of the paper and plastic is important and some contamination is allowed, then running the air classifier at an air speed of 2000 ft/min would result in 100% recovery of the paper and plastic fraction. If the objective is to produce a pure metal (aluminum and steel) fraction as the reject (drop to the bottom), running the classifier at 2000 ft/min would result in a pure aluminum/steel mixture, but fully 50% of the aluminum would be lost to the extract.

This test gives the theoretical settling velocities of the particles if the drop distance is great relative to the distance needed to accelerate to terminal settling velocity. For such materials as paper and sheet plastic, this is not a problem. With materials that have high velocities, however, the acceleration distance might be significant, and a simple drop test might give misleading results. Hasselriis has shown that it is possible to calculate the settling velocity from drop-test data using the following relationship.[15]

$$y = \frac{v_s^2}{g} \ln\left(\frac{1}{2}\right)\left[e^{v_s^{gt}} - e^{v_s^{gt}}\right]$$

where

$\quad y$ = distance traveled, ft
$\quad v_s$ = terminal settling velocity, ft/sec
$\quad g$ = acceleration due to gravity, 32.2 ft/sec^2
$\quad t$ = time to travel distance y, sec

This equation requires an iterative solution, since it cannot be readily solved for the terminal settling velocity.

Figure 5-21 is a plot showing the theoretical performance of an air classifier as estimated from a drop test compared to the actual performance. In general, the actual performance should be worse than the theoretical best, which should be viewed as an ideal goal.

In the absence of a drop test, the terminal velocities for various materials can be calculated using an empirical equation:[16]

$$v_s = 1.9 + 0.092\rho_s + 5.8A$$

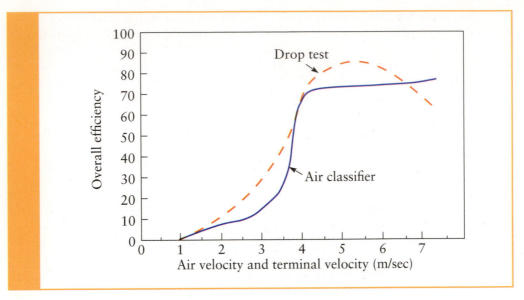

Figure 5-21 Theoretical air classification based on a drop test versus actual air-classifier performance.

where

v_s = terminal (falling) velocity, ft/sec
ρ_s = particle density, lb/ft³
A = particle area (e.g., for a plate, A = length × width)

The size limits to this equation are between 0.0625 and 1.00 in.², which generally is considerably smaller than shredded MSW.

If the area function in the preceding equation is eliminated, little loss in accuracy results.[17,32] In other words, the terminal velocity seems to be most significantly affected by density and only slightly by area and other variables (within the limits used in the experiments). The model, using only density as the independent variable, would then be

$$v_s = 1.91 \rho_s^{1/2}$$

This model shows reasonable agreement with steel, aluminum, paper, and balsa wood as the materials.

Air Classifier Performance

Table 5-1 shows some typical results of air classifier performance with shredded refuse as feed.

The fuel characteristics of the extract in Table 5-1 are shown in Table 5-2. Also tabulated are the fuel properties of two other fuels prepared from MSW by air classification. The first is fuel composed of cubettes that were prepared from air-classified MSW, and the second is a proprietary product known as Eco-Fuel, which is air classifier extract chemically treated to make it biologically stable. Additional information on the properties of refuse-derived fuel is found in Chapter 7.

Table 5-1 Typical Air Classification Results for Shredded MSW

	Percent by Weight	
	Shredded Refuse	Extract from Air Classifier
Noncombustible		
Rocks and dirt	0.3	0
Ferrous metal	7.8	0.08
Nonferrous metal	1.0	0.05
Glass and other	7.8	1.82
Total	**16.9**	**1.95**
Combustible		
Paper	52.2	78.8
Wet garbage	11.8	0.1
Yard and garden	6.7	8.6
Other	12.2	10.6
Total	**82.9**	**98.1**

Source: Murray, D. L., and C. L. Liddell. 1977. "The Dynamics, Operation and Evaluation of an Air Classifier." *Waste Age* (March).

Table 5-2 Fuel Characteristics of Air Classification Shredded and Processed Refuse

	I Air-Classified MSW	II Cubettes	III Eco-Fuel
Moisture (%)	15	15	10
Ash (%)	8.3	6	11
HHV, Btu/lb	6930	6800	6900

Sources: I: [16], II: [17], III: [19]

Paper and cardboard seem to be insensitive to particle size, whereas the heavy fraction is somewhat more sensitive, as shown in Table 5-3. Moisture content does not seem to influence greatly the recovery of lights, although a drop of perhaps 5% in recovery is expected when the moisture content doubles.[18]

Air classifier performance also can be expressed in terms of the air:solids ratio, as shown in Figure 5-22. As the air flow is increased for a given solids feed rate, higher recoveries are experienced. Conversely, lower feed rates for a given air velocity enhance recovery.[20] The capacity of a classifier should be sufficient to withstand sudden surges.[18]

Although the effect of feed rate on air classifier performance has not been fully studied, it seems reasonable that performance would deteriorate with increased feed rate. If a light particle is fed into the throat section, it must accelerate to attain the speed of the air stream. If heavier particles are present, these interfere with this acceleration and, hence, increase further the congestion within the throat.

Table 5-3 Effect of Particle Size on Recovery in an Air Classifier

	Nominal Particle Size, in. (90% passing)		
	1.9	3.1	24.3
Moisture	27.0	32.1	24.3
Paper and cardboard, percent recovered	92.3	90.4	92.8
Grit, percent recovered	74.8	87.7	86.2
Plastic, percent recovered	73.2	70.6	92.7
Percent heavies in extract	1.5	1.7	3.5

Source: Murray, D. L., and C. L. Liddell. 1977. "The Dynamics, Operation and Evaluation of an Air Classifier." *Waste Age* (March).

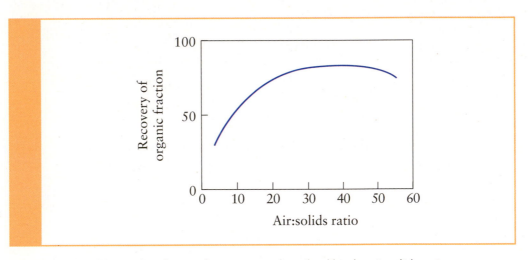

Figure 5-22 Air classifier performance as described by the air:solids ratio.

It is reasonable that the feed rate and average residence time of particles within the throat section are related as

$$t = \left(\frac{V}{Q}\right)^{n}$$

where

t = average residence time of particles
V = volume within throat section
Q = feed flow rate
n = factor that accounts for the congestion within the throat; $n > 1.0$

It is possible to achieve performances that are better than theoretical by taking advantage of the same principles that are used in the operation of the jig. The air classifier is a float/sink separator, and it functions because some particles have a lower aerodynamic velocity and others a higher aerodynamic velocity. To achieve separation on the basis of density only and not on the aerodynamic velocity, air classifiers have been modified to force the particles to continually accelerate

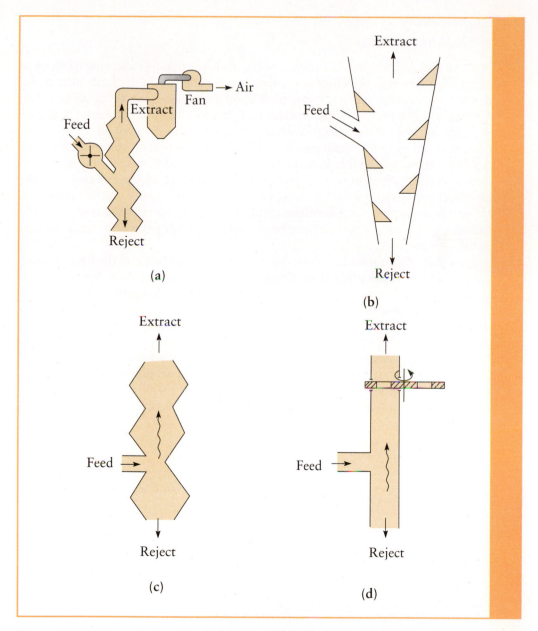

Figure 5-23 Various air classifiers: (a) zig-zag, (b) baffled, (c) constricted, and (d) pulsed flow.

and decelerate. Four designs have been tested to take advantage of this possibility. These are shown in Figure 5-23.

a. Zig-zag classifier

b. Baffled classifier

c. Constricted air classifier

d. Pulsed flow classifier

All four of these modifications use the principle of never allowing particles to attain their terminal velocities and enhancing separation on the basis of material density. In addition, all four create high turbulence within the air classifier throat section, and this promotes the separation of particles that might be stuck together.

The *zig-zag air classifier* consists of a column with a series of 90 or 60° turns. Laboratory work with smoke tracers has demonstrated that at nominal air velocities (flow/cross-sectional area of tube) of 2.5 m/sec (500 ft/min) a central air core is formed with turbulent vortices at the corners that touch the central core.[21] As the air velocity is increased to 3.5 m/sec (750 ft/min), the corner vortices disappear altogether, indicating fully turbulent conditions. The fraction of material reporting as heavies and lights obviously can be altered by changing the flow characteristics in the classifiers.

Within the turbulent vortices, the clumps are broken up, and the light particles are transferred to the upward air stream. The heavy particles drop from vortex to vortex until they exit at the bottom. The dropping action also helps break apart any agglomerated particles. The heavy particles tend to slide down the lower sides of the throat until they are hit by the upward rush of air at the corner (Figure 5-24). If the downward velocity of the particle is sufficiently great relative to the airstream velocity, the particle takes path C, as shown in the figure. This allows it to continue its downward movement on the wall of the next segment. If, however, the air-stream velocity is sufficiently great, the particle may be caught within the air stream and experience trajectory A. If the particle moves along trajectory B, it has a 50% chance of going up or down.[22]

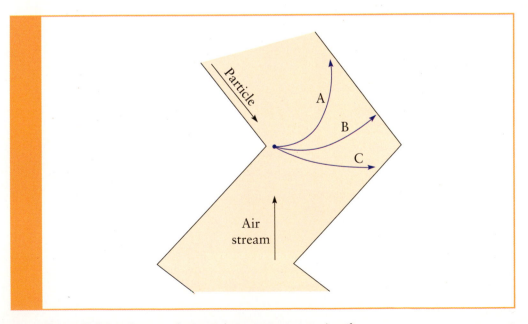

Figure 5-24 Behavior of a particle in a zig-zag air classifier.

A creative alternative is to make the air classifier behave like a jig by pulsing the fluid and causing the particles to constantly be accelerating and decelerating. The *pulsed flow air classifier* has a straight throat, but the flow of air to the classifier is pulsed so that the particles never are allowed to attain their terminal settling velocities.[3]

All four types of modified classifiers—zig-zag, baffled, constricted, and pulsed flow—have the added advantage that the turbulence caused by the air flow helps to break up the clumps of refuse that have been stuck together because of mechanical accidents (e.g., pieces of metal stuck into paper) or adhesive forces (e.g., grease and oil acting as glue).

Removing the Separated Particles from the Air Stream

Air classifiers can operate successfully only if the extracted materials are removed from the air stream once they have been separated. This operation can be accomplished by using a large chamber, but such settling chambers are inadequate both from the standpoint of efficiency and because of the large space requirements. A more efficient means of removing the suspended particles from the air stream is to use a *cyclone*.

The operation of the cyclone is illustrated in Figure 5-25. The air and solid particles enter the cyclone chamber at a tangent, setting up a high-velocity rotational air movement within the chamber. The solid particles (having greater mass) move outward toward the inside wall, are slowed down on contact, and eventually drop to the bottom of the chamber under the influence of gravity. The air exits through the central tube, free of solids.

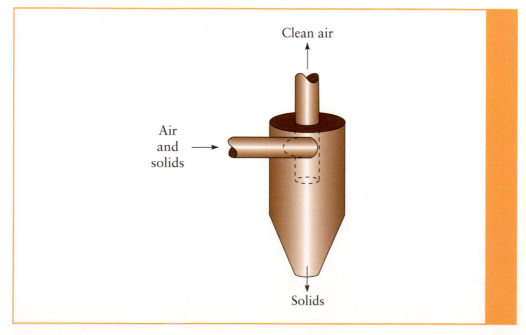

Figure 5-25 Cyclone used for air cleaning.

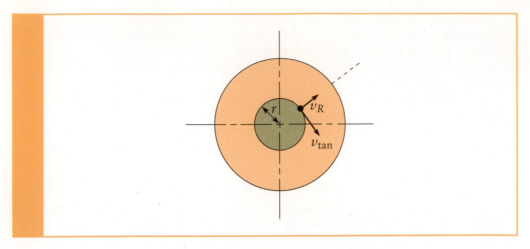

Figure 5-26 Radial movement of a particle within a cyclone.

The objective of a cyclone is to move particles to the outside by centrifugal action. The radial velocity of a particle, v_R is important, as illustrated in Figure 5-26. Assuming laminar conditions (a bad assumption), the particle terminal radial velocity is governed by Stokes's law, except that the gravitational acceleration must be replaced by centrifugal acceleration, defined as

$$a = r\omega^2$$

where

   a = centrifugal acceleration, m/sec^2
   r = radius, m
   ω = rotational velocity, rad/sec

so that

$$v_R = \frac{d^2(\rho_s - \rho)\omega^2 r}{18\mu}$$

where

   d = particle diameter, m
   ρ = density of air, kg/m^3
   ρ_s = density of particle, kg/m^3
   μ = air viscosity, N-s/m^2

Recognizing that $r\omega^2 \times v_{tan}^2/r$, where v_{tan} is the tangential velocity, the radial velocity can be expressed as

$$v_R = v_s \frac{v_{tan}^2}{gr}$$

which is the basic design equation for cyclones. The objective is (for any feed into the cyclone) to have the highest possible v_R, and this can be achieved by having a high tangential velocity (v_{tan}), a high particle settling velocity (v_s), and a small r. The latter observation leads to using banks of small-diameter cyclones instead of a large single unit. For a given volume, the bank of small-diameter cyclones is far more efficient in removing particles from the air stream. Unfortunately, with refuse-derived fuel, a small diameter will result in clogging, and large diameters are absolutely essential.

This analysis may be somewhat simplified by assuming that $\rho_s \gg \rho$. Thus

$$v_R \approx \frac{d^2 \rho_s \omega^2 r}{18\mu}$$

At any radius r, the tangential velocity is

$$v_{tan} = r\omega$$

and

$$\omega = \frac{v_{tan}}{r}$$

where ω is the rotational velocity in rad/sec. The time t needed for one rotation is thus

$$t = \frac{r}{2\pi v_{tan}}$$

The distance traveled by a particle during one rotation is

$$S \approx v_R t \approx \frac{d^2 \rho_s \omega^2 r}{18\mu} \left(\frac{r}{2\pi v_{tan}} \right)$$

where S is the radial distance traveled by a particle during one rotation. (This is necessarily approximate, since the radius and hence the v_{tan} changes as the particle moves.) Since

$$v_{tan} = \omega r,$$

$$S \approx \frac{d^2 \rho_s v_{tan}}{36\mu\pi}$$

the objective is to increase S (the radial distance traveled by the particles during a single rotation). The faster the particle collides with the wall, the faster it is removed from the air stream. From this equation, S can be increased by increasing the particle size d, increasing its density ρ_s, increasing the tangential velocity v_{tan}, and decreasing the fluid viscosity μ. Experience with cyclones for the removal of the shredded light fraction has shown that large factors of safety are needed.

It is useful to review this analysis of separation in a cyclone. It starts with the assumption that Stokes's law holds, which is patently false in the highly turbid environment of a cyclone. When this analysis is used in air pollution control, such as for the removal of fine particulates (about 5 μm in diameter) from combustion, the particles are so small that they are not influenced by the turbulence and tend to obey Stokes's law even in the presence of macroturbulence. In theory, the boundary layer around the particles is still laminar due to their very small size.

This, of course, does not occur when the cyclone is used to separate out shredded paper from the air-classified waste stream. The paper particles, which tend to have dimensions in the centimeter range instead of in micrometers, are most certainly behaving in the turbulent flow regime.

Second, the use of the Stokes equation assumes that each particle is behaving as if it were not affected by other surrounding particles. This assumption is clearly not true for air-classified paper in a cyclone where the feed is a high concentration of paper and other light materials.

This is not to say that this analysis is untrue. It simply must be applied with great care and understanding. Research into the processing of light material (such as shredded paper) is greatly needed.

5-4-4 Other Float/Sink Devices

Three additional options using float/sink technology have been used for MSW processing. The *heavy-liquid separators* substitute a denser liquid for water. For example, a mixture of tetrabromoethane and acetone has been used for the separation of aluminum from heavier materials. This liquid has a specific gravity of about 2.4, and the sink fraction of shredded, air-classified, and screened refuse (with the ferrous fraction removed) can produce a fairly high concentration of aluminum. Two major problems with the use of such heavy liquids are the inability to readily vary the specific gravity as needed and the cost of the chemicals. A significant fraction can be lost during the operation as a result of adsorption to the waste material. Another commercially successful heavy liquid is pentachlorethane (C_2HCl_5: specific gravity 1.67 at 20°C),[17] which has been used for coal cleaning.

Heavy-media separators differ from heavy liquids in that the specific gravity is varied by adding colloidal solids. For example, a mixture of ferrosilicon and water (85:15) with a surface-active agent can be used to attain specific gravities over 3.0. In one study,[23] aluminum was removed by first sinking it in a fluid with a specific gravity of 1.4 and then floating it in a liquid of specific gravity at 3.0. This method also suffers from the problem of capture and retention of the fluid, resulting in higher operating costs for replacement, as well as potential wastewater treatment problems. But it has advantages. The principal advantage of heavy-media separation can be demonstrated by the following argument.[24] Suppose that a mixture is composed of two solids, **a** and **b**, with **a** being denser than **b**. Because of size differences, the larger **b** particles may have higher settling velocities than the small **a** particles, and hence separation by settling cannot be complete. The range of sizes that can be separated can be calculated by recognizing that, at equal settling velocities (assuming laminar flow and Stokes's law),

$$v_a = \frac{d_a^2 g(\rho_{s_a} - \rho)}{18\mu} \quad \text{and} \quad v_b = \frac{d_b^2 g(\rho_{s_b} - \rho)}{18\mu}$$

If the two velocities are equal,

$$\left(\frac{d_a}{d_b}\right)^2 = \frac{\rho_{s_b} - \rho}{\rho_{s_a} - \rho}$$

and separation is possible if

$$\frac{d_a}{d_b} > \left(\frac{\rho_{s_b} - \rho}{\rho_{s_a} - \rho}\right)^{1/2}$$

If it is possible to choose a fluid with a density very nearly equal to one of the solid materials, the classification can be made more efficient. For example, if the fluid density approaches that of particle **b**, the ratio d_a/d_b approaches zero, meaning that very large particles can be separated from even the very small a particles. If the fluid density is greater than the density of **b** and less than that of **a**, complete separation is possible.

The third method of achieving float/sink separation is to use an *upflow separator* with water as the fluid. Effective specific gravities of between 1.1 and 2.0 have been used in commercial devices. Such upflow devices have been used to separate heavy organics (leather, plastics, textiles, etc.) from metals and glass in the heavy fraction of air-classified refuse.[17]

A process that, strictly speaking, is a float/sink process but has a special twist is *flotation*. In this process, the selected solids are floated to the surface of the slurry by means of attached gas bubbles. The key to successful flotation is the selective adhesion of air bubbles to the material that is to be floated. The actual separation, after the material has been made lighter by air-bubble attachment, can be by frothing or by a simple gravity separator such as a shaking table. The usual separation method in resource recovery operations is froth flotation, and the common application is the removal of glass from ceramics and other contaminants.

5-5 MAGNETS AND ELECTROMECHANICAL SEPARATORS

5-5-1 Magnets

Magnets are used to separate ferrous materials from the rest of refuse. The code obviously is the magnetic property of ferrous materials, such as steel. Typical magnet arrangements are shown in Figure 5-27. The usual process uses a belt magnet installed above a conveyor belt, with the belt on the magnet moving across the conveyor carrying the refuse. Steel cans and other ferrous materials are pulled from the refuse and adhere to the underside of the belt covering the electromagnet. The belt moves these materials off to the side where they cease to be under the influence of the magnetic field and drop off the conveyor. The effectiveness of magnets depends on several variables, such as the height of the magnet above the conveyor belt carrying the refuse. The closer the magnet is to the refuse, the better the effectiveness of ferrous removal (Figure 5-28). The greater the magnetic force employed, the greater will be the recovery of ferrous material.[25] The speed of the conveyor also affects the recovery, with higher speeds showing reduced recovery, as would be expected. The magnetic field simply does not have enough time to act on the ferrous material and pull it out of the refuse. Finally, the greater the burden depth on the conveyor belt, the lower the recovery, as shown in Figure 5-29. This relationship is exponential, because the buried ferrous material has an exponentially increasing force on it due to the material above it. Thus, burden depth and the ability to maintain this at some shallow and even depth before it runs under the magnet are the main concerns in designing magnetic separation operations.[25]

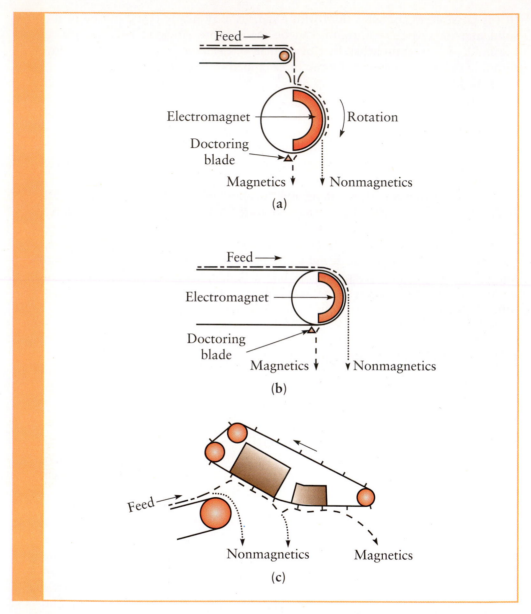

Figure 5-27 Typical magnets used in removal of ferrous materials: (a) drum holding magnet, (b) belt holding magnet, (c) suspended type magnetic separator.

A problem faced with the use of long conveyors prior to magnetic separation is that the jiggling of the material on the conveyor is not unlike a gold-miner's pan, and separation by density occurs, with the steel cans migrating to the bottom of the charge on the conveyor belt. When the refuse is then passed under the magnet, the ferrous material must be able to move through the refuse piled on top to reach the underside of the magnet—an often impossible task. Good engineering requires

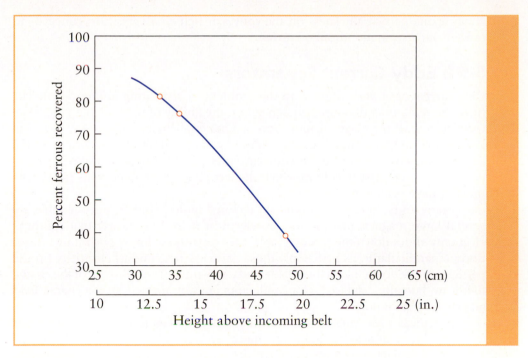

Figure 5-28 Height of a magnet above incoming refuse affects ferrous recovery.
Source: [25]

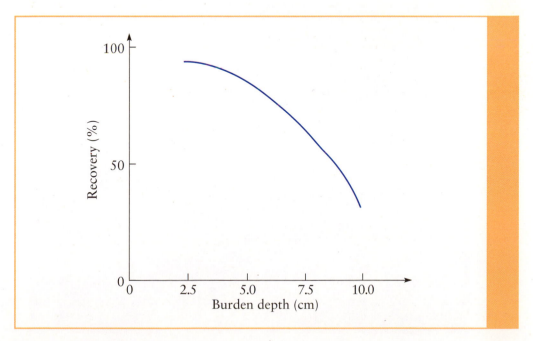

Figure 5-29 Greater burden depth reduces the recovery of ferrous material from a
conveyor belt. Source: [26]

either a one-particle thickness on the conveyor belt or a very short conveyor to prevent such separation from occurring.

5-5-2 Eddy Current Separators

Eddy current separators respond to the problem of separating nonferrous metals from the remainder of refuse and depend on the ability of metals to conduct electrical current. If the magnetic induction in a material changes with time, a voltage is generated in that material, and the induced voltage will produce a current, called an eddy current. The feed to an eddy current separator might be the reject component from air classifiers from which the ferromagnetic components (steel cans mostly) have been removed.

Many eddy current separators are inclined tables. Underneath the table are several large magnets that produce an electrical field. If a particle that conducts electricity slides down the inclined table, the electrostatic forces push it in a direction perpendicular to its path. Only those particles that conduct electricity (as the particles move down the inclined table and come under the influence of the charge field) are laterally displaced. Nonconductors are not affected by the charge field and drop straight down.

A typical eddy current separator is shown in Figure 5-30. This device is a modification of a linear induction motor in that it generates a sine wave of magnetic intensity, which travels down the length of the motor with alternating

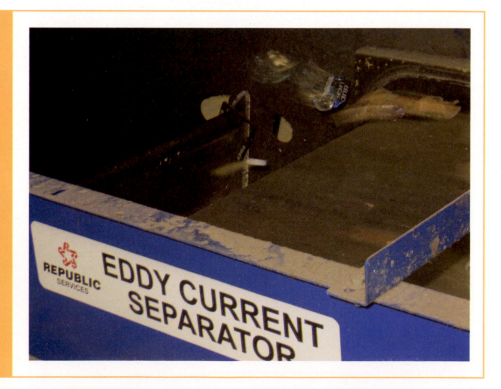

Figure 5-30 Eddy current separator. (Courtesy of William A. Worrell)

Table 5-4 Eddy Current Separation of Shredded Refuse from Which Ferrous Has Been Removed

| | | Aluminum Cans and | Percent by Weight | | |
	Weight (kg)	Other Aluminum	Aluminum Foil	Other Inorganics	Organics
Extract	14.5	88.8	0.2	5.5	5.5
Middlings	7.4	60.2	2.1	11.6	25.1
Reject	85.7	85.7	7.0	3.0	86.6
Feed	107.5	21.9	2.6	3.9	71.5

Source: [27]

north- and south-pole components. As the metal-rich concentrate passes over the linear induction motor, eddy currents are induced in an electrical conductor that appears on the surface of the table. The induced magnetic fields associated with the eddy currents in the metals interact with the moving field generated by the motor, which pushes the conductors (nonferrous metals) along the linear motor. All that is necessary to achieve removal is to orient the motor transverse to the direction of the feed, so as to repel the metal away from the main direction of travel. The mixed material is fed to one end of the nonmagnetic belt, which travels over the linear induction motors positioned on the underside of the belt. Recovery of the metal concentrate is on the top side of the belt, where the material to be removed is ejected by the linear induction motor against the retaining wall and into the extract area. The rejects are not affected by the eddy currents and therefore flow along the lower portion of the belt area. Table 5-4 shows the performance of a typical linear induction motor in separating aluminum from a shredded refuse from which ferrous material has been removed.[27]

Using the numbers in Table 5-4, the eddy current separator was able to achieve a recovery of only little more than 50% with a purity of only 89%—not generally acceptable to secondary materials dealers. Thus, a hand removal of contaminants or additional screening is required after the eddy current separator has been used.[28,29]

5-5-3 Electrostatic Separation Processes

Charged particles under the influence of electrostatic forces obey laws of attraction and repulsion similar to those for magnets. Certain materials can be coded by being electrically charged; then they are separated by being attracted to the opposite-charged electrode or by being repelled from a like-charged electrode.

Separating plastics has been a problem, because the particle properties (the codes) are so similar for various types of plastics. One method of achieving separation is to allow the shredded or granulated plastic particles to acquire an electrical charge by friction as the particles rub against each other. Different plastics have a markedly different ability to take on electrostatic charge, as seen by the charging progression shown in Figure 5-31. The charge effectively codes the particles, and

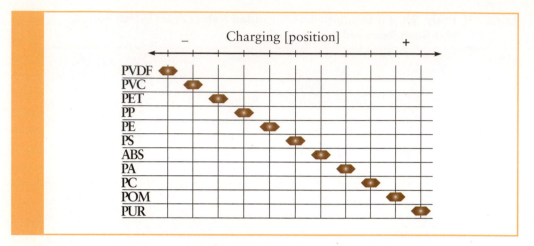

Figure 5-31 Triboelectric charging progression. (Courtesy Steinert)

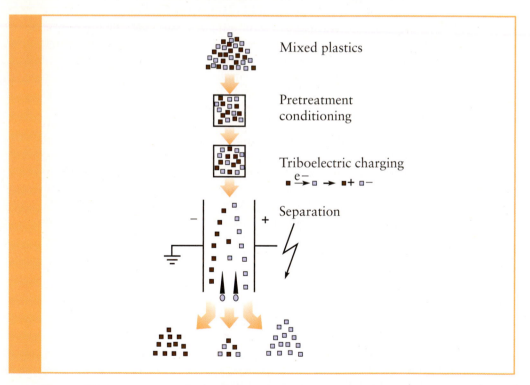

Figure 5-32 Separation of triboelectrically charged plastic. (Courtesy Steinert)

they then can be separated by allowing them to fall freely through a high-voltage field and into separate compartments, as shown in Figure 5-32. Such a process is called a *triboelectric charging* process, since the charge is placed on the particles due to friction. Each type of plastic has a different potential for acquiring a charge, and thus, separation can be made.[30]

5-6 OTHER DEVICES FOR MATERIALS SEPARATION

Listed here are some devices that may someday prove useful in MSW processing.

Stoners, also called pneumatic tables, differ from jigs only in that air is substituted for water as the pulsating fluid. The general principles of operation are the same as for jigs using water. Most commercial models, in addition to pulsating air, use shaking tables, thus providing two forms of the energy input. These devices have been applied for the recovery of aluminum from shredded and screened waste, although operating data have not been made public. Stoners have been used for many years in the agricultural industry for removing stones and other impurities from peanuts, beans, and so on. They are most frequently employed where the two-part separation into light and heavy fractions involves a minor fraction of heavies, with a density difference between the two of at least 1.5:1.[31]

Inclined tables can be used to separate particles of various densities and sizes. For example, coal can be washed by separating slate and other heavy materials from the raw ore by "washing" the mixture down an inclined table and removing the heavy contaminants through ports located along the incline.

Shaking tables differ from simple inclined tables only in that the table is shaken with a differential movement in the direction perpendicular to fluid flow. In addition, all modern shaking tables, such as the **Wilfrey Table**, are equipped with riffles, which are long slots in the table also perpendicular to the flow.

Optical sorting attempts to respond to a major problem with the recovery of glass. Waste glass is of many colors, and such mixtures have low market value. Clear glass alone has substantial value, while contamination by even 5% amber glass makes it essentially unmarketable, since it cannot be used to produce clear glass products. At this time, the only technique, other than hand sorting, for color-sorting glass seems to be by the use of the wavelength of transmitted or reflected light from the glass.

A popular optical sorter, which has in the past been used for the separation of diamonds in diamond mines and rotten corn and peanuts in agriculture, is shown in Figure 5-33, which illustrates the basic principle. The individual particles are moved by means of a high-speed pulley from a bin and flung through a light-detection box or sensor. Within this detector are three lights and three photocells. Each light is bounced off a background reference slide into a photocell, and the current produced is sensed electronically. Should the particle falling through the sensor be of the same color (some average of reflected and transmitted light) as the background slide, the photocell will not detect any difference and nothing happens. These particles are thus passed through the device as *extract*. Should a particle be lighter or darker than the background slide, however, the current change will trigger a compressed-air ejector located immediately below the sensor box and the *reject* particle is blown into a different bin. Dividing a feed of glass particles into three categories—flint (clear), green, and other—requires two passes, because the color sorters on the market are inherently binary devices. In such a case, the first sorting would be based on light intensity reaching the photocells, passing the flint particles, and ejecting the colored particles. In the second sorting of the colored particles, green filters might be placed over the photocells and the electronics tuned so that green particles are perceived as *light* and are passed;

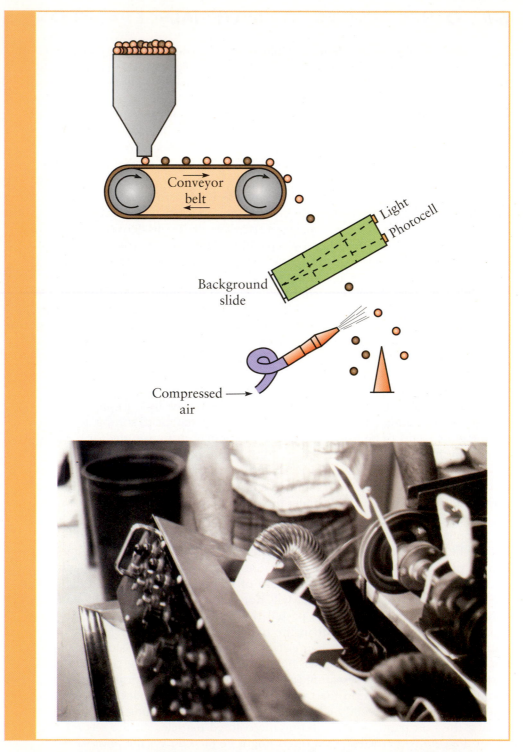

Figure 5-33 Optical sorting. (Courtesy P. Aarne Vesilind)

amber (and other colors on the red end of the spectrum) particles are perceived as *dark* and are ejected.

This device works well when the particles are uniform in size and the electronics can be set so as to make the device reject even suspicious particles. For example, of 100 peanuts entering the sensor, two might be rotten. The device can be adjusted to sense these two, but since it is so finely tuned, it may also eject two other peanuts which may be perfectly good but may have been oriented during their trajectory so as to produce a shadow and thus be rejected. The peanut farmer accepts this margin of safety and gladly discards two good peanuts as long as he is reasonably sure that he has removed all rotten ones. With glass separation, however, the mix is not colored pieces and 98 clear pieces out of 100. Instead, a 50/50 mix is common, and this presents problems of efficiency of operation. To date, it has not been possible to achieve the required purity of product that would make this device a practical alternative for glass sorting. In addition, a color sorter requires a relatively narrow size distribution in the feed, and this requirement places considerable constraints on size-reduction processes.

Optical sorting has also been applied to other commodities in a MRF. For example, it is possible to sort plastic bottles by resin type by using a full spectrum near infrared (NIR) sensor to analyze and classify polymers. Plastic bottles are first dropped on a vibratory feeder (Figure 5-34) and then transported to the optical sorter (Figure 5-35). The optical sorter uses a sensor to identify the polymer and then short air puffs of air to sort the material into 2 or 3 categories.

Figure 5-34 Material being fed to an optical sorter. (Courtesy of William A. Worrell)

Figure 5-35 Optical sorter for plastic sorting. (Courtesy of William A. Worrell)

Bounce and adherence separators have been used in Europe and in one facility in the United States. The bounce and adherence separator is a short conveyor on which the angle and speed can be adjusted. The material falls onto the middle of an angled conveyor, and as the belt moves the material upward, some types of material—such as paper, cardboard and film plastic—lie on (adhere to) the belt, travel up the conveyor, and fall off the upper end. Other materials—such as bottles and cans—roll down the conveyor and fall off the lower end. The optimum efficiency of this separation must be achieved by trial and error as the angle and speed of the conveyor belt are adjusted.

The adherence material can be transported to a paper picking area, and the bounce material would proceed to a ferrous magnet, eddy current separator, or glass picking area. The advantage of this separation is that by using a first (if crude) cut, the process loading on each subsequent unit operation is reduced, thus enhancing the effectiveness of separation.

5-7 MATERIALS SEPARATION SYSTEMS

Each of the unit operations in the previous section accomplishes one separation, and its performance thus far has been analyzed in isolation. Obviously, the placement of the unit operations in series will affect the final products. While there is no "best" series for any given feed, experience over the years has shown that certain combinations seem to perform well.

For example, Figure 5-36 shows a system for processing mixed MSW. In this process, the MSW is stored on a floor where unacceptable items—such as large appliances, old lawn mowers, etc.—are removed and sent to the landfill. The first separation step is a trommel screen, which also acts as a bag opener. Both the reject and extract from the trommel screen are sent to magnets for removal of ferrous products (mostly cans), which are then sent to a can flattener and then to storage. The extract from the trommel (undersize) is sent to the landfill. The reject (oversize) goes to a hand-sorting operation where corrugated cardboard, PETE plastic bottles, HDPE plastic, and aluminum cans are removed. The first three go to balers and then storage, while the aluminum cans are run through a magnet to remove any stray steel cans, and the product is then sent to storage. Following hand sorting, the remaining material is shredded and air-classified. Providing hand sorting ahead of the shredder is an excellent idea, because it minimizes the chance of explosive materials finding their way to the shredder. The shredded material is then air-classified to produce a refuse-derived fuel, which is stored for transportation to a power plant, and the reject is again sent to the landfill.

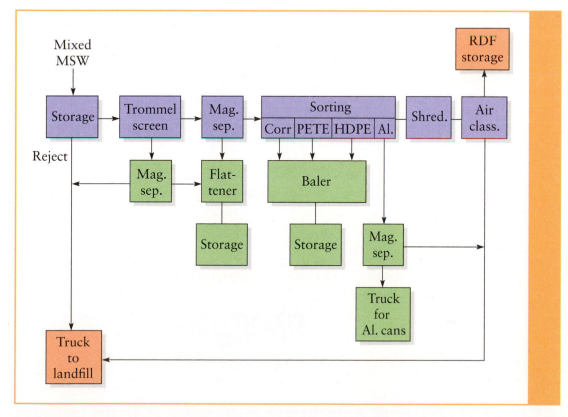

Figure 5-36 A typical dirty materials recovery facility for mixed waste.

Depending on the level of separation performed at the household level, source-separated material collected at curbside or at materials recycling centers can be processed in various combinations. Consider, for example, a system where the householder is asked to sort out all papers, regardless of type. Since mixed paper has a lower market value, a simple separating process consisting of hand sorting would be appropriate, as shown in Figure 5-37. The mixed paper is received at the plant, and a conveyor belt presents the papers to the pickers, who make decisions based on types of paper to be separated, such as office paper, newspaper, and magazines. Unwanted material, such as plastics, would be rejected.

If the separated material consisted of bottles and cans, the processing facility might function like the setup shown in Figure 5-38. Here the aluminum cans are removed from the collection vehicle and run through a magnet for the removal of stray steel cans. Unless the separation has been done by a person on a truck, an operator has to be assigned to the operation to prevent glass bottles and plastics from entering the magnet.

The role of the solid waste engineer, when employed by a municipality or waste authority, is to understand the overall performance of such facilities, and to relate the performance to the available markets. A knowledge of the quality and quantity, as well as dependability, of the waste supply is critical. These engineers must also understand the operational costs and the details of the financing, including sale of the products.

Detailed design of a materials recovery facility is often done by consulting firms, and engineers practicing as consultants must have a workable knowledge of the individual unit operations. These engineers have to understand the separation process as well as such ancillary requirements as traffic flow, power needs, and the structural aspects of the building in which the equipment is to function.

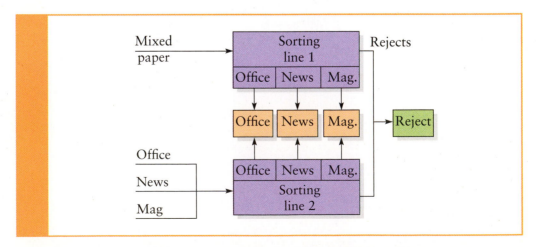

Figure 5-37 A materials recovery facility for previously separated paper waste.

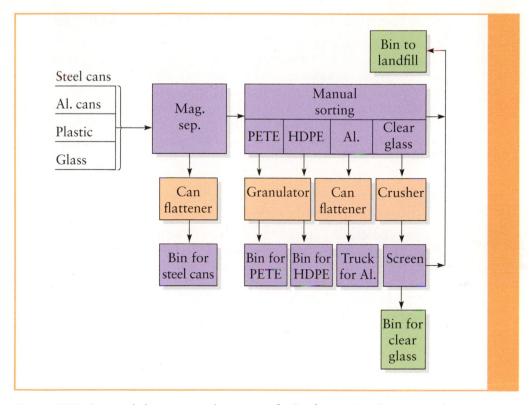

Figure 5-38 A typical clean materials recovery facility for previously separated bottles and cans.

Finally, engineers employed by the manufacturers of the equipment must understand how their equipment will fit into a materials recovery system and what it will be expected to do. These engineers are as critical to the success of the facility as the other two types of engineers. They have a clear conflict, however, since they are employed by the manufacturer to sell equipment, but they have to make sure that this equipment works. They cannot, for example, say that the capacity of a trommel screen is 5 tons/hour, when they know full well that other facilities have found that it can function well only at less than 3 tons/hour. The problem is that there are many trommel screen manufacturers, and the tendency is to oversell the product's performance to make the sale.

In summary, there are three levels of engineering responsibility:

- Engineers who must understand the overall nature of the system, including the nature of the feedstock and the markets for the products.

- Engineers who must understand the system and how each unit operation is to perform within the materials recovery system.

- Engineers who must understand each unit operation and who must be careful to apply such equipment appropriately.

5-7-1 Performance of Materials Recovery Facilities

The potential performance of all such systems, regardless of the arrangement of the unit operations, can be summarized using a notation originally devised by Hasselriis.[15] The system is based on the idea that each unit operation rejects some fraction of the feed and extracts the remaining, and that these fractions of reject and extract are the same regardless of where the unit operation is placed in the process train. Define f as the *split* or the fraction of material rejected by any unit operation, and thus $1 - f$ is the fraction of material extracted by the unit operation. The use of this idea is illustrated by the following example.

EXAMPLE 5-4

Consider a processing operation consisting of an air classifier, a trommel screen, a magnet, and a disc screen placed in various sequences. The feed is assumed to contain four materials: paper, glass, ferrous, and aluminum. The f-value (fraction of material rejected) for each of the unit operations for each of the materials is shown here:

	Air Classifier	Trommel Screen	Magnet	Disc Screen
Paper	0.1	0.9	1.0	0.9
Glass	0.8	0.1	1.0	0
Ferrous	1.0	1.0	0	1.0
Aluminum	0.9	0.9	0.9	0.9

The assumption is that the air classifier will reject 10% of the paper and accept 90% of the paper ($f = 0.1$ and $1 - f = 0.9$). The trommel screen, however, will reject 90% of the paper.

Assume further that the feed rates of the mixed materials are as follows.

Material	Feed, lb/h
Paper	70
Glass	10
Ferrous	10
Aluminum	10

SOLUTION

Consider now option 1, the placement of the unit operations as

```
        extract          extract
           ↑                ↑
feed → air classifier → magnet → trommel → reject
                                    ↓
                                 extract
```

Begin by placing the feed quantities in the form of a vertical array (A) followed by an array of the *f*-values (f).

$$
\begin{matrix}
A & & f \\
\begin{bmatrix} 70 \\ 10 \\ 10 \\ 10 \end{bmatrix} & \times &
\begin{bmatrix} 0.1 & 1.0 & 0.9 \\ 0.8 & 1.0 & 0.1 \\ 1.0 & 0 & 1.0 \\ 0.9 & 0.9 & 0.9 \end{bmatrix}
\end{matrix}
$$

Then calculate the reject from the air classifier by multiplying the *f*-value times the feed, or 70 × 0.1 for the paper, 10 × 0.8 for the glass, and so on. The resulting reject (B) (or that material that moves on to the next unit operation) is thus 7 for paper, 8 for glass, and so on. The entire array looks like the following:

$$
\begin{matrix}
A & & f & & B \\
\begin{bmatrix} 70 \\ 10 \\ 10 \\ 10 \end{bmatrix} & \times &
\begin{bmatrix} 0.1 & 1.0 & 0.9 \\ 0.8 & 1.0 & 0.1 \\ 1.0 & 0 & 1.0 \\ 0.9 & 0.9 & 0.9 \end{bmatrix} & = &
\begin{bmatrix} 7 & 7 & 6.3 \\ 8 & 8 & 0.8 \\ 10 & 0 & 0 \\ 9 & 8.1 & 7.3 \end{bmatrix}
\end{matrix}
$$

For example, the fraction of paper rejected by the air classifier is 0.1, so that 70(0.1) = 7. This reject then moves to the second unit operation, the magnet, where all of it is rejected, or 1.0(7) = 7. Moving on to the trommel screen, 0.9 is rejected, or 7(0.9) = 0.63, which is the reject from the trommel. This result can be used to calculate the recovery and purity of various products. For example, the recovery of paper (that fraction of paper entering the system that is actually recovered as the extract from the air classifier) is calculated by first noting that the paper in the extract from the air classifier is 70 − 7 = 63. The recovery is then

$$R_{paper_{air\ classifier}} = \frac{63}{70}100 = 90\%$$

The purity of the paper is calculated by noting than in addition to the 63 of paper, the air classifier extract includes 10 − 8 = 2 of glass, no ferrous, and 10 − 9 = 1 of aluminum for a total of 66. The purity is then

$$R_{paper_{air\ classifier}} = \frac{63}{66}100 = 95\%$$

Similarly, the recovery of glass (as the trommel extract)

$$R_{glass_{trommel}} = \frac{8 - 0.8}{10}100 = 72\%$$

and its purity is

$$R_{glass_{trommel}} = \frac{7.2}{0.7 + 0.8 + 7.2}100 = 83\%$$

Suppose the unit operations are now arranged as follows:

```
                          extract         extract
                            ↑               ↑
feed → trommel screen → magnet → air classifier → reject
             ↓
          extract
```

The array would look like this:

$$
\begin{array}{cccc}
A & f & & B
\end{array}
$$

$$
\begin{bmatrix} 70 \\ 10 \\ 10 \\ 10 \end{bmatrix} \times \begin{bmatrix} 0.9 & 1.0 & 1.0 \\ 0.1 & 1.0 & 0.8 \\ 1.0 & 0 & 1.0 \\ 0.9 & 0.9 & 0.9 \end{bmatrix} = \begin{bmatrix} 63 & 63 & 63 \\ 1 & 1 & 0.8 \\ 10 & 0 & 0 \\ 9 & 8.1 & 7.3 \end{bmatrix}
$$

Using the product array, we can calculate the recovery of paper by first noting that the extract from the air classifier is $63 - 6.3$ or about 56.7. Thus the recovery is

$$
R_{paper_{air\,classifier}} = \frac{56.7}{70} 100 = 81\%
$$

but now the purity of the paper is calculated by noting that the extract includes, in addition to the paper, $1 - 0.8 = 0.2$ of glass, 0 of ferrous, and $8.1 - 7.3 = 0.8$ aluminum, for a total of 57.7. The purity is then $(56.7/57.7)100 = 98.2\%$, which is far better than with the first option. Clearly, if the primary objective of the materials recovery facility is to produce a paper product of high purity (say, greater than 98% purity), then the second option would be the better arrangement of the unit operations (even though only 81% of the paper would be recovered).

A caution that ought to be obvious: We assume in the foregoing analysis that the fraction of material extracted or rejected is the same regardless of what unit operation precedes the separation step, so that the quality of the feed does not matter in the recovery or purity of the product. This is not true, of course, but as a means of arriving at rough calculations and analyses for separation, the process is useful.

From the previous discussion, it is clear that there is a multitude of ways we can construct materials recovery facilities. No attempt has been made in this chapter to present an exhaustive list of the possibilities. Handbooks such as Taggart's *Handbook of Mineral Dressing*[32] and manuals such as Susan Kinsella and Richard Gertman's *Single Stream Recycling Best Practices Manual* present an impressive compilation of machines and processes and include some techniques used in MRFs. An emerging field such as resource recovery has to borrow from many fields and select those devices that show most promise. With thousands of facilities throughout the United States and around the world processing solid waste, recyclables, and yard waste, the different designs and combinations of processing equipment are almost endless. The "perfect" system has yet to be designed, and given the variable nature of solid waste, it never will be designed. Each facility has a unique set of requirements, and the engineer must factor all of these into the design.

5-8 FINAL THOUGHTS

In the United States and many other countries, environmental decision making has become increasingly public and consultative. A decision to construct a power station, a prison, or a waste-to-energy plant affects many people. These people rightly have a say in the planning process through formal and informal participation, including consultation, lobbying, and public hearings. The opportunity for class action suits and the relative ease with which United States law grants standing to concerned individuals in environmental cases provide additional access for public involvement. Despite all of this representation, one large group of people receives no hearing. These are the people yet to be born.

The effects of today's management decisions will be felt for years to come. Indeed, many decisions have no effects until decades later. For example, it may take generations for waste containers to corrode, for their contents to leach, for the leachate to migrate and pollute groundwater, and for toxic effects to occur. The only persons to be adversely affected by a management decision may be the only persons who have no say in the decision. They cannot speak up for themselves, and, unlike children, they have no one legally responsible for representing their interests. In fact, we often deliberately ignore the interests of future generations. For decades, we have produced (and continue to produce) long-lived radioactive wastes without any agreement about how to keep them safe in the future or even whether they can be kept safe at all. If we were serious about protecting future generations, we wouldn't generate the waste until we have figured out how to manage it safely.

Instead, there is no clear long-term responsibility for hazardous waste management. United States legislation is much more comprehensive and, in principle, tougher than that of most countries. The main statute, the Resource Conservation and Recovery Act of 1976, requires waste facility operators to provide "perpetual care" for depositories. "Perpetual care," however, is defined as a period of only 30 years. In terms of the interests of future generations, when we say "perpetual" we should *mean* perpetual—not just a few decades.

References

1. Rietema K. 1957. "On the Efficiency in Separating Mixtures of Two Components." *Chemical Engineering Science* 7:89.

2. Worrell, W. A. and P. A. Vesilind. 1980. "Evaluation of Air Classifier Performance." *Resource Recovery and Conservation* 4.

3. Stessel, R. 1982. *Pulsed Flow Air Classification*. MS Thesis, Department of Civil and Environmental Engineering. Durham, North Carolina: Duke University.

4. Hering, R. and S. A. Greely. 1921. *Collection and Disposal of Municipal Refuse*. New York: McGraw-Hill Book Company.

5. Engdahl, R. B. 1969. *Solid Waste Processing*. EPA OSWMP. Washington, D.C.

6. Henstock, M. 1978. Discussion at Engineering Foundation Conference, Rindge, New Hampshire (July).

7. Trezek, G. J. and G. Savage. 1976. "MSW Component Size Distribution Obtained from the Cal Resource Recovery System." *Resource Recovery and Conservation* 2:67.

8. Matthews, C. W. 1972. "Screening." *Chemical Engineering* (July 10).

9. Rose, H. E. and R. M. E. Sullivan. 1957. A *Treatise on the Internal Mechanics of Ball, Tube and Rod Mills*. London: Constable and Co.

10. Nelson, W., A. Kruglack, and M. Overton. 1975. *Trommel Screening*. Duke Environmental Center, Duke University: Durham, North Carolina.

11. Makar, H. V. and R. S. DeCesare. 1975. "Unit Operations for Nonferrous Metal Recovery." *Resource Recovery and Utilization*. ASTM STP 592, pp. 71–88.

12. Beardsley, J. B. 1937. *From Wheat to Flour;* pp. 28–29. Chicago: Wheat Flour Institute.

13. Worrell, W. A. 1978. *Testing and Evaluation of Three Air Classifier Throat Designs*. Duke Environmental Center. Durham, North Carolina: Duke University.

14. Richardson, J. F. and W. N. Zaki. 1954. "Sedimentation and Fluidisation." *Transactions* Institute of Chemical Engineering 32:35.

15. Hasselriis, F. 1984. *Refuse-Derived Fuel Processing*. Boston: Butterworths.

16. Boettcher, R. A. 1972. *Air Classification of Solid Waste*. EPA OSWMP, SW–30c. Washington, D.C.

17. Salton, K., I. Nagano and S. Izumin. 1976. "New Separation Technique for Waste Plastics." *Resource Recovery and Conservation* 2:127.

18. Murray, D. L. and C. L. Liddell. 1977. "The Dynamics, Operation and Evaluation of an Air Classifier." *Waste Age* (March).

19. Arthur D. Little, Inc. Personal communication.

20. Midwest Research Institute. 1979. *Study of Processing Equipment for Resource Recovery Systems*. EPA Contract 68-03-2387, Final Report, Washington, D.C.

21. Sweeney, P. J. 1977. *An Investigation of the Effects of Density, Size and Shape Upon the Air Classification of Municipal Type Solid Waste*. Report CEEDO-TR-77-25, Tyndall Air Force Base, Florida Civil and Environmental Engineering Development Office.

22. Senden, M. M. G. and M. Tels. 1978. "Mathematical Model of Air Classifiers." *Resource Recovery and Conservation* 2:129.

23. Michaels, E. L., K. L. Woodruff, W. L. Fry-berger, and H. Alter. 1975. "Heavy Media Separation of Aluminum from Municipal Solid Waste." *Transactions* Society of Mining Engineering 258:34.

24. Foust, A. S., et al. 1960. *Principles of Unit Operations*. New York: John Wiley & Sons, Inc.

25. Parker, B. L. 1983. *Magnetic Separation of Ferrous Materials from Shredded Refuse*. MS Thesis. Durham, North Carolina: Duke University.

26. Spence, P. 1977. *Magnetic Separation—A Laboratory Study*. Duke Environmental Center (April). Durham, N.C.: Duke University.

27. Abert, J. G. 1977. "Aluminum Recovery—A Status Report." NCRR

Bulletin 7, no. 2. Washington, D.C.: National Center for Resource Recovery.

28. Zavestski, S. 1992. "Reynolds Aluminum." *Waste Age* (October): p. 67.

29. Malloy, M. G. 1996. "Pouring It On in Ohio." *Waste Age* (March): p. 61.

30. Steiner. "Separating Plastics with the Same Density by Means of a Free Fall Separating Plant." Promotional literature. Koln, Germany.

31. Drobney, N. L., H. E. Hull, and R. F. Testin. 1971. *Recovery and Utilization of Municipal Solid Waste*. EPA OSWMT SW–10c. Washington, D.C.

32. Taggart, A. F. 1974. *Handbook of Mineral Dressing*. New York: John Wiley & Sons.

Abbreviations Used in This Chapter

HDPE = high-density polyethylene
PETE = polyethylene terephthalate

MRF = materials recovery facility
RDF = refuse-derived fuel

Problems

5-1. Suppose that a proposal is made to hand-sort all of the refuse from a town of 100,000 people. Estimate the personnel and cost required, and describe the problems involved in implementing such a program.

5-2. Estimate the critical speed and horsepower requirements for a trommel screen, 3 m in diameter, processing refuse so as to run 25% full.

5-3. If the angle of a trommel screen were increased (steeper), how would this affect the critical speed, horsepower, and efficiency?

5-4. Aluminum chips of uniform diameter 1.0 in. are to be separated from glass of 100% less than 0.5 in. by use of a single-deck reciprocating screen. The total feed rate is 1 ton/h, consisting of 0.1 ton/h of aluminum and 0.9 ton/h of glass. It is required to produce an aluminum fraction that is 99% pure (by weight). Find the area and size of screen required.

5-5. Using the aluminum/glass mixture specified in Problem 5-4, assume that the smallest glass particle of significance is 0.05 in. in diameter. Estimate the air-classifier size and throat velocity required to separate the glass from the aluminum.

5-6. If the pressure drop through a cyclone is assumed to be 1 in. of water, estimate the diameter of cyclone required for various air-flow rates to remove 1 in. diameter aluminum chips.

5-7. Estimate the air velocity required in an air classifier to suspend a piece of glass of 0.5 in. in diameter. Suppose that a 0.4 volume fraction of the throat section were occupied by glass. What would the required air velocity be to suspend this glass?

5-8. What density of fluids is needed to separate glass, aluminum, and plastics into the three individual components? What specific fluids might be used for this purpose?

5-9. Two air-classifier manufacturers report the performance given in Table P5-9 for their units.

Table P5-9

Manufacturer A:

    Recovery of organics—80%
    Recovery of inorganics—80%

Manufacturer B:

    Overall recovery—60%
    Purity of light fraction—95%

    Assume a feed that consists of 80% organics. Compare the performance of the two units.

5-10. A shredded refuse is to be screened to recover the glass fraction. For the purposes of this analysis, the waste to be screened is divided into "glass" and "nonglass" fractions. The material to be screened has 25% by weight of "glass" and 75% by weight of "non-glass." The size distribution of each fraction is shown in Table P5-10.

Table P5-10

| Particle Size (in.) | Percent Finer Than | |
	Glass	Nonglass
2	100	20
1	100	10
0.5	80	8
0.25	20	5
0.125	5	0

    a. Would it be possible to select a screen size such that all (100%) of the glass can be recovered (as screen extract)?

    b. If a glass product of minimum 90% pure is to be obtained from the screening in Problem 5-10, what size holes must be used, and what is the recovery at this purity?

5-11. A 2.4-m-diameter trommel screen rotating at 4 rpm is 7.6 m long and set at an angle (β) of 15°. The feed rate is 10 tonnes per hour of raw MSW.

    a. What is the critical speed?

    b. At 4 rpm, would you expect the MSW to be cascading, cataracting, or centrifuging?

5-12. A drop test was conducted for paper and plastics. The results are shown in Figure 5-39. In this test, the pieces were dropped from a height and their terminal settling velocities calculated (Table P5-12). The paper is to be separated from the plastic using an air classifier with at least 95% purity of the paper. What air velocity should be used in the air classifier to obtain the maximum recovery of paper at the required purity?

Table P5-12

| Terminal Settling Velocity (m/sec) | Percent of Particles with Terminal Settling Velocities Less Than | |
	Paper	Plastic
0.5	0	0
1.0	15	0
1.5	80	25
2.0	100	68
2.5	100	95
3.0	100	100

5-13. A center-fed sink/float apparatus is 2 meters deep and has a fluid with a density of 1.2 g/cm³ and a viscosity of 0.015 poise. Three plastics are to be separated, with the results given in Table P5-13.

Table P5-13

Plastic	Feed (kg/h)	Density (g/cm³)	Overflow (kg/h)
A	42	1.4	5
B	38	1.1	35
C	20	0.9	18

    a. What is the recovery of plastic B in the overflow?

    b. What is the purity of plastic B in the overflow?

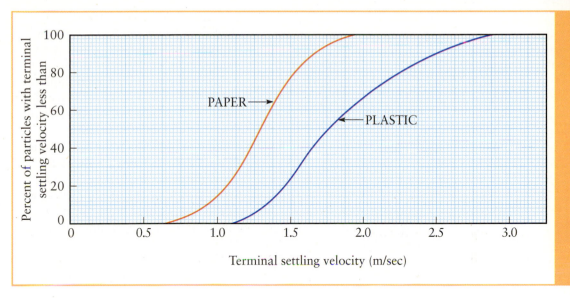

Figure 5-39 Results of drop test with paper and plastic pieces. See Problem 5-12.

c. If plastic B is made of spheres 0.5 mm in diameter, how long would the spheres take to reach the top?

5-14. Two materials, A and B, are to be separated using two unit operations, 1 and 2. The feed has 10 tons/h A and 4 tons/h B. The split (fraction of material rejected by each operation) is given in Table P5-14.

Table P5-14

Material	Split (fraction rejected) by Unit Operation	
	1	2
A	0.2	0.5
B	0.4	0.8

a. Which sequence of operations, (1→2) or (2→1), will yield the greatest recovery of material A?
b. What will be the purity of material A using that sequence?

c. What will be the efficiency of separation for the entire process train with regard to material A using the Worrell-Stessel equation?

5-15. Plot a curve showing the fraction of feed exiting as extract in a trommel screen versus the speed of the trommel. (Extract is what falls through the holes.)

5-16. The specific gravity of aluminum is about 1.6, and the viscosity of air is about 5×10^{-7} lb-s/ft². If a velocity of 1500 feet per minute is required to suspend shredded pieces of an aluminum can, what would be the aerodynamic diameter?

5-17. What feed characteristics and operational variables will make the eddy current separator less effective than it might theoretically be?

5-18. An air classifier receives a feed of organics and inorganics, and the two are to be separated. A preliminary test is conducted; the results are given in Table P5-18.

Table P5-18

Organics in the Feed (lb/h)	Inorganics in the Feed (lb/h)	Air Velocity (ft/sec)	Organics in Extract (lb/h)	Inorganics in Extract (lb/h)
18	20	20	5	0.5
18	20	40	8	1
18	20	50	12	2
18	20	60	17	3.5
18	20	80	18	6

a. If an extract of minimum 85% is required, what is the highest recovery possible?

b. Using the Worrell-Stessel efficiency definition, what air velocity yields the highest efficiency?

5-19. Three separation unit operations—trommel screen, magnet, and air classifier—are available for use in separating a feed made up of paper, ferrous, and glass. The splits (fraction of each material being rejected in each operation) are as given in Table P5-19.

Table P5-19

Material	Feed in tons/ hour	Trommel Screen	Magnet	Air Classifier
		Fraction Rejected		
Paper	40	0.9	0.9	0.1
Ferrous	10	0.5	0.05	0.95
Glass	10	0.1	1.0	0.5

Using the array notation presented in the last section of this chapter, determine the order and combination of any three of the unit operations if the objective is to produce the highest recovery and purity of ferrous material. You can use one, two, or three unit operations.

5-20. A community collects mixed cans—aluminum and steel. Design a materials separation facility for separating aluminum cans from steel cans. Show the floor plan, including all conveying equipment, storage bins, etc. Analyze the robustness of your facility by assuming that each of the unit operations (including conveyors) goes down—one at a time. Can your facility still function?

5-21. A shredded MSW is to be air-classified in a straight-wall air classifier with throat dimension of 18 in. × 18 in. What should be the maximum capacity of the fan? Assume the air is incompressible and make all other necessary assumptions.

5-22. Explain why a pulsed-flow air classifier would have better performance characteristics than a straight-vertical-tube air classifier. Start your explanation by first defining what is meant by "better performance."

5-23. Figure 5-40 is a sketch of a clean materials recovery facility. The flow rates of the various materials streams are indicated by the numbers—all having units of tons/day.

a. What percent of steel entering the plant is recovered by the magnet?

b. What is the purity of the paper produced by the air classifier?

c. What is the fraction of incoming MSW that ends up in the landfill?

d. Which of the unit operations is a polynary separator?

5-24. Design a materials recovery facility for a mixed municipal solid waste that will have as the end product the following materials: clear class, colored glass, aluminum, ferrous material, and corrugated cardboard. Draw a plan for such a facility with unit operations clearly labeled. Include on this plan the receiving station and all methods of material transport between unit operations, and indicate the fate of material not collected.

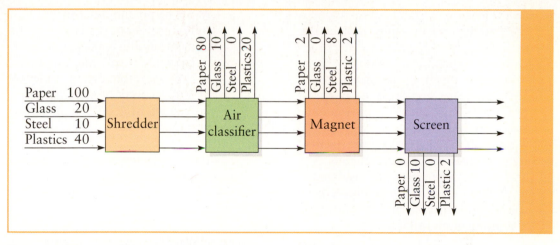

Figure 5-40 Flow of four materials through a materials recovery facility. See Problem 5-23

5-25. Construct a matrix array using appropriate splits for the materials separation system consisting of a trommel screen, a magnet, and an air classifier. If the objective is to produce clean ferrous (expressed as both the recovery and purity), which configuration would be chosen? Assume three materials make up the mixed feed: paper, ferrous, and glass. Table P5-25 provides the feed tonnages and the splits (a split is defined as the fraction continuing on, or rejected by, the unit operation):

Table P5-25

| Material | Tons per Hour | Split | | |
		Trommel Screen	Magnet	Air Classifier
Paper	40	0.90	0.90	0.10
Ferrous	10	0.5	0.05	0.95
Glass	10	0.10	1.00	0.50

5-26. Shredded solid waste was run through a set of sieves with the results given in Table P5-26.

Table P5-26

Sieve Size (inches)	Fraction of Feed Retained on Sieve
10	0.05
8	0.20
4	0.32
2	0.18
1	0.20

What is the characteristic size of this shredded material? (You may wish to use the Rosin-Rammler paper.), as shown in Figure 4-15.

5-27. What is an *eddy current separator* and how does it work?

5-28. Explain in your own words how a jig works.

5-29. Which would separate a mixture of steel and aluminum cans, a boance and adherence separator or a ferrous magnet? Explain why.

5-30. Using your knowledge of trommel screen behavior, plot a typical curve showing the percent recovery (extract) versus trommel (a) speed, (b) slope, and (c) feed rate. Start by thinking of the extremes. For example, the speed at zero rpm is one extreme.

5-31. Design a facility for separating aluminum cans from steel cans. Show all of the storage points, unit operations, and means of conveyance. Submit a floor plan and an elevation.

5-32. An industrial materials recovery facility is to separate out two materials, A and B. Air classification is considered one option for such separation. A drop test is performed on the materials; the results are given in Table P5-32.

Table P5-32

Terminal Settling Velocity, m/sec	Percent of Particles with Falling Velocities Less than the Given Velocity	
	A	B
0.25	10	0
0.50	30	5
0.75	85	10
1.00	100	50
1.25	100	80
1.50	100	100

a. Will it be possible to achieve 100% recovery of material A as the extract from the air classifier?

b. What would be the highest purity of material A that can be achieved?

c. At 50% recovery of material A as the extract, what will be the purity of the extract with respect to material A?

5-33. According to Figure 5-20, is it possible to completely separate one of the four materials from the other three? Explain your answer.

5-34. Using Figure 5-27, what is the advantage of the suspended type magnet over the other two types of magnets?

CHAPTER 6

Biological Processes

Municipal refuse contains about 75% organic material, which can be converted to useful energy by combustion, as discussed in the next chapter, or to useful products by biological processes, which is the topic of this chapter.

The three components of MSW of greatest interest in the bioconversion processes are garbage (food waste), paper products, and yard wastes. The garbage fraction of refuse varies with geographical location and season. Dietary habits, of course, affect its composition and quantity, as does the standard of living. Kitchen garbage grinders in more affluent communities transfer much of the putrescible waste from the refuse stream to the sewerage system, and the reduction of the garbage fraction is a continuing trend in the United States and in many other countries.[1]

The garbage fraction also has by far the highest moisture content of any constituent in MSW, but the moisture is rapidly transferred to absorbent materials (such as newspapers) as soon as contact is made. Garbage also tends to be well mixed in MSW, and therefore it is often difficult to find identifiable bits of garbage in mixed refuse other than the large pieces, such as orange peels or apple cores. Garbage is even better distributed in MSW if the waste is shredded.

The fraction of paper in MSW tends to remain fairly stable throughout the year. However, the total amount of paper discarded has been decreasing.

Yard waste generation is seasonal.[48] For example, in some climate zones there is essentially no yard waste generated during the winter months.

In addition, the amount of yard waste can vary from week to week, often reflecting the weather. An important factor impacting compostability of yard waste is the amount of lignin because lignin compounds do not easily decompose in biological processes. As the percentage of woody biomass increases in yard waste, the lignin content increases. In addition, in different tree species the lignin content varies from 15 to 40%.[47]

The two methods of biological conversion discussed here make use of the organic fraction of refuse. The methods (*anaerobic digestion* and *composting*) are *broad-spectrum processes*, where the specific organisms responsible for the bioconversion are not identified or isolated, and the processes are described by empirical data. Because composting and anaerobic digestion do not begin with raw material composed of only one chemical, the specific biochemical reactions involved in these processes are numerous, and therefore it is not possible to approach them as one would describe the hydrolysis of cellulose.

6-1 METHANE GENERATION BY ANAEROBIC DIGESTION

When organic matter decays under anaerobic conditions (absence of free oxygen), the end products include gases such as methane (CH_4), carbon dioxide (CO_2), small amounts of hydrogen sulfide (H_2S), ammonia (NH_3), and a few others. The recognition long ago that methane is an excellent fuel prompted wastewater treatment plant design engineers to digest (decompose) waste solids and capture this gas for use in heating buildings and running machinery in the treatment plant. While the quantity of methane generated in a wastewater treatment plant is not sufficient to consider its conversion to pipeline gas, the potential for producing pipeline gas from decomposing refuse has a lot of merit.

Ideally, the production of methane and carbon dioxide can be calculated using the following equation:

$$C_aH_bO_cN_d + \left(\frac{4a - b - 2c + 3d}{4}\right)H_2O \rightarrow$$

$$\left(\frac{4a + b - 2c + 3d}{8}\right)CH_4 + \left(\frac{4a - b + 2c + 3d}{8}\right)CO_2 + dNH_3$$

Example 6-1 illustrates how this equation can be used if the chemical composition of a material is known.

EXAMPLE 6-1

Estimate the production of CO_2 and CH_4 during the anaerobic decomposition of glucose.

SOLUTION

The general formula for glucose is $C_6H_{12}O_6$; hence by the equation above, $a = 6$, $b = 12$, $c = 6$, and $d = 0$.

$$C_6H_{12}O_6 + \left(\frac{24 - 12 - 12}{4}\right)H_2O \rightarrow \left(\frac{24 + 12 - 12}{8}\right)CH_4 + \left(\frac{24 - 12 + 12}{8}\right)CO_2$$

$$C_6H_{12}O_6 \rightarrow 3CH_4 + 3CO_2$$

Note that the equation balances. The molecular weights are $180 \rightarrow 3(16) + 3(44)$; hence 1 kg of glucose produces 0.73 kg of CO_2 and 0.27 kg of CH_4. Recalling that 1 gram molecular weight of a gas at standard temperature and pressure occupies 22.4 liters, the production of CO_2 and CH_4 from 1 kg of glucose is 746 liters each of methane and carbon dioxide.

Unfortunately, the chemical composition of MSW is difficult, if not impossible, to determine, although some attempts have been made to do so. The best approximation is that the organic fraction of refuse can be described by the chemical formula $C_{99}H_{149}O_{59}N$. With this formula, the previous equation estimates that the production of methane from a landfill is 257 liters of methane per kilogram of wet refuse (total, organic plus inorganic, assuming wet refuse is 50% biodegradable organic). In using this equation, note that the only carbon that can participate in the production of gas is from decomposable materials, such as food waste and paper. Other organics, most importantly plastics, do not decompose to produce gas.

The two ways of generating methane are to capture the gases produced in landfills or to digest organics in an anaerobic digester using tanks similar to those used in wastewater treatment plants or another structure, such as a horizontal or vertical cylinder. Methane generation both in anaerobic digesters and in landfills is discussed in this section, although a more complete presentation of landfill gas production is found in Chapter 8. Much of the following anaerobic decomposition theory applies to both processes, however.

6-1-1 Anaerobic Decomposition in Mixed Digesters

The two basic metabolic pathways for the decomposition or degradation of wastes are *aerobic* (with oxygen) and *anaerobic* (in the absence of oxygen). While an aerobic system might be generally represented as

[complex organics] + oxygen $\rightarrow CO_2 + H_2O + NO_3^- + SO_4^{-2}$ + other products the anaerobic decomposition of organics can be described as

$$[\text{complex organics}] + \text{water} \rightarrow CO_2 + CH_4 + H_2S + NH_4^+$$

The end products in aerobic decomposition are all stable, possessing no additional energy to be used by decomposing organisms (they are at their highest oxidation state). The products of anaerobic decomposition, on the other hand, still contain energy. Ammonia and hydrogen sulfide could be still further oxidized, and methane contains considerable energy.

The microorganisms responsible for anaerobic decomposition can be divided into two broad categories:

1. *Acid formers* that ferment the complex organic compounds to more simple organic forms, such as acetic and propionic acids. These hardy organisms can be either facultative or strict anaerobes.

2. *Methane formers* that convert the organic acids to methane. These organisms are strict anaerobes and have very slow growth rates—two characteristics that cause considerable problems in anaerobic processes in wastewater treatment and will similarly plague anaerobic decomposition of refuse. Methane formers are very sensitive to various environmental factors. They are strict anaerobes and quite sensitive to temperature changes. Two different groups of methane formers seem to exist: one group (*mesophilic*) operating best around 30 to 38°C (85 to 100°F) and a second group (*thermophilic*) operating best around 50 to 58°C (120 to 135°F). The methane formers also require stable and neutral pH. Sufficient alkalinity (resistance to pH drop) should be present to prevent the pH from falling below 6.8. Finally, methane formers are sensitive to the presence of toxic materials, such as heavy metals and pesticides.

During the acid-forming stage, the first step in the process involves extracellular enzymes produced by acid formers, which break down the large complex organic molecules. For example, the enzymes cellobiase and cellulase break down cellulose to glucose, and lipase breaks fat to shorter-chained fatty acids. This process is energy consuming.

Other bacteria then metabolize the glucose and other products into organic acids, mostly acetic and propionic acid. These simple organic acids then serve as substrate for methanogens. This methane formation is performed by a number of organisms that have specific substrates and roles in the overall reaction. The two reactions

$$CH_3COOH \rightarrow CH_4 + CO_2$$

$$4CH_3CH_2COOH + 2H_2O \rightarrow 7CH_4 + 5CO_2$$

for acetic and propionic acids, respectively, are actually the net results of a large number of steps. The resulting gas varies in composition but averages about 60% methane with a heating value between 500 and 700 Btu/ft³ (4700 and 6500 kJ/m³).[3]

The total amount of gas theoretically available from the anaerobic digestion of MSW is considerably more than has been captured to date in pilot plant facilities. About 54% of the volatile solids have been found to pass through the digester[4] and have not been converted to CO_2 and CH_4.

Anaerobic digestion systems are characterized as either low-solids or high-solids systems. While there is no definitive division between these two designations, low-solids systems are usually under 10% solids and are considered wet processes. High-solids systems are over 20% solids and are considered dry processes.

In a low-solids system, the digestion of refuse involves hardware and a flow diagram not unlike the anaerobic digestion used in wastewater treatment. In the case of MSW, the organics are first separated from the refuse and are slurried with sewage sludge or some other suitable liquid. The resulting mixture is digested in a heated and enclosed tank. The gas is captured either under a floating cover or in a separate tank. The residual of the digestion process is a dark, odoriferous slurry that must be properly managed.

Pilot plant studies have shown that the total gas produced is strongly influenced by detention time in the digester and the digester temperature, as shown in Figure 6-1. Note that the 45°C result is lower than 40°C, suggesting the existence of both a mesophilic (40°C) and a thermophilic (60°C) operating range.

In addition to temperature and residence time, other variables are important in this process, such as the maintenance of total anaerobiosis. It is also necessary to maintain a neutral pH level—never below 6.2, at which point methane production ceases. Since this is a biological process, the provision of adequate nutrients (such as nitrogen) is required. If the C/N ratio of the waste is not sufficient for full decomposition, another source of N is needed, such as sewage sludge rich in nitrogen. The C/N ratio for typical MSW has been reported as 24:1 with some values as high as 40:1,[5] while a ratio of 20:1 as a minimum is required for active anaerobic decomposition. Raw primary sludge[6] has a C/N ratio of about 16:1.

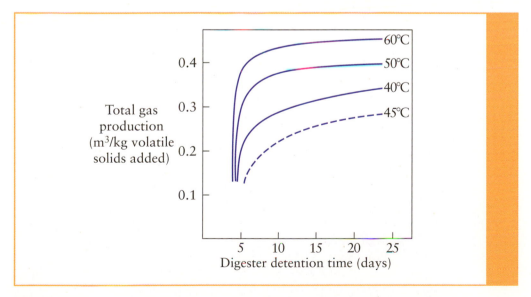

Figure 6-1 Gas production from anaerobic digestion of MSW. Source: Based on Pfeffer, J. T., and J. C. Liebman, "Energy from Refuse by Bioconversion Fermentation and Residue Disposal Processes," *Resource Recovery and Conservation* 1: 295.

Finally, toxic materials can be detrimental to anaerobic digestion, and these must be controlled by removing potential toxins before they get to the digester. If present in the digester they must be removed by adsorption or precipitation. The latter method has been successfully applied in order to remove metals with sulfide in wastewater treatment plants.[7]

MSW digestion might be described as the decay or reduction in volatile (organic) matter as

$$\frac{dS}{dt} = -K_d S$$

where

S = concentration of the biodegradable material (measured as volatile suspended solids, or a specific material if the system feed is controlled), mg/liter at time t

K_d = decay constant, days^{-1}

t = time, days

This is simply a first-order decay equation, stating that the rate of decay is proportional to the organics remaining, a reasonable assumption if the process rate is not time dependent. After integration,

$$\frac{S}{S_0} = e^{-K_d t}$$

where S_0 is the original organic solids concentration, at the starting time $t = 0$, given in mg/liter.

The materials balance within a completely mixed continuous digester would be

[rate of input] − [rate of output] + [rate of positive or negative accumulation] = [rate of net change]

If the digester is operating at steady state, the net change is zero, and

$$\frac{QS_0}{V} - \frac{QS}{V} - K_d S = 0$$

where

Q = flow rate through digester, m^3/day

V = volume of digester, m^3

The accumulation term is negative because the organic material is being destroyed. The hydraulic residence time is $\bar{t} = Q/V$ or

$$\bar{t} = \frac{S_0 - S}{K_d S}$$

Hence, if K_d is known, the required residence time for any reduction in solids can be calculated. Batch laboratory experiments can be used to obtain values of K_d by plotting the values of log S/S_0 versus time and measuring the slope as $(K_d/2.303)$. Values of K_d for refuse slurries have not been reported.

The process kinetics also may be described in terms of the gas produced instead of the volatile matter destroyed. Using a similar mass balance, Pfeffer[8] found that it was possible to describe the reactor performance by the model

$$\frac{G_0 - G}{G} = K_g \bar{t}$$

where

G_0 = maximum gas production attainable, estimated at 0.547 liter total gas per gram volatile solids in the reactor

G = daily gas production, liters/g volatile solids

$\bar{t}$ = hydraulic residence time, days

K_g = rate constant, days^{-1}

K_g seems to have two distinct values. The initial rate is rapid and lasts between 5 and 10 days followed by a significantly lower rate. Table 6-1 is a listing of the K_g values. At 45°C there is a substantial drop in K_g from 40°C, indicating again the existence of mesophilic and thermophilic regimes in anaerobic digestion.

The quantity of gas generated can be estimated by entering a plot (such as Figure 6-1) at the calculated $\bar{t}$ (hydraulic residence time) and reading off the gas production. Because of the heterogeneous nature of the waste and the fact that not all the organics decompose, any theoretical calculations probably would be fruitless. Therefore, laboratory studies to determine kinetic constants for a particular waste are necessary.

6-1-2 Potential for the Application of Anaerobic Digesters

After the 1973 and 1979 worldwide oil crises, when many communities were considering waste-to-energy plants, anaerobic digestion was viewed as a lower-cost, more environmentally friendly alternative method of generating energy. At that time, much of the research was focused on mixed waste low-solids anaerobic digesters. However, for a variety of reasons, no large anaerobic digestion plants were built.

The process is plagued by potential problems. There is no way to ensure the removal of toxic materials before the waste goes into the digesters, and "sour" digesters (such as those encountered in wastewater treatment plants) are a definite possibility.

The problem of mixing a paper slurry has continued. Even pilot plant scale mixing with fairly dilute slurries has been found to be a problem. The desired solids concentration for these digesters is at 10%, which is a highly viscous and

Table 6-1 Rate Constants, K_g for Gas Production in Anaerobic Digesters

Temperature, °C	Rate Constant (day^{-1})	
	Initial	Final
35	0.055	0.003
40	0.084	0.043
45	0.052	0.007
50	0.117	0.030
55	0.623	0.042
60	0.990	0.040

Source: [8]

thixotropic slurry. In wastewater treatment practice, where solids concentrations normally range from 3 to 5%, mixing has always been a problem. Tracer studies have shown that typical primary digesters seem to have only 25% of their volume mixed—the remaining being dead space.[10] Such problems will surely plague refuse-digestion facilities as well. A demonstration project in Florida continued to break shafts on mixers because of the high fibrous content of the waste.

Large land areas are required by the digesters, a minimum 12 acres (5 ha) for a 1000 ton/day (900 tonne/day) plant.[9] This can be a problem where transport costs prohibit long-range refuse movement and if the treatment facility must be located on expensive urban land. And finally, the problem of what to do with the effluent and residue has not been solved. The sludge does seem to dewater readily (as it should) with all the fiber in it, but its ultimate disposal is an additional problem in the application of this process.

In the last few years there has been more interest in low-solids (less than 10% total solids) anaerobic digestion of organic material. Toronto, Canada built a pilot plant and is constructing a second plant (Figure 6-2). Sacramento, California also has a small facility. These plants are designed to process a source separated organic waste that is suitable for a low-solids facility. For example, food waste is a desirable feed stock, but yard waste is not.

To manage all organics, European companies are building high-solids (greater than 20% total solids) anaerobic digesters. While some plants process only source separated organics (such as food waste and green waste), others process

Figure 6-2 Low-solids anaerobic digester. (Courtesy William A. Worrell)

Figure 6-3 Anaerobic digestion tunnel. (Courtesy William A. Worrell)

organics that include a significant fraction of nonorganic. A high-solids digestion system is designed to feed and process "dry" material. Unlike a low-solids system, which generates effluent, a high-solids system recycles the effluent to inoculate the incoming feed material. Some designs have no mixing after feeding, and the organics are placed into long, air-tight tunnels (see Figure 6-3). Other designs slowly transport the material through a vertical or horizontal fermentation drum; this results in some mixing. Figure 6-4 illustrates the process schematics of three of the largest European manufacturers.

Figure 6-5 shows a Kompogas anaerobic digestion plant near Zurich, Switzerland. This plant accepts source separated organics, including food waste

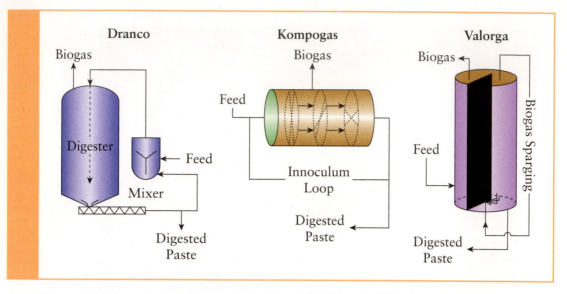

Figure 6-4 Schematic drawing of various dry-solids anaerobic digesters. Source: Adapted from Vandevivere, P., L. De Baere, and W. Verstraete, "Types of anaerobic digesters for solid wastes, in *Biomethanization of the Organic Fraction of Municipal Solid Wastes*, J. Mata-Alvarez, Editor. 2002, IWA Publishing: Barcelona. pp. 111–140, as used in Current Anaerobic Digestion Technologies Used for Treatment of Municipal Organic Solid Waste, California Integrated Waste Management Board, March 2008. © 2008, 2010 by the California Department of Resources Recycling and Recovery (CalRecycle). All rights reserved. Used by permission.

Figure 6-5 Kompogas anaerobic digestion plant in Switzerland. (Courtesy William A. Worrell)

in compostable bags. The waste is shredded and then placed in a horizontal fermentation drum for two to three weeks. This anaerobic process results in the generation of methane for use as a fuel. The fermented material (digestate) is then composted aerobically. The products are methane gas and a good quality soil amendment.

In addition to digesters being either wet or dry, they also can be single-stage, two-stage, or batch facilities. Single-stage digesters are simple to design, build, and operate and operate as a continuous feed system. A two-stage digester separates the initial hydrolysis and acid-producing fermentation from the methanogenesis. These systems are also continuous feed systems. In Europe, most of the systems are single-stage systems. Finally, there are also batch digesters that are fed initially and then allowed to operate. The advantages and disadvantages of the various systems are shown on Table 6-2.

Table 6-2 **Advantages and Disadvantages of Various Anaerobic Digestion Systems**

	Criteria	Advantages	Disadvantages
Single-Stage, Wet System	Technical	Derived from well-developed wastewater treatment technology Simplified material handling and mixing	Short-circuiting Sink and float phases Abrasion with sand Complicated pre-treatment
	Biological	Dilution of inhibitors with fresh water	Sensitive to shock as inhibitors spread immediately in reactor VS lost with removal of insert fraction in pre-treatment
	Economic and Environmental	Less expensive material handling equipment	High consumption of water and heat Larger tanks required
Single-Stage, Dry System	Technical	No moving parts inside reactor Robust (insert material and plastics need not be removed) No short-circuiting	Not appropriate for wet (TS < 5%) waste streams
	Biological	Less VS loss in pre-treatment Larger OLR (high biomass) Limited dispersion of transient peak concentrations of inhibitors	Low dilution of inhibitors with fresh water Less contact between microorganisms and substrate (without inoculation loop)
	Economic and Environmental	Cheaper pre-treatment and smaller reactors Very small water usage Smaller heat requirement	Robust and expensive waste handling equipment required
Two-Stage System	Technical	Operational flexibility	Complex design and material handling
	Biological	Higher loading rate Can tolerate fluctuations in loading rate and feed composition	Can be difficult to achieve true separation of hydrolysis from methanogenesis
	Economic and Environmental	Higher throughput, smaller footprint	Larger capital investment

Table 6-2 **Advantages and Disadvantages of Various Anaerobic Digestion Systems** (*continued*)

	Criteria	Advantages	Disadvantages
Basic System	Technical	Simplified material handling Reduced pre-sorting and treatment	Compaction prevents percolation and leachate recycling
	Biological	Separation of hydrolysis and methanogenesis Higher rate and extent of digestion than landfill bioreactors	Variable gas production in single-reactor systems
	Economic and Environmental	Low cost Appropriate for landfills	Less complete degradation of organics (leach bed systems)

Source: [44]

6-1-3 Methane Extraction from Landfills

Landfills are very large anaerobic digesters. However, unlike the previously discussed facilities, landfills are not optimized for gas production. Some communities have tried to create a bioreactor landfill to increase the rate of stabilization and gas production. A bioreactor landfill is operated to rapidly transform and degrade organic waste. The increase in waste degradation and stabilization is accomplished through the addition of liquid to enhance microbial processes. Liquid must be added to almost the field capacity of the landfill. Field capacity of the landfill is the point at which the landfill is saturated with water prior to any percolation and can range from 35 to 60% moisture. To achieve field capacity in waste starting at 10 to 20% moisture requires between 40 and 80 gallons per cubic yard of waste.[45] The extraction and use of gas from landfills is discussed in detail in Chapter 8.

6-2 COMPOSTING

Composting differs from the previously discussed anaerobic process in that it is an aerobic process, and the end product is the partially decomposed organic fraction. Composting is often promoted as a "natural" process of solid waste treatment. One reason for this reputation is that compost piles can be readily constructed in the backyard, and the product is a useful soil conditioner. It is little wonder, therefore, that municipal engineers and city councils are besieged by citizens groups urging that composting be initiated in their community in place of alternative solid-waste disposal schemes such as landfilling and combustion, which many people view as a waste of money and natural resources.

6-2-1 Fundamentals of Composting

Aerobic microorganisms extract energy from the organic matter through a series of exothermic reactions that break down the material to simpler materials. The basic aerobic decay equation holds:

$$[\text{complex organics}] + \text{oxygen} \rightarrow CO_2 + H_2O + NO_3^- + SO_4^{2-}$$
$$+ [\text{other less complex organics}]$$
$$+ [\text{heat}]$$

Table 6-3 Destruction of Some Common Pathogens and Parasites during Composting

Salmonella typhosa	No growth beyond 46°C; death within 30 min at 55–60°C and within 20 minutes at 60°C; destroyed in a short time in compost environment
Salmonella sp.	Death within 1 h at 55°C and within 15–20 min at 60°C
Escherichia coll	Death for most within 1 h at 55°C and within 15–20 min at 60°C
Shigella sp.	Death in 1 h at 55°C
Entamoeba histolytica cysts	Death within a few minutes at 45°C
Trichinella spiralis larvae	Quickly killed at 55°C
Brucella abortus or Br. suis	Death within 3 min at 62°C and within 1 h at 55°C
Streptococcus pyogenes	Death within 10 min at 50°C
Mycobacterium tuberculosis var. hominis	Death within 15–20 min at 66°C or after momentary heating at 67°C
Corynebacterium diphtheriae	Death within 45 min at 55°C
Ascaris lumbricoides eggs	Death in less than 1 h at 50°C

Source: [12]

During this decomposition, the temperature increases to about 70°C (160°F) in most well-operated composting operations. As the reaction develops, the early decomposers are mesophilic bacteria followed after about a week by thermophilic bacteria, actinomycetes, and thermophilic fungi.[12] Above 70°C, spore-forming bacteria predominate. As the decomposition slows, the temperature drops, and mesophilic bacteria and fungi reappear. Protozoa, nematodes, millipedes, and worms are also present during the later stages. The concentration of dead and living organisms in compost can be as high as 25%.[12]

The elevated temperatures destroy most of the pathogenic bacteria, eggs, and cysts. Some of the more common pathogens and their survival at elevated temperatures are shown in Table 6-3. The product of thermophilic composting is essentially free of pathogens. All potential pathogens, including resistant parasites such as Ascaris eggs and cysts of *Entamoeba histolytica*, are destroyed.[13]

A critical variable in composting is the moisture content. If the mixture is too dry, the microorganisms cannot survive, and composting stops. If there is too much water, the oxygen from the air is not able to penetrate to where the microorganisms are, and the mixture becomes anaerobic. Typically compost should have a moisture content of between 40% to 60%. The right amount of moisture, whether wastewater sludge or other sources of water, that needs to be added to the solids to achieve just the right moisture content can be calculated from a simple mass balance:

$$M_p = \frac{M_a X_a + 100 X_s}{X_s + X_a}$$

where

M_p = moisture in the mixed pile ready to begin composting, as percent moisture

M_a = moisture in the solids, such as the shredded and screened refuse, as percent moisture

X_a = mass of solids, wet tons

X_s = mass of sludge or other source of water, tons (This assumes that the solids content of the sludge is very low, a good assumption if waste activated sludge is used, which is commonly less than 1% solids.)

EXAMPLE 6-2

Ten tons of a mixture of paper and other compostable materials has a moisture content of 7%. The intent is to make a mixture for composting of 50% moisture. How many tons of water or sludge must be added to the solids to achieve this moisture concentration in the compost pile?

SOLUTION

$$M_p = \frac{M_a X_a + 100 X_s}{X_s + X_a} = \frac{(10 \times 7) + (100 \times X_s)}{10 + X_s} = 50$$

Solving for X_s yields 8.6 tons of water or sludge.

If the water to be added is expressed in gallons instead of tons, the water balance equation reads

$$M_p = \frac{M_a X_a + 0.416 W_s}{X_a + 0.00416 W_s}$$

where W_s = water or sludge to be added in gallons. The other variables are as defined previously.

All biochemical conversion processes (such as composting) are in essence two-step operations. The first step is the decomposition of complex molecules of waste materials into simpler entities. If there is no nitrogen available, this is the full extent of the process. If nitrogen is available, however, the second step is the synthesis of the breakdown products into new cells. These new microorganisms contribute to the process, and the system operates in balance.

Because of the high rate of microbial activity, a large supply of nitrogen is required by the bacteria. If the reaction were slower, the nitrogen could be recycled, but since many reactions are occurring concurrently, a sufficient nitrogen supply is necessary. The requirement for nitrogen can be expressed as the C/N ratio, as before.

A C/N of 20:1 is the ratio at which nitrogen is not limiting the rate of decomposition. Above a C/N of 80:1, thermophilic composting cannot occur, because the nitrogen severely limits the rate of decomposition. Most systems operate between these extremes. Some researchers recommend an optimal C/N ratio of 25:1.[14] A C/N ratio higher than this can increase the time to maturity. Nitrogen can become limiting at a C/N ratio greater than about 40:1.[15] At higher pH levels, the nitrogen will be lost into the atmosphere as ammonia gas if the C/N ratio exceeds 35:1.

The C/N ratio generally decreases during the composting process as the organic carbon is converted to carbon dioxide. Unless there is a concurrent loss of ammonia (which is possible as the pH becomes more basic), the nitrogen content remains fairly constant, resulting in a decrease in the C/N ratio.[15]

The C/N ratios for various materials used in composting are shown in Table 6-4.

The calculation of carbon and nitrogen levels and the C/N ratio is straightforward and based on mass balances. If two materials such as shredded refuse and sewage sludge are mixed, the carbon of the mixture is calculated as

$$C_p = \frac{C_r X_r + C_s X_s}{X_r + X_s}$$

where

C_p = carbon concentration in the mixture prior to composting, as percent of total wet mass of mixture

C_r = carbon concentration in the refuse, as percent of total wet refuse mass

C_s = carbon concentration in the sludge, as percent of total wet sludge mass

X_s = total mass of sludge, wet tons per day

X_r = total mass of refuse, wet tons per day

The pH of the compost pile varies with time, showing an initial drop, then increasing to between 8.0 and 9.0, and finally leveling off between 7.0 and 8.9.[16] If the compost heap becomes anaerobic, however, the pH continues to drop due to the action of the anaerobic acid formers. As long as the pile stays aerobic, there is sufficient buffering within the compost to allow the pH to stabilize at an alkaline

Table 6-4 Carbon/Nitrogen Ratios for Various Materials

	C/N
Food waste	
Raleigh, NC	15.4
Louisville, KY	14.9
MSW (including garbage)	
Berkeley, CA	33.8
Savannah, GA	38.5
Johnson City, TN	80
Raleigh, NC	57.5
Chandler, AZ	65.8
Sewage sludge	
Waste activated	6.3
Mixed digested	15.7
Wood (pine)	723
Paper	173
Grass	20
Leaves	40–80
Sawdust	511

Source: [15]

level. For educational purposes, the progression of pH and temperature in a compost pile can be readily demonstrated by laboratory-scale apparatus.[17]

The pH also affects nitrogen loss, because ammonia (NH_3) escapes as ammonium hydroxide is formed above a pH value of 7.0, which decomposes to water and gaseous NH_3. Thus, efficient compost operations, which operate around a pH of 8.0, cannot retain nitrogen at a greater concentration than C/N of about 35:1.[12]

The time required for a compost pile to mature depends on such factors as the putrescence of the feed, the insulation and aeration provided, the C/N ratio, the particle size, and other conditions. Usually, two weeks is considered the minimum time for the adequate composting of shredded municipal refuse in windrows. Mechanical composting plants, using inoculation of previously composted materials, can accomplish decomposition in 2 or 3 days. This material is still quite active, however, and usually requires further stabilization.[12]

The completion of composting is judged primarily on the basis of a slight drop in temperature and a dark brown color. A more accurate measure is the determination of starch concentration in the compost. Starch is readily decomposable, and thus, its disappearance is a good indicator of mature compost. A simple laboratory method for measuring starch in compost is available, although the technique yields only qualitative information.[18,19] This technique also can be applied to a composting demonstration project for the classroom.[17] A more rigorous measure of the end point is the drop in the C/N ratio to perhaps 12:1. Higher C/N ratios will result in continued decomposition of the compost after it is applied, and the subsequent robbing of nitrogen from the soil.

Recognizing the difficulties involved in the processing of a heterogeneous material such as municipal solid waste, Golueke[15] suggests that the viability of any biochemical process be judged on the basis of the organisms employed. Regardless of what biochemical process is used, the organisms should have the following characteristics:

> *Not Fastidious.* They will work under adverse conditions (e.g., wide temperature range) and be tolerant of environmental change.
>
> *Ubiquitous.* They should exist in nature, since pure stock cultures degenerate with time and rarely stay pure.
>
> *Persistent.* They must grow in the environment without special assistance.
>
> *Not Picky.* They should be able to use a broad spectrum of substrates.

If these criteria are used to judge the efficacy of composting, the process would pass with flying colors. Composting is able to handle many organic wastes and seems to be insensitive to changes in flow rates and feed characteristics. From a purely fundamental and biochemical perspective, composting makes a great deal of sense. However, it is important not to forget the old saying: "Garbage in, garbage out." Thus what is being composted is very important.

6-2-2 Composting Organic Waste

The organic fraction of MSW can be composted, and the products from such facilities may have significant environmental and cash value. Organic wastes are now composted in thousands of communities throughout the world.

For example, in Canada, over half of the households participate in some form of composting.[46]

Composting on a municipal scale is an uncomplicated process. At its simplest, a passively aerated compost system consists of shredded and/or screened source-separated organic waste placed in long parallel piles, called windrows. Typically, windrows are laid in long rows of about 4 to 6 ft (1.2 to 2 m) high (Figure 6-6). Moisture content is maintained near 50% by adding water and/or sludge as needed. Because the reaction is aerobic, oxygen must be made available to the microorganisms, and this is done by turning the pile with a specially constructed agitator (Figure 6-7). While this is the least expensive and simplest method of composting, it is not always used. Other factors such as regulations, odor problems, space availability, organic waste composition, and climate (rain, snow, or desert) may require different composting methods.

A more sophisticated system uses an actively aerated system. Windrows are formed on top of perforated PVC pipes (Figure 6-8) or concrete floors (Figure 6-9) with an air distribution system and air blowers for forced aeration. (Figures 6-10 and 6-11). Instead of turning the windrow periodically to aerate the windrow, air is forced through the windrows. The air blowers supplying the air may be continuous or intermittent. Enough air must be supplied to support the aerobic composting. Without enough oxygen, the pile will become anaerobic and cause odor problems. On the other hand, too much air will cool the pile and stop the composting activity.

Figure 6-6 Windrow composting system. (Courtesy William A. Worrell)

Figure 6-7 Mobile aerator for windrows. (Courtesy William A. Worrell)

Figure 6-8 Perforated pipe for air distribution. (Courtesy William A. Worrell)

Figure 6-9 Air distribution channels. (Courtesy William A. Worrell)

Figure 6-10 Air blowers for an aerated static pile. (Courtesy William A. Worrell)

Figure 6-11 Air blowers for a covered indoor aerated static pile. (Courtesy William A. Worrell)

In an aerated static pile composting system, if the air is sucked into the windrow, the system is referred to as a negative aeration system. If the air is blown into the windrow, then it is a positive aeration system. Each method has advantages and disadvantages. With a negative aeration system, the air passing through the pile is sucked into the pipe at the bottom of the pile by creating a vacuum. This air can then be discharged into an air treatment system, such as a biofilter or scrubber, to minimize odors. However, a negative aeration system uses more power and tends to compact the compost pile. The biggest disadvantage of a positive aeration system is that the air is blown through the pile and exits into the atmosphere. Thus odors are not controlled. To control odors a layer of finished compost or a synthetic cover can be placed on top of the pile. Thus the air must pass through the cover, which will remove some odors (see Figure 6-12).

Other options provide even more control of the compost process. For example, the aerated composting may be done under a permanent roof (see Figure 6-13), which prevents run-off from rain during the composting operation. Another option is to compost inside a building (see Figure 6-14) so that there will be no run-off and air can be collected and treated. Finally, aerated composting can be done inside fully enclosed tunnels (see Figure 6-15) for the highest level of control of temperature, moisture, and odor.

Aerobic decomposition results in a dark brown earthy-smelling material that has low nutrient value but is an excellent soil conditioner.

Figure 6-12 Plastic cover on a windrow compost pile. (Courtesy William A. Worrell)

Figure 6-13 Aerated composting under a roof. (Courtesy William A. Worrell)

Figure 6-14 Aerated composting inside a building. (Courtesy William A. Worrell)

Figure 6-15 Aerated composting tunnel doors. (Courtesy William A. Worrell)

The ranges of nitrogen, phosphorus, and potassium in finished compost from MSW are shown in Table 6-5.

Table 6-5 Fertilizer Value of Compost from MSW

Nutrient	Range, as Fraction of Total Solids in %
Nitrogen	0.4 to 1.6
Phosphorus	0.1 to 0.4
Potassium	0.2 to 0.6

Source: [12]

6-2-2 Composting Municipal Solid Waste

In the past, some processes were designed to compost MSW without the removal of the inorganic fraction, which is a practice referred to as *mixed waste composting*. The practical application of mixed waste composting is limited by three serious problems:

1. Lack of markets for the finished product from MSW.
2. Small reduction in the total refuse volume requiring disposal.
3. Environmental factors of composting plants, specifically odor.

The first is the most serious of the three problems. The majority of mixed waste composting plants in the United States have closed because they could not sell (or even give away) their product. There is very little use for compost that includes contaminates such as glass, plastic, and potentially toxic items. While the level of undesirable contaminants in compost made of yard waste and leaves is surprisingly low,[23] the compost produced from this raw material can be used in gardens without threat of contamination from pesticides or heavy metals.

Metals in compost manufactured from MSW, by contrast, can be at levels that present significant public health concern.[24] Lead, for example, is typically at a level of 800 ppm in compost made from MSW.[25] When composting plants are designed, however, the economic analysis invariably includes a profit from the sale of compost.

While there may be some advantage to using mixed waste composting prior to landfilling or incineration, it will be costly and is most likely driven by regulation. For example, if regulations prevent the landfilling of untreated waste, then it may be necessary to compost the waste before it is landfilled.

The second problem, the limited reduction in the volume of waste, usually has technical solutions because the process can be changed to actually capture items such as glass, metals, and so on. In fact, composting may be considered simply another process within a complete materials recovery plant. One pilot-scale process has been run to produce a putrescible material from the screening (7-mesh) of the light fraction of air-classified refuse.[20] This material consisted of 6% of the raw refuse and contained 74% volatile solids. Although in this study the material was digested for methane production, it also could be readily composted.

The fraction of putrescible material in refuse has been steadily declining over the years and will no doubt continue to do so. As the organic fraction is reduced, the compostable fraction similarly drops, and it is unlikely that more than 50% of refuse could ever be recovered as compost. The remaining 50% must be disposed of by normal means, and this cost must be added to the total cost of the composting facility.

The last problem is mostly one of odor production. Although some people insist that the odor from compost piles is pleasant, these people are in the distinct minority. Compost plants do smell—there is no doubt of that—and thus, the plants must be located fairly far from residential areas. This requirement adds another cost—transportation.

MSW composting plants are paid a tipping fee to accept MSW. If the plant cannot sell the compost, it must continue to accept incoming MSW to maintain a cash flow. This results in an ever-increasing pile of compost on-site. As the pile builds, it eventually is too big to be turned, and it becomes anaerobic, causing odor problems.

In conclusion, it seems unlikely that mixed waste composting will be used on a large scale, at least in more developed countries. Whenever composting of refuse on a municipal scale is suggested, the decision makers should always bear in mind the dismal record of past mixed waste composting efforts. All of these plants were built under an aura of optimism and enthusiasm, and all of the economic analyses promised success. The engineer and decision maker should ask what is so different in the proposed new compost plant that would allow it to succeed while all the others have failed.

While composting MSW presents many problems, composting of some feed stocks has real benefit. Bioremediation can be used to treat soil and groundwater contaminated with oil and other petroleum products, solvents, and pesticides. Microbes, such as bacteria, use contaminants as a source of food and energy. The advantages of composting in cleaning up the contamination seems well established.[28]

Various types of industrial and agricultural wastes also can be readily composted. A combination such as sawdust and chicken manure, for example, produces a superior compost, high in nitrogen.

Sludges from sewage treatment plants have been used for many years as sources of nutrients in composting.[29] The problem with composting sludge always has been that the solids compact too tightly, leaving no spaces for air to enter the pile. This problem can be solved by mixing the sludge with wood chips and composting the mixture. The wood chips are composed of poorly decomposable cellulose and lignin and can be readily removed from the stabilized material by simple screening and then can be reused. Raw sludges can be composted to a passable soil conditioner or a high-grade topsoil.

6-3   FINAL THOUGHTS

Engineers working for municipalities—either directly or indirectly—must have some degree of autonomy. Professionals correctly believe that professional autonomy is beneficial to the welfare of the public. If the government starts telling physicians how to treat people, telling preachers what to preach, or telling engineers how to build things, then the public loses. Accordingly, these professions have jealously guarded their autonomy in the name of the public good. The engineering profession recognizes that, if engineering is to maintain its professional autonomy, the public has to trust engineers, and it is very much to the advantage of engineering and the public at large to maintain this trust.

Such autonomy can be taken away by the state, of course, as witnessed in nations that have totalitarian governments (such as the former Soviet Union) in which once-proud and independent engineers became tools of the communist government and had little say in technical decisions. The inability of those engineers to voice their concerns about projects that were counterproductive, wasteful in resources, and harmful to the public was in great part responsible for the eventual downfall of the Soviet empire.[39]

Engineers, as all other professionals, must work to maintain public trust. To promote trust in the engineering profession, professional engineering societies have all drafted statements that express the values and aspirations of the profession—statements commonly referred to as a "Code of Ethics."

One of the earliest codes of ethics in the United States was adopted in 1914 by the American Society of Civil Engineers (ASCE). Based in spirit on the original Code of Hammurabi,[40] the 1914 ASCE Code addressed the interactions between engineers and their clients and among engineers themselves. Only in the 1963 revisions did the ASCE Code include statements about the engineer's responsibility to the general public, stating as a fundamental canon the engineer's responsibility for the health, safety, and welfare of the public.

In 1997, ASCE modified its code to include the commitment of engineers to sustainable development. The term *sustainable development* was first popularized by the World Commission on Environment and Development (also known as the Brundtland Commission), which is sponsored by the United Nations. Within this report, sustainable development is defined as "development that meets the needs of the present without compromising the ability of future generations to meet their own needs."[41] Sustainable development can be defined in a number of ways—and indeed the Brundtland Report itself includes ten different definitions—while a report for the United Kingdom Department of the Environment contains thirteen pages of definitions.[42]

Although the original purpose of introducing the idea of sustainable development was to recognize the rights of the developing nations in using their resources, sustainable development has gained a wider meaning and now includes educational needs and cultural activities, as well as health, justice, peace, and security.[43] All of these are possible if the global ecosystem is to continue to support the human species. We owe it to future generations, therefore, not to destroy the earth they will occupy. Using biological processes to produce useful products such as methane and recovery nutrients is increasing. We recognize that the use of non renewable fossil fuels for our energy use is not in keeping with the principles of sustainable development and contributes to the impact of greenhouse gas. Composting also contributes to nutrient recovery. For example, phosphorous is a valuable limited resource and should be reused for agricultural applications and not lost in landfills.

References

1. Alter, H. 1989. "The Origins of Municipal Solid Waste: The Relationship Between Residues from Packaging Materials and Food." *Waste Management and Research* 7:103–114.

2. Bell, J. M. 1964. "Characteristics of Municipal Refuse." *Proceedings* National Conference on Solid Waste Research, American Public Works Association, February.

3. Pfeffer, J. T. 1974. *Reclamation of Energy from Organic Refuse.* 670/2-74–016. Cincinnati, Ohio.

4. Pfeffer, J. T. and J. C. Liebman. 1976. "Energy from Refuse by Bioconversion Fermentation and Residue Disposal Processes." *Resource Recovery and Conservation* 1:295.

5. Hagerty, D. J., J. L. Pavoni, and J. E. Heer. 1973. *Solid Waste Management.* New York: Van Nostrand Reinhold Co.

6. Vesilind, P. A. 1979. *Treatment and Disposal of Wastewater Sludge.* Ann Arbor, Michigan: Ann Arbor Science Publishers.

7. Lawrence, A. W. and P. L. McCarty. 1965. "The Role of Sulfide in Preventing Heavy Metal Toxicity in Anaerobic Treatment." *Journal of the Water Pollution Control Federation* 37:113.

8. Pfeffer, J. T. 1974. "Temperature Effects on Anaerobic Fermentation of Domestic Refuse." *Biotechnology and Bioengineering* 16:77.

9. Hille, S. J. 1975. *Anaerobic Digestion of Solid Waste and Sewage Sludge to Methane.* EPA OSWMP SW-15g, Washington, D.C.

10. Monteith, H. D. and J. P. Stephenson. 1997. "Mixing Efficiencies in Full-scale Anaerobic Digesters by Tracer Methods." *Proceedings* Symposium on Sludge Treatment. Waste-water Technology Centre, Burlington, Ontario, Canada.

11. Bowerman, F. 1979. "Methane Generation from Deep Landfills." *Proceedings* Engineering Foundation Conference on Resource Recovery, Henniker, New Hampshire (July).

12. Golueke, C. G. 1972. *Composting*. Emmaus, Pennsylvania: Rodale Press, Inc.

13. Wilby, J. S. 1962. "Pathogen Survival in Composting Municipal Wastes." *Journal Water Pollution Control Federation* 34:80.

14. Barktoll, A. W. and R. A. Nordstedt. 1991. "Strategies for Yard Waste Composting." *BioCycle* 32, no. 5:60–65.

15. Golueke, C. 1978. "State of the Art of Byconversion Processes." *Proceedings Engineering Foundation Conference on Resource Recovery, Rindge, New Hampshire* (July).

16. *An Analysis of Composting as an Environmental Remediation Technology*. 1998. EPA 530-R-98–008.

17. Vesilind, P. A. 1973. "A Laboratory Exercise in Composting." *Compost Science* (September-October).

18. Lossin, R. D. 1971. "Compost Studies." *Compost Science* (March-April).

19. Vesilind, P. A. 1973. *Solid Waste Engineering Laboratory Manual*. Durham, North Carolina: Department of Civil Engineering, Duke University.

20. Diaz, L. F., F. Kurz, and G. J. Trezek. 1976. "Methane Gas Production as Part of a Refuse Recycling System." *Compost Science*.

21. Jones, K. H. 1991. "Risk Assessment: Comparing Composting and Incineration Alternatives." *MSW Management* 1, no. 3:29–32, 36–39.

22. Walter, R. 1971. "How to Compost Leaves." *American City* (June): p. 116.

23. Roderique, J. O. and D. S. Roderique. 1995. Quoted in *Decision Maker's Guide to Solid Waste Management*. EPA 530-r-95–023. Washington, D.C.

24. Oosthnoek, J. and J. P. N. Smit. 1987. "Future of Composting in the Netherlands." *BioCycle* (July).

25. Richard, T. L. and P. Woodbury. 1993. *Strategies for Separating Contaminants from Municipal Solid Waste*. Cornell University Waste Management Institute.

26. Cole, M. A., L. Zhang, and X. Liu. 1995. "Remediation of Pesticide Contaminated Soil by Planting and Compost Addition." *Compost Science and Unitization* 3:20–30.

27. Savage, G., L. Diaz, and C. G. Golueke. 1985. "Disposing of Hazardous Wastes by Composting." *BioCycle* 26:31–34.

28. McKinley, V. 1984. *Microbial Activity in Composting Sewage Sludge*. Ph.D. Dissertation, University of Cincinnati.

29. Wilson, G. B. and J. M. Walker. 1973. "Composting Sewage Sludge: How?" *Compost Science* (September-October).

30. Trezek, G. J. and C. G. Golueke. 1978. "Availability of Cellulosic Wastes for Chemical or Biochemical Processing." *Proceedings Energy, Renewable Resources and New Foods, AIChE Symposium Series* N. 72, p. 158.

31. Span, L. A., J. Medeiros, and M. Mandel. 1976. "Enzymatic Hydrolysis of Cellulosic Wastes to Glucose." *Resource Recovery and Conservation* 1.

32. Goldstein, I. S. 1974. *The Potential for Converting Wood into Plastics and Polymers or into Chemicals for the Production of These Materials*. Raleigh, North Carolina: School of Forest Resources, North Carolina State University.

33. Hajny, G. J. and E. T. Reese (Eds.). 1969. *Cellulases and Their Applications*, Advances in Chemistry Series 95. Washington, D.C.: American Chemical Society.

34. Callahan, C. and C. E. Dunlop. 1974. *Construction of a Chemical-Microbial Plant for Production of Single Cell Protein from Cellulosic Wastes*. EPA, SW 24C. Washington, D.C.

35. Rogers, C. J., E. Coleman, D. F. Sino, T. C. Purcell, and P. V. Scoparius. 1972. "Production of Fungal Proteins from Cellulose and Waste Celluloids." *Environmental Science and Technology* 8:715.

36. Cookson, A. and G. Froeisdorph. 1973. *The Nitrite Accelerated Photochemical Degradation of Cellulose as a Pre-treatment for Microbiological Conversion to Protein.* EPA. 670-2-73-052. Washington, D.C.
37. Porteus, A. 1972. "WP Disposal Process Turns Cellulose Material into Alcohol." *Paper Trade Journal* (February 7).
38. McAbee, M. K. 1974. "Japan Pushes Alcohol for Protein Process." *Chemical Engineering News* (December 9): p. 11.
39. Graham, L. 1996. *The Ghost of the Executed Engineer.* Cambridge, Massachusetts: Harvard University Press.
40. Hope, R. F. (Ed). 1994. "Hammurabi, King of Babylonia." *The Code of Hammurabi, King of Babylonia*, about 2250 B.C. Holmes Beach, Florida: Wm. W. Gaunt.
41. Herkert, J. R., A. Farrell, and J. Winebrake. 1996. "Technology Choice for Sustainable Development." *Technology and Society*, IEEE 15, no. 2.
42. Pearce, D., A. Markanya, and E. B. Barber. 1989. *Blueprint for a Green Economy.* Report for the UK Department of the Environment. London: Earthscan Publication.
43. World Commission on Environment and Development. 1987. *Our Common Future.* Oxford: Oxford University Press.
44. California Integrated Waste Management Board. Current Anaerobic Digestion Technologies Used for Treatment of Municipal Organic Solid Waste. March 2008.
45. Waste Management, Inc. 2000. "The Bioreactor Landfill: the next generation of landfill management," white paper from WMI-2000.
46. Statistics Canada, July 2013, "Composting by Households in Canada," Publication Number 16-002-x.
47. Sarkanen K.V., Ludwig C.H., 1971, Lignins: *Occurrence, Formation, Structure and Reactions.* John Wiley and Sons, New York.
48. Denafas, G., Ruzgas, T., Martuzevičius, D., Shmarin, S., Hoffmann, M., Mykhaylenko, V., Ogorodnik, S., Romanov, M., Neguliaeva, E., Chusov, A., Turkadze, T., Bochoidze, I., and Ludwig, C., 2014. "Seasonal Variation of Municipal Solid Waste Generation and Composition in Four East European Cities." *Resour. Conserv. Recy.*, 89:22–30.

Abbreviations Used in This Chapter

ASCE = American Society of Civil Engineers
HDPE = high-density polyethylene
MSW = municipal solid waste
PVC = polyvinyl chloride

TS = total solids
VS = volatile solids
OLR = Organic loading rate

Problems

6-1. Assume that refuse has a C/N ratio of 24:1. If raw sludge with a C/N of 16:1 is to be added to refuse to reach a required C/N of 20:1, how much of each—refuse and sludge—is needed? (*Note:* Use a wet weight basis.)

6-2. Assume that a city of 50,000 people uses on the average 20 million m^3 of natural gas per year. What fraction of this demand could be met by digesting the refuse from this community?

6-3. Using the approximate chemical analyses of refuse—$C_{99}H14_9O5_9N$—what is the empirical formula for the organic fraction of refuse? Estimate the theoretical production of methane from this hypothetical compound.

6-4. Describe how your community manages its organic waste.

6-5. Suppose you are a city engineer. An advertisement appears in a trade journal for a new in-vessel composting system that produces methane gas as a valuable end product. Your city manager, who has no technical training, asks you to find out more about this system. You decide to write a letter to the company, Bioscam Inc., requesting more information. What questions would you ask? Write such a letter.

6-6. Assume a rough estimate of MSW composition is as follows:

	% By Weight
Paper (assume all cellulose, $C_6H_{10}O_5$)	50
Glass	10
Steel	10
Aluminum	10
Food waste (assume sugar $C_6H_{12}O_6$)	20

If 10,000 metric tonnes of this material is placed in a landfill, how much methane gas would theoretically be produced by anaerobic decomposition?

6-7. Suppose a wet mixture of old bread, leaves, and soil is placed in a pile 1 m tall, and a long thermometer is stuck in the middle of the pile. Every other day, the pile is aerated by throwing the mixture into a new pile. Draw a graph showing the temperature in the pile during the first two weeks and explain why these temperature changes occurs.

6-8. An industry operates a landfill into which an organic waste having the formula $C_4H_8O_4$ decays anaerobically.

a. How much methane gas will be generated if the decomposition is complete? Express your answer as "lb gas/lb waste."

b. Realistically, given the nature of the anaerobic reaction, how much methane gas might actually be produced (approximately)? Why only this much?

6-9. Estimate the production of CO_2 and CH_4 during the anaerobic decomposition of ethanol, C_2H_6O.

6-10. What would be three good uses for methane collected from a landfill?

6-11. Describe how a Kompogas in-vessel composting system works.

CHAPTER 7

Thermal Processes

Municipal solid waste is high in organic content and may also carry infectious and hazardous materials. Thus, the combustion of MSW at high temperatures, sufficient to kill infectious agents and produce energy and inorganic ash, has been attractive to communities that either have no landfills or where landfilling is too costly.

The burning of dumps is, of course, ancient history, as this is no longer permitted in most countries due to air pollution and water pollution problems. Engineered MSW combustion facilities, called *incinerators*, were first used in Europe and later adopted by many larger municipalities in the United States.

For the most part, these early incinerators were poorly designed and operated and produced not only horrendous air pollution but also were inefficient in volume reduction and disinfection.[1]

The technology of incineration advanced with time, especially in Europe, and with strict air pollution control and better management of solid waste input stream, incineration remained an option for many municipalities. There was always the notion among combustion engineers that incineration was an unfortunate operation. It was, and is, a method of wasting energy. High-energy material is introduced as the feed and low-energy material is produced. Why can't this energy be recovered?

---

[1]The incinerator constructed in the 1940 in Durham, NC, was derisively called the "Durham toaster."

Such thinking led engineers in the 1940s to design incinerators with heat recovery systems.[38] Initially these were hot air systems where the heated air was used for the heating of nearby buildings, but the absence of demand during warm weather made these systems inefficient. The next designs were adaptations of coal-fired power plants where water was converted to steam and the steam used for generating electricity. Again, it took some time to develop the appropriate technology since MSW is quite different from coal in how it behaves in a combustor. It is more corrosive, it has higher water content, and often it is unpredictable.

The next step in the development of MSW combustion was a better control and processing of the feed. This feed material can also be combined with coal or other fossil fuels and burned in so-called **co-fired** combustion units.

After the two oil embargos in the 1970s, many communities embraced waste-to-energy as a solution to managing their solid waste and also for producing renewable energy. In 1980, the United States combusted 2.7 million tons of MSW. By 1990, 29.7 million tons of MSW were combusted. During this decade, there were other factors that contributed to the rapid growth of new plants, such as favorable tax treatment, tax exempt financing, and lucrative power-purchase agreements from utilities.

However, after this decade of expansion, very few new plants have been built. In fact, 22 years after 1990, the amount of MSW combusted has decreased to 29.3 million tons. In the 1980s, California had planned for numerous waste-to-energy plants from San Francisco to San Diego. In reality, only three were built in the late 1980s with all of the other planned plants being abandoned. The last planned California waste-to-energy plant was defeated by the San Diego County Board of Supervisors in 1991. There are many reasons why new plants have not been built in the United States, such as the inability to site new plants and the high capital and operating cost to build a plant compared to the cost of landfilling.

Nevertheless, in the year 2012, 11.7% of the MSW produced in the United States was combusted in facilities that included energy recovery.[40] Unlike the United States, in Europe, modern plants have continued to be built. The data for different European countries is shown in Figure 2-2 in Chapter 2. Theoretically, the combustion of refuse produced by a community is sufficient to provide about 5 to 15% of the electrical power needs for the residents. Refuse therefore is not an insignificant source of power.

While some would argue that waste-to-energy technology's time has passed, the renewed focus on domestically produced renewable energy may cause communities to once again consider waste-to-energy. The nuclear catastrophe in Fukushima has lighted the discussions about the options for alternative energy generation. Principally, just burying the wastes in landfills is a loss of materials and energy, and therefore, countries which mainly landfill their wastes have a high potential to increase their share in renewable energy from waste. Internationally, waste-to-energy (WTE) or energy-from-waste (EfW), as it is called in Europe, continues to be a widely used alternative to landfilling.

While data about global installations are scarce, a recent report states:

> Today, almost 2,200 WTE plants are active worldwide. They have a disposal capacity of around 270 million tons of waste per year. More than 200 thermal treatment plants with a capacity of over 60 million annual tons were constructed between 2009 and 2013. We estimate almost 500 new plants with a capacity of about 160 million annual tons to be constructed by 2023.[42]

Many of the international plants incorporate advanced features that have been developed in the last decades.

7-1   HEAT VALUE OF REFUSE

One of the earliest measures of heat energy still widely used by American engineers is the *British thermal unit* (Btu), which is defined as that amount of energy necessary to heat one pound of water one degree Fahrenheit. The internationally accepted unit of energy is the *Joule*. Other common units for energy are the *calorie* and *kilowatt-hour* (kWh)—the former used in natural sciences, the latter in engineering.

Another expression of heat value, used by the oil industry's International Energy Agency (IEA) and Organization for Economic Co-operation and

Development (OECD), is *the tonne-oil-equivalent* (*toe*), which is based on the metric ton (or tonne).

A *toe* is a useful unit where huge quantities of energy are involved and is defined as the amount of energy released by burning one tonne of crude oil. It is equivalent to approximately 42 gigajoules, although different crude oils have different energy values. The *toe* unit is of little value in the design of MSW combustion systems but needs to be understood when world or national energy policy and needs are reviewed. Table 7-1 shows the conversion factors for all of these units (emphasizing the fact that all are measures of energy and are interchangeable).

The amount of energy or heat value in an unknown fuel can be estimated by ultimate analysis, compositional analysis, proximate analysis, and calorimetry.

7-1-1 Ultimate Analysis

Ultimate analysis uses the chemical makeup of the fuel to approximate its heat value. The most popular method using ultimate analysis is the DuLong equation, which originally was developed for estimating the heat value of coal:

$$Btu/lb = 145\ C + 620\left(H - \frac{1}{8}O\right) + 40\ S$$

where C, H, O, and S are the weight percentages (dry basis) of carbon, hydrogen, oxygen, and sulfur, respectively. The DuLong formula is cumbersome to use in practice, and it does not give acceptable estimates of heat value for materials other than coal.[1]

Table 7-1 Energy Conversions

To Convert	To	Multiply By
Btu	Calories	252
	Joules	1054
	kWh	2.93×10^{-4}
	toe	2.52×10^{-8}
Calories	Btu	3.97×10^{-3}
	Joules	4.18
	kWh	1.16×10^{-6}
	toe	1.00×10^{-10}
Joules	Btu	9.49×10^{-4}
	Calories	0.239
	kWh	2.78×10^{-7}
	toe	2.39×10^{-11}
Kilowatt-hours	Btu	3413
	Calories	8.62×10^{5}
	Joules	3.60×10^{6}
	toe	8.60×10^{-5}
Toe	Btu	3.97×10^{7}
	Calories	1.00×10^{10}
	Joules	4.18×10^{-10}
	kWh	1.16×10^{4}

Another equation for estimating the heat value of refuse using ultimate analysis is[2]

$$Btu/lb = 144\ C + 672\ H + 6.2\ O + 41.4\ S - 10.8\ N$$

where C, H, O, S, and N are again the weight percentages (dry basis) of carbon, hydrogen, oxygen, sulfur, and nitrogen, respectively, in the combustible fraction of the fuel. That is, the sum of all of these percentages must add to 100%.

Other approaches are used to estimate the heat value of wastes considering its elemental composition and a correction term for the moisture content. The empirical formula according to Boie[43] as given in source [44] gives the net calorific value (NCV), which is another term used for the Low Heat Value (LHV) as follows:

$$NCV\ [MJ/kg] = 34.8\ C + 93.9\ H + 10.5\ S + 6.3\ N - 10.8\ O - 2.5\ W$$

where: C = carbon content [kg/kg], H = hydrogen content [kg/kg], S = sulfur content [kg/kg], N = nitrogen content [kg/kg], O = oxygen content [kg/kg], and W = water content [kg/kg].

As mainly organics contribute to the energy content, only the major elements typically found in organic and biochemical compounds are considered in this formula.

The elemental analysis can be conducted using standard methods such as those published by the American Society for Testing and Materials (ASTM).[3] The elements C, H, N, and S are measured in one procedure by thermally decomposing the sample at high temperatures (typically $> 1000°C$) in a catalytic reactor. The introduction of oxygen produces the gases NO_2, CO_2, H_2O, and SO_2. These are separated in a gas-chromatograph (GC) and analyzed by thermo-conductivity. Moisture is usually determined by thermo-gravimetric (TG) measurements, and oxygen can be measured by pyrolysis of the sample in helium gas where oxygen is tied up as carbon monoxide. The CO can then be measured with the GC. Table 7-2 lists some typical values found in MSW.

7-1-2 Compositional Analysis

Ultimate analysis provides a theoretical maximum of the heat value but does not take into account any inefficiencies or chemical interactions that might occur during combustion. An alternative to ultimate analysis is *compositional analyses*, which are based on the actual content of the MSW sample. One such formula is[5]

$$Btu/lb = 49R + 22.5(G + P) - 3.3W$$

Table 7-2 Typical Ultimate Analyses of MSW

Constituent	Average of Three Different Samples of MSW
Carbon	51.9
Hydrogen	7.0
Oxygen	39.6
Sulfur	0.37
Nitrogen	1.1

Source: [4]

Note: The numbers in this table have been adjusted for zero inerts and zero moisture.

Table 7-3 Typical Heat Values of MSW Components

Component	Heat Value, Btu/lb Dry Weight
Food waste	2000
Paper	7200
Cardboard	7000
Plastics	14,000
Textiles	7500
Rubber	10,000
Leather	7500
Garden trimmings	2800
Wood	8000
Glass	60
Nonferrous metals	300
Ferrous metals	300
Dirt, ashes, other fines	3000

Source: Modified from [7]

where

R = plastics, percent by weight of total MSW, on dry basis
G = food waste, percent by weight of total MSW, on dry basis
P = paper, percent by weight of total MSW, on dry basis
W = water, percent by weight, on dry basis

Using regression analysis and comparing the results to actual measurements of heat value, an improved form of a compositional model is suggested:[6]

$$Btu/lb = 1238 + 15.6R + 4.4P + 2.7G - 20.7W$$

where

R = plastics, percent by weight, on dry basis
P = paper, percent by weight, on dry basis
G = food wastes, percent by weight, on dry basis
W = water, percent by weight, on dry basis

An even more exact compositional analysis is possible if the composition of the MSW or processed fuel is known. The heat value of the complex fuel can be calculated by using typical heat values of its components, as listed in Table 7-3. The calculation is shown in Example 7-1.

EXAMPLE 7-1

A processed refuse-derived fuel has the following composition:

Component	Fraction by Weight, Dry Basis
Paper	0.50
Food waste	0.10
Plastics	0.30
Glass	0.10

SOLUTION

Estimate the heat value based on the typical values in Table 7-3.

0.50 (7200) + 0.10 (2000) + 0.30 (14,000) + 0.10 (60) = 8006 Btu/lb

Some of the components of refuse may not always be tuned finely enough in such a compositional analysis. Paper, for example, comes in many categories, and each of these categories has its own heat value, as shown in Table 7-4.

If the waste composition is known in terms of the wet weight, the fraction of moisture has to be subtracted from the composition fractions. Typical moisture concentrations of MSW components are shown in Table 7-5. The calculation of heat value for a wet sample is shown in Example 7-2.

Table 7-4 Heat Value of Various Types of Paper

Type of Paper	Heat Value, Btu/lb Dry Weight
Newspaper	7520
Cardboard	6900
Kraft	6900
Beverage and milk boxes	6800
Boxboard	6700
Tissue	6500
Colored office paper	6360
White office paper	6230
Envelopes	6150
Treated paper	6000
Glossy paper	5380

Source: [8]

Table 7-5 Typical Moisture Contents of MSW

Component	Typical Moisture, Percent
Food waste	70
Paper	6
Cardboard	5
Plastics	2
Textiles	10
Rubber	2
Leather	10
Garden trimmings	60
Wood	60
Glass	2
Nonferrous metals	2
Ferrous metals	3
Dirt, ashes, other fines	8

Source: Adapted from [9]

EXAMPLE 7-2

Using the refuse-derived fuel composition shown in Example 7-1 on a wet weight basis, calculate the heat value of this mixture.

SOLUTION

As shown in Table 7-5, 6% of the wet weight of paper is water that does not contribute to the heat value. Hence, the first term from the calculation in Example 7-1 is multiplied by $1.0 - 0.06 = 0.94$, and so on.

$$0.50 \ (7200) \ 0.94 + 0.10 \ (2000) \ 0.30 + 0.30 \ (14,000) \ 0.98$$
$$+ \ 0.10 \ (60) \ 0.98 = 7570 \ \text{Btu/lb}$$

7-1-3 Proximate Analysis

In proximate analysis, it is assumed that the fuel is composed of two types of materials: volatiles and fixed carbon. The amount of volatiles can be estimated by loss of weight when the fuel sample is burned at some elevated temperature, such as 600 or 800°C, and the fixed carbon is estimated by the weight loss when the sample is combusted at 950°C. A commonly used proximate analysis equation for estimating the heat value of refuse is

$$\text{Btu/lb} = 8000A + 14,500B$$

where

A = volatiles, fraction of all dry matter lost at 600°C
B = fixed carbon, fraction of all dry matter lost between 600°C and 950°C

Another form of a proximate analysis equation is

$$\text{Btu/lb} = 2500D - 330W$$

where

D = fraction volatile material, dry basis, defined as weight loss at 800°C
W = fraction water, dry basis

Depending on whether the moisture is included or not, the results are presented as Btu/lb as received or as Btu/lb on dry basis. Table 7-6 lists some representative moisture, volatiles, and fixed carbon fractions for several refuse components.[10]

7-1-4 Calorimetry

Calorimetry is the referee method of determining the heat value of mixed fuels. Figure 7-1 shows a schematic sketch of a *bomb calorimeter*. The bomb is a stainless-steel ball that screws apart. The ball has an empty space inside into which the sample to be combusted is placed. A sample of known weight (such as a small piece of coal) is placed into the bomb, and the two halves are screwed shut. Oxygen under high pressure is then injected into the bomb, and the bomb is placed in

Table 7-6 Typical Proximate Analysis of MSW Components

Component	Fraction by Weight			
	Moisture	Volatile	Fixed	Ash
Mixed paper	0.102	0.759	0.084	0.054
Yard waste	0.752	0.186	0.045	0.016
Food waste	0.783	0.170	0.036	0.010
Polyethylene	0.002	0.985	0.001	0.012
Wood	0.200	0.697	0.113	0.008

Source: [10]
Note: If the values in Table 7-6 are to be used in the proximate analysis equation, the fractions have to be recalculated on the basis of *dry* matter.

an adiabatic water bath with wires leading from the bomb to a source of electrical current. By means of a spark from the wires, the material in the steel ball combusts and heats the bomb, which in turn heats the water. The temperature rise in the water is measured with a thermometer and recorded as a function of time. Figure 7-2 shows the trace of a typical calorimeter curve.

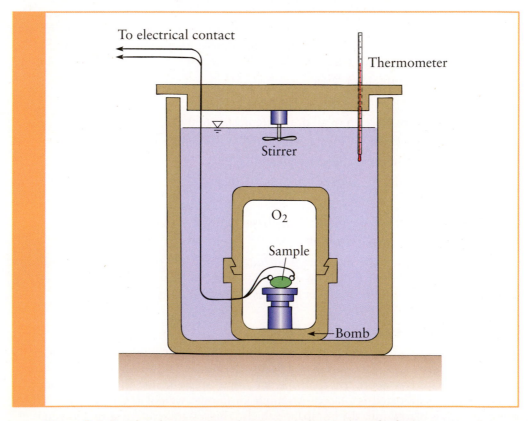

Figure 7-1 Bomb calorimeter used to measure heat value of a fuel.

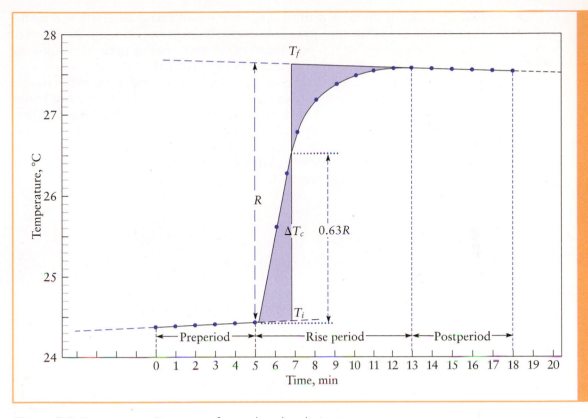

Figure 7-2 Temperature/time trace from a bomb calorimeter.

More accurately, Figure 7-2 is a plot of temperature (T) versus time (t) and is called a *thermogram*. Because the initial (preperiod) and final (postperiod) slopes of the thermogram may be different and because the temperature rise due to a chemical reaction is usually nonlinear, it is necessary to decide on a procedure to estimate the true net temperature rise (ΔT_c) in the experiment. A common method is to extrapolate the postperiod slope backward in time and the preperiod slope forward in time (the dotted lines in Figure 7-2). The net temperature rise (ΔT_c) can be measured from the difference of these two extrapolated lines at some intermediate time in the reaction period. The exact location of this intermediate time depends on the calorimeter and on the reaction being studied. A frequently used criterion for estimating this time is to measure ΔT at the point where the shaded areas in the figure are equal. It can be shown that these shaded areas are equal when the temperature increase is approximately $0.63\Delta T$. From the thermogram, the difference between the initial temperature (T) and the final temperature (T_f) is the temperature rise (ΔT) of the reaction.

The water container of a bomb calorimeter is well insulated, so no heat is assumed to escape from the system. All of the energy liberated during the combustion is used to heat the water and the stainless steel bomb. The heat energy is calculated as the temperature increase of the water times the mass of the water plus the bomb. Since one calorie is defined as the amount of energy necessary to raise the temperature of one gram of water one degree Celsius, and knowing the grams of water in the calorimeter, it is possible to calculate the energy in calories. Knowing the weight of the sample, its heat value can be calculated.

Each calorimeter is different and must be standardized using a material for which the heat of combustion is known precisely. Plus, the heat generated by the combustion of the ignition wire must be taken into account if accurate analyses are required. Typically, benzoic acid is used as the standard. A benzoic acid pellet specially manufactured for this purpose is combusted in a bomb calorimeter, and the temperature rise (ΔT) is determined from the *thermogram* and used to calculate the heat capacity of the calorimeter C_v. That is,

$$C_v = \frac{UM_b}{\Delta T}$$

where

C_v = heat capacity of the calorimeter, cal/°C
U = heat of combustion of benzoic acid, cal/g, °C
M_b = mass of benzoic acid pellet, g
ΔT = rise in temperature from thermogram, °C

For very accurate determinations, the heat generated by the combustion of the fuse wire must be included in the calculations.

$$C_v = \frac{[6318M_b + 1643M_w]}{\Delta T}$$

where

C_v = heat capacity of the calorimeter, cal/°C
M_b = mass of benzoic acid tablet used, g
M_w = mass of ignition wire used, g
ΔT = temperature rise, °C

Note that 6318 is the heat of combustion of benzoic acid expressed as cal/g, and 1643 is the heat of combustion of nickel chromium wire, cal/g.

EXAMPLE 7-3

A benzoic acid pellet weighing 5.00 g is placed in a bomb calorimeter along with 0.20 g fuse wire. The benzoic acid is ignited, and the temperature rise is 3.56°C. What is the heat capacity of this calorimeter?

SOLUTION

$$C_v = \frac{[6318(5.00) + 1643(0.2)]}{3.56} = 8966 \, \text{cal/°C}$$

Knowing the heat capacity of the calorimeter (C_v), the heat generated by the combustion of a material of unknown heat value may be determined by measuring the temperature rise that occurs upon the combustion of the material:

$$U = \frac{C_v \Delta T}{M}$$

where

U = heat value of unknown material, cal/g
ΔT = rise in temperature from thermogram,
M = mass of the unknown material, g
C_v = heat capacity of the calorimeter, cal/°C

EXAMPLE 7-4

A 10 g sample of a refuse-derived fuel (RDF) is combusted in a calorimeter that has a heat capacity of 8966 cal/°C. The detected temperature rise is 4.72°C. What is the heat value of this sample?

SOLUTION

$$U = \frac{C_v \Delta T}{M} = \frac{8966 \times 4.72}{10.00} = 4231\,\text{cal/g}$$

The conversion from cal/g to Btu/lb is 1.78 × cal/g = Btu/lb, so that this fuel has a heat value of 7531 Btu/lb.

An important aspect of calorimetric heat values is the distinction between *higher heating value* and *lower heating value*. The higher heating value (HHV) is also called the *gross calorific energy*, while the lower heating value (LHV) is also known as the *net calorific energy*. The distinction is important in the design of combustion units.

In a calorimeter, as organic matter combusts, the products of combustion are (ideally) carbon dioxide and water. The water produced is in vapor form. As the bomb cools, however, this water condenses, yielding heat that is measured as part of the temperature rise. The HHV is calculated by including the contribution due to this *latent heat of vaporization* (the heat required to produce steam from water). But such condensation does not occur in a large furnace. The hot flue gases carry the water vapor outside the furnace, and condensation cannot take place. The LHV is then the HHV minus the latent heat of vaporization that has occurred in the bomb calorimeter. For design purposes, the LHV is a much more realistic number, but American engineers usually express heat values in terms of HHV. When heat values of refuse or other fuels are specified, therefore, it must be clear whether the numbers are expressed as HHV or LHV.

In many engineering practices, heat value refers only to HHV. In this text, unless otherwise specified, heat value refers to HHV. Although calorimetry is the reference method of measuring heat value of a fuel, it does not actually simulate the behavior of that fuel in a full-scale combustor. There are at least three reasons why the HHV number overestimates the actual heat value in combustion: the presence of metals, the incomplete combustion of organics, and the water leaving in a gaseous state.

Some metals, most notably aluminum, will oxidize at sufficiently high temperatures to yield heat. The oxidation of aluminum can be described as

$$4Al + 3O_2 \rightarrow 2(Al_2O_3) + heat$$

This reaction is highly exothermic, yielding 13,359 Btu/lb (31,070 kJ/kg). If an MSW sample contains a significant amount of aluminum, the measured heat value in the calorimeter will reflect this exothermic reaction because the temperature of combustion in the calorimeter is so high. In a typical full-scale combustor, however, the temperature is not sufficiently high to force the oxidation of aluminum.[11]

The second problem with the use of calorimetric heat values is that while all organic material will oxidize in a calorimeter, this will not occur in a full-scale combustor. The amount of unburned organics can vary from 2 to 5%, depending on the effectiveness of the operation. In one laboratory study, samples of RDF were combusted for different times in various furnace temperatures, and the resulting ash was analyzed for calorific value. The results, pictured in Figure 7-3, show that

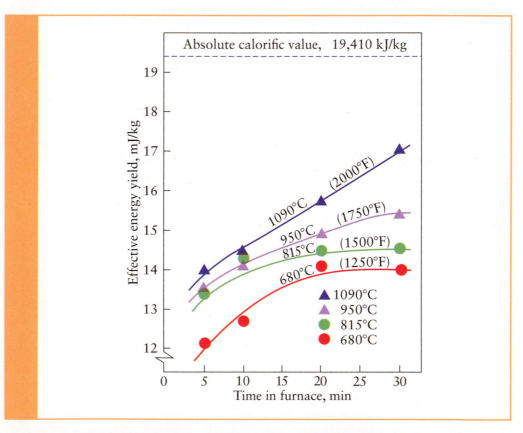

Figure 7-3 MSW combustion requires time to achieve full energy recovery. Source: [25]

even at reasonable residence times and temperatures, a significant fraction of the fuel is unburned.

A third reason why HHV normally overestimates the energy which can be recovered from the wastes in an WTE plant is the fact that a large fraction of the water leaves the WTE plant as gaseous water. It is not common to install condensers to recover this energy, which is also known as "latent heat." This difference mainly reflects the difference between LHV and HHV. Therefore, if no such condenser units are implemented, it is more appropriate to use LHV instead of HHV values to calculate the potentially recoverable energy from wastes. Such an approach assumes that the water vapor produced when the fuel is burned is released via the exhaust. In this context it is important to verify which assumptions have been used to calculate the overall energy efficiency of a process. The result will be different using LHV or HHV to calculate the process's efficiency. In most cases the values are given for HHV; however, in the literature this is not always done in a consistent manner.

Because MSW is such a heterogeneous and unpredictable fuel, engineers often need to have "rules of thumb" for estimating the heat values. For MSW, one rule of thumb is that one ton of MSW produces 5000 lb of steam, and this steam produces 600 kW of electricity. This value, however, depends on the combustion technology and the heating value of the MSW.

7-2 MATERIALS AND THERMAL BALANCES

7-2-1 Combustion Air

The energy from the sun is stored using the process of photosynthesis in organic molecules, and this energy is slowly released as the organic materials decompose. Animals, of course, receive their energy from plants by eating them and extracting energy for their maintenance needs as well as cell growth. In simplest of terms, the photosynthesis process is

$$x CO_2 + \text{sunlight} + \text{nutrients} + x H_2O \rightarrow \text{nutrients} \cdot (H_2CO)_x + x O_2$$

The $(H_2CO)_x$ represents an infinite variety of carbohydrates. The degradation of the high-energy organics is then

$$\text{nutrients} \cdot (H_2CO)_x + O_2 \rightarrow x CO_2 + x H_2O + \text{nutrients} + \text{heat energy}$$

Combustion of the organic fraction of refuse is simply a very rapid decomposition process, which is strongly exothermic. The end products from the combustion of the hydrocarbons are essentially the same as in slower aerobic biochemical decomposition that reaches the final low-energy products.

Of interest to engineers is the amount of heat produced in the combustion operation. The heat value of pure materials can be estimated from thermodynamics. For example, the combustion of pure carbon is a two-stage reaction:

$$C + O \rightarrow CO + 10{,}100 \text{ J/g}$$

and

$$CO + O \rightarrow CO_2 + 22{,}700 \text{ J/g}$$

although it is usually expressed as a single reaction:

$$C + O_2 \rightarrow CO_2 + 32{,}800 \text{ J/g}$$

The amount of oxygen necessary to oxidize some hydrocarbon is known as *stoichiometric oxygen*. Consider the simple combustion of carbon:

$$C + O_2 \rightarrow CO_2 + \text{heat}$$

That is, one mole of carbon combines with one mole of molecular oxygen. The molecular weight of carbon is 12 and of oxygen 16, so it takes two oxygens or 32 grams of oxygen to react with 12 grams of carbon. The stoichiometric oxygen is then 2.67 g O_2/g C.

EXAMPLE 7-5

Calculate the stoichiometric oxygen required for the combustion of methane gas.

SOLUTION

The equation is

$$CH_4 + 2O_2 \rightarrow CO_2 + 2H_2O$$

That is, it takes 16 grams of methane (12 + 4) to react with 2 × 2 × 16 = 64 grams of oxygen. (The molecular weight is 16, there are two oxygens, and there are two moles.) Thus, the stoichiometric oxygen required for the combustion of methane is 64/16 = 4 g O_2/g CH_4.

Normally, refuse is not burned using pure oxygen, however, and air is used as the source of oxygen. A correction factor therefore is needed to recognize that the air is 23.15% oxygen by weight. Dividing the stoichiometric oxygen by 0.2315 yields *stoichiometric air*.

EXAMPLE 7-6

Calculate the stoichiometric air required for the combustion of methane gas.

SOLUTION

Since the stoichiometric oxygen from Example 7-5 is 4 g O_2/g CH_4, the stoichiometric air requirement is 4/0.2315 = 17.3 g air/g methane.

7-2-2 Efficiency

Figure 7-4 illustrates how a power plant operates. Water is heated to steam in a boiler, and the steam is used to turn a turbine, which drives a generator. This diagram can be simplified to a simple energy balance where *energy in* has to equal *energy out* (energy wasted in the conversion + useful energy) plus the energy accumulated in the box. Expressed as an equation:

$$\begin{vmatrix} \text{rate of} \\ \text{energy} \\ \text{ACCUMULATED} \end{vmatrix} = \begin{vmatrix} \text{rate of} \\ \text{energy} \\ \text{IN} \end{vmatrix} - \begin{vmatrix} \text{rate of} \\ \text{energy} \\ \text{OUT} \end{vmatrix} + \begin{vmatrix} \text{rate of} \\ \text{energy} \\ \text{PRODUCED} \end{vmatrix} - \begin{vmatrix} \text{rate of} \\ \text{energy} \\ \text{CONSUMED} \end{vmatrix}$$

Of course, energy is never produced or consumed in the strict sense; it is simply changed in form.

Just as processes involving materials can be studied in their ***steady-state*** condition, defined as no change occurring over time, energy systems also can be in steady state. If there is no change over time, there cannot be a continuous accumulation of energy, and the equation must read

[rate of energy IN] = [rate of energy OUT]

or if some of the energy out is useful and the rest is wasted,

$$\begin{vmatrix} \text{rate of} \\ \text{energy} \\ \text{IN} \end{vmatrix} = \begin{vmatrix} \text{rate of} \\ \text{energy} \\ \text{USED} \end{vmatrix} + \begin{vmatrix} \text{rate of} \\ \text{energy} \\ \text{WASTED} \end{vmatrix}$$

If the input and useful output from a black box are known, the efficiency of the process can be calculated as

$$E = \frac{\text{energy USED}}{\text{energy IN}} \times 100$$

where E = efficiency (%).

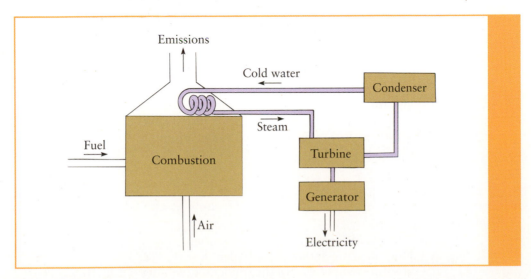

Figure 7-4 Operation of a fossil-fuel power plant.

EXAMPLE 7-7

A coal-fired power plant uses 1000 Mg (megagrams, or 1000 kg, commonly called a metric tonne) of coal per day. The energy value of the coal is 28,000 kJ/kg (kilojoules/kilogram). The plant produces 2.8×10^6 kWh of electricity each day. What is the electrical efficiency of the power plant?

SOLUTION

Energy In = (28,000 kJ/kg $\times$ 1000 Mg/day) $\times$ (1 $\times$ 10³ kg/Mg)
          = 28 $\times$ 10⁹ kJ/day

Useful Energy Out = 2.8 $\times$ 10⁶ kWh/day $\times$ (3.6 $\times$ 10⁶) J/kWh $\times$ 10⁻³ kJ/J
                  = 10.1 $\times$ 10⁹ kJ/day

Efficiency (%) = (10.1 $\times$ 10⁹)/(28 $\times$ 10⁹) $\times$ 100 = 36%

One of the most distressing problems in the production of electricity from MSW—or from any fossil fuel for that matter—is that the power plants are less than 40% efficient. The reason for such low efficiency is that the waste steam must be condensed to water before it again can be converted to high-pressure steam. Using the previous energy balance:

$$0 = Q_0 - Q_U - Q_W$$

where

Q_0 = energy flow in
Q_U = useful energy out
Q_W = wasted energy out

Note that energy flow can be in any number of terms such as kJ/sec or Btu/hr. The efficiency of this system, as previously defined, is

$$E(\%) = (Q_U)/(Q_0) \times 100$$

From thermodynamics, it is possible to prove that the greatest efficiency (least wasted energy) can be achieved with the *Carnot heat engine*, and this efficiency is determined by the absolute temperature of the surroundings. The efficiency of the Carnot heat engine is defined as

$$E_c(\%) = \frac{T_1 - T_0}{T_1} \times 100$$

where

E_c = Carnot efficiency, %
T_1 = absolute temperature of the boiler, °K = °C + 273
T_0 = absolute temperature of the condenser (cooling water), °K

Since this is the best possible situation, any real-world system must be less efficient, or

$$E \leq E_c$$

$$\frac{Q_U}{Q_0} \leq \frac{T_1 - T_0}{T_1}$$

Modern coal boilers generate steam with temperatures as high as 600°C, while environmental restrictions limit condenser water temperature to about 20°C. Thus, the best expected efficiency is

$$E_c(\%) = \frac{(600 + 273) - (20 + 273)}{(600 + 273)} \times 100 = 66\%$$

A power plant also experiences energy losses due to hot stack gases, evaporation, friction losses, etc.; these losses typically reduce the efficiency from the theoretical value of 66% to a value in the range of 47–38% for plants of the size in the range 20 to 0.1 MW, respectively. Typically larger plants of the same technology tend to be more efficient. As a comparison, the efficiency of nuclear power stations in Switzerland ranges from 30 to 35%. In contrast, a gas-fired combined-cycle power station, which uses methane as fuel, can reach efficiencies as high as 58%.[46]

Modern WTE plants can recover heat and/or power. Heat, in the form of steam, can be used for district heating or process heating. District heating is typically only needed during the winter months while process heating is needed year round. However, if the process heat is for a specific industrial operation, there is always the risk that the industry could reduce operating hours or close during the operating life of the WTE plant. Power, in the form of electricity, can usually be sold into an electrical grid continuously. However, the price paid for the electricity could fluctuate depending on the demand.

By using an extraction turbine which has openings in its casing for extraction of a portion of the steam at some intermediate pressure before condensing the remaining steam, it is possible to generate both steam and electricity. Thus the amount of steam produced can be adjusted based on the demand for steam.

Since the production of electric power from the steam represents an energy loss, the efficiency of producing electricity will be lower than the efficiency for producing steam. Figure 7-5 shows the heat and power efficiencies for 29 WTE plants in Switzerland in 2006. To determine overall efficiency, the power and heat efficiency numbers are added together. Thus, those plants that only produce power (electricity) have an efficiency of 13 to 23% (heat efficiency = 0). On the other hand, those plants that produce heat (steam) and power (electricity) have a much higher efficiency.

In summary, the efficiencies for power generation are different from plant to plant. Whereas an older WTE plant would only convert about 20% of the energy contained in MSW to power, the currently most efficient WTE plant in Amsterdam converts more than 30% of the net energy content to power; that is, about 850 kWh per ton of waste. It is therefore known as Waste Fired Power Plant (WFPP). The concept and principles for this high-efficiency plant are described in source [45].

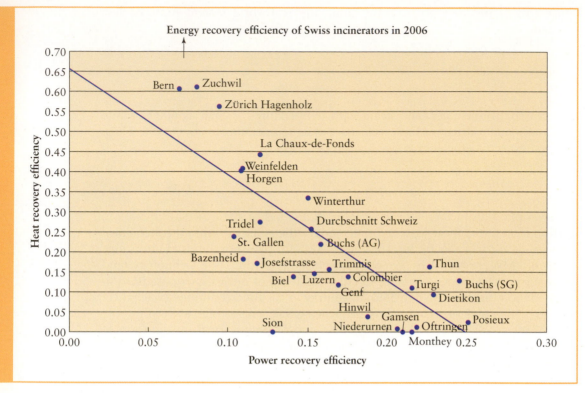

Figure 7-5 Heat and power recovery efficiencies of 29 WTE plant plants in Switzerland. The diagonal line indicates the minimal efficiency to obtain subsidies in order to cover the true costs for power generation; that is, KEV subsidies. Adapted from source [51].

7-2-3 Benefit/Cost

Fossil fuel power plants buy fuel and produce power. The hope is that the cost of the fuel (plus cost of operation) is less than the income from the sale of the power. This definition of efficiency does not apply to solid waste heat recovery facilities, however, since the fuel has a negative cost. People actually pay the plant to take fuel (solid waste). The effectiveness of a MSW energy recovery facility is therefore calculated as a *benefit/cost* ratio where the cost (capital cost, operating cost, and cost of residue disposal) is compared to the benefit (sale of energy and tipping fees charged to the users).

If such a calculation produces a number less than 1.0, meaning that the operation loses money, the tipping fees have to be increased. Increased tipping fees could cause users to go to other facilities, which would result in even lower income from tipping fees. As an alternative, there could be subsidies provided by government programs. In some cases this subsidy takes the form of increased revenues from the sale of the energy. For example, in Switzerland 50% of the energy content in MSW is accounted as "renewable," which allows the operators of WTE plants to sell this energy at a higher rate.

7-2-4 Thermal Balance on a Waste-to-Energy Combustor

Energy (of whatever kind), when expressed in common units, can be pictured as a quantity that flows, and thus it is possible to analyze energy flows using the same concepts used for materials flows and balances. As before, a black box is any process or operation into which certain flows enter and others leave. If all of the flows can be correctly accounted for, then they must balance.

A thermal balance on a large combustion unit is difficult because much of the heat cannot be accurately measured. For instructional purposes, a black box can be used to describe the thermal balance, as shown in Figure 7-6. Assume that in this facility the heat is recovered as steam.

The heat input to this black box is from heat value in the fuel and the heat in the water entering the water-wall pipes. The output is the sensible heat in the stack gases, the latent heat of water, the heat in the ashes, the heat in the steam, and the heat lost due to radiation. If the black box is thought of in dynamic terms, the balance would be

$$
\begin{vmatrix} \text{rate of} \\ \text{heat} \\ \text{accumulated} \end{vmatrix} = \begin{vmatrix} \text{rate of} \\ \text{heat in} \\ \text{the fuel} \end{vmatrix} + \begin{vmatrix} \text{rate of} \\ \text{heat in} \\ \text{the water} \end{vmatrix}
$$

$$
- \begin{vmatrix} \text{rate of} \\ \text{heat out} \\ \text{in the} \\ \text{stack gases} \end{vmatrix} - \begin{vmatrix} \text{rate of} \\ \text{heat out} \\ \text{in the} \\ \text{steam} \end{vmatrix}
$$

$$
- \begin{vmatrix} \text{rate of heat out} \\ \text{as latent} \\ \text{heat of vaporization} \end{vmatrix} - \begin{vmatrix} \text{rate of} \\ \text{heat out} \\ \text{in the ash} \end{vmatrix} - \begin{vmatrix} \text{rate of} \\ \text{heat loss due} \\ \text{to radiation} \end{vmatrix}
$$

If the process is in a steady state, the first term (accumulation) is zero, and the equation can be balanced. The best way to illustrate this is with an example.

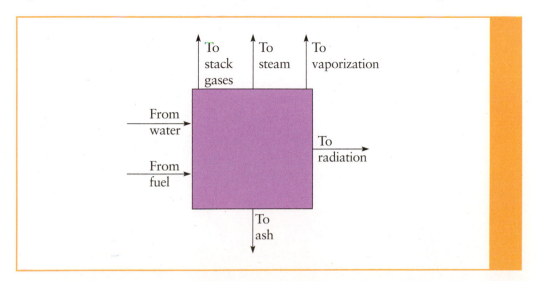

Figure 7-6 Black box showing energy flow in a combustor.

EXAMPLE 7-8

A combustion unit is burning RDF consisting of 85% organics, 10% water, and 5% inorganics (inerts) at a rate of 1000 kg/h. Assume the heat value of the fuel as 19,000 kJ/kg on a moisture-free basis. Assume the combustion unit is refractory lined with no water wall and no heat recovery. Assume that the air flow is 10,000 kg/h and that the under- and overfire air contributes negligible heat. Assume further that 5% of the heat input is lost due to radiation and that 10% of the fuel remains uncombusted in the ash, which exits the combustion chamber at 800°C. The specific heat of ash is 0.837 kJ/kg/°C, and the specific heat of air is 1.0 kJ/kg/°C. What is the temperature of the stack gases?

SOLUTION

(a) Heat from the combustion is

1000 kg/h × 0.85(organics) × 0.9 × 19,000 kJ/kg = 14,535,000 kJ/h

(b) The heat due to water vaporization is

1000 kg/h × 0.10 × 2575 kJ/kg = 257,500 kJ/h

where 0.10 is the 10% water in the RDF and 2575 kJ/kg is the latent heat of vaporization for water.

(c) The heat loss due to radiation is

$0.05 \times 16.15 \times 10^6 = 807,500$ kJ/h

(d) The heat lost in the ashes is both the sensible heat in the ashes as well as the unburned fuel. The total amount of ash is the inerts plus the unburned organics or

[1000 kg/h × 0.05 (inorganics)] + [1000 kg/h × 0.85 (organics)
    × 0.1] = 135 kg/h

If the temperature of the ash is 800°C, the heat loss due to the hot ash is

800°C × 135 kg/h × 0.837 kJ/kg/°C = 90,396 kJ/h

where 0.837 kJ/kg/°C is the specific heat of ash.

(e) The heat lost in the stack gases is then calculated by subtraction, or

0 = 14,535,000 − 257,500 − 807,000 − 90,396 − X

or $X \approx 13,400,000$ kJ/h. If the total air is 10,000 kg/h, and if the specific heat of air is 1.0 kJ/kg/°C, the temperature of the stack gases is

$$\frac{13,400,000\,\text{kJ/hr}}{10,000\,\text{kg/hr}\,1.0\,\text{kJ/kg/°C}} = 1340°C$$

which is about 2,500°F—quite high.[31] These hot gases could be run through a heat exchanger to produce steam as a valuable product of the combustion process.

When purchasing a waste-to-energy plant, some municipalities have tried to have the manufacturer guarantee the electricity output per ton of waste. In such cases, the manufacturer then requires the municipality to have a certain minimum heat value of its solid waste. No municipality can guarantee the heat value of its solid waste, and thus, process efficiencies must be analyzed on the basis of a thermal balance. But the problem with this approach is that calculating the thermal balance is very difficult with such a variable fuel as solid waste, and the cost of monitoring and sampling the waste is significant. One alternative is to require the manufacturer to meet two criteria:

1. Ash must not exceed a percent combustible level.
2. Exhaust gas from the boiler must be within a predetermined temperature range.

These two criteria ensure complete combustion of the solid waste and recovery of the heat, and both criteria can be easily monitored.

7-3   COMBUSTION HARDWARE USED FOR MSW

In years past when refuse was burned without recovering energy, the units were known as *incinerators*, a name no longer used by the industry because of the sorry record of these facilities. Poor design, inadequate engineering, and inept operation combined to produce an ash still high in organics and smoke that (even in the days of little industrial air-pollution control) caused many communities to shut down the incinerators.

Without energy recovery, the exhaust gas from these units was very hot. Typical air pollution control for particulates consisted of a dry cyclone. As the requirement for particulate control increased, electrostatic precipitators (ESPs) were required for control of particulates, but the hot exhaust gas exceeded the acceptable inlet temperature for ESP. If a wet cyclone were used prior to the ESP to cool the gas, any moisture carryover would cause severe corrosion problems in the ESP. It became very difficult to upgrade old incinerators, and most of them were shut down.

7-3-1 Waste-to-Energy Combustors

Modern combustors combine solid waste combustion with energy recovery (Figure 7-7). Such combustors have a *storage pit* for storing and sorting the incoming refuse (Figure 7-8), a *crane* for charging the combustion box Figure 7-9, a *combustion chamber* consisting of bottom grates on which the combustion occurs (Figure 7-10), the *furnace* or *combustion chamber*, the *heat recovery system* of pipes in which water is turned to steam (Figure 7-11), the *ash-handling system*, and the *air-pollution control system*. The units take their name from the fact that the combustion chamber is lined with refractory bricks that are heat resistant, very much like the firebricks in the home fireplace. These units produce steam in a *boiler* located at the top of the combustion chamber.

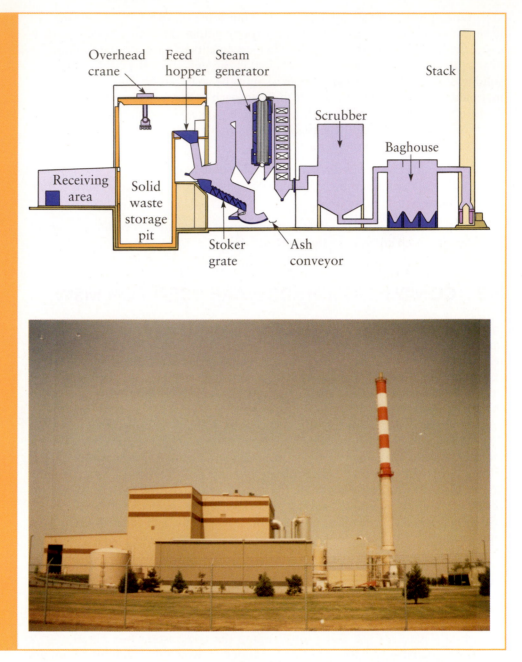

Figure 7-7 A typical municipal solid waste combustor. (Courtesy William A. Worrell)

Figure 7-8 Storage pit with water misters to control dust. (Courtesy William A. Worrell)

The heart of the combustion process is the chamber in which the combustion occurs. In most units, the refuse is moved through the combustion chamber on a moving grate, and the design and operation of these grates often determines the success or failure of the entire process. The functions of the grates are to provide tumbling so that the MSW can be thoroughly burned, to move the refuse down and through the combustion chamber, and finally, to provide *underfire air* to the refuse through openings in the grates. The underfire air both assists in the combustion and cools the grates. Some newer grates are also water cooled. The control of underfire air is also an important variable in maintaining a desired operating temperature in the combustion chamber. Most refuse combustors operate in the range of 1800 to 2000°F (980 to 1090°C), which ensures good combustion and elimination of odors.

The temperature within the combustion chamber is critical for successful operation. If it is too low, say below 1400°F (770°C), then many of the plastics will not burn, resulting in poor combustion. Above 2000°F (1090°C) slagging of metals in the ash could become a problem. Thus, the window for effective operation is not large, and close control needs to be kept on the charge to the combustion chamber and the amount of overfire air and underfire air.

As the amount of excess air is increased, the temperature drops. The relationship between the availability of air and the temperature in the combustion chamber is shown in Figure 7-12. Note that at stoichiometric air quantities the temperature increases to intolerable levels. As shown in Figure 7-12, maintaining a temperature of 2000°F (1090°C) in the combustion chamber requires about 100% excess air. Also note that the temperature drops if the unit is operated in the

Figure 7-9 Crane for moving MSW. (Courtesy William A. Worrell)

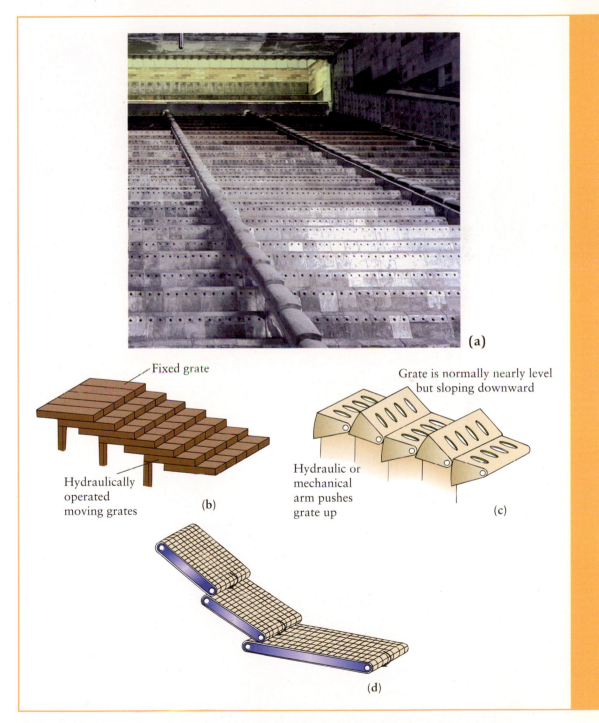

Figure 7-10 Grates in an MSW combustor. The underfire air is blown through the holes in the grates. (Photo courtesy Hitachi, Zosen Inova) The drawings show three types of grates: (b) reciprocating, (c) rocking, (d) traveling. (Courtesy P. Aarne Vesilind)

Figure 7-11 Boiler tubes for steam generation. (Courtesy William A. Worrell)

starved-air mode, which provides less than stoichiometric air to the combustion unit. Such combustors are discussed next.

The air blown into the combustion chamber above the refuse is logically *overfire air*, and its purpose is to provide the oxygen necessary for combustion as well as to enhance the turbulence in the combustion chamber.

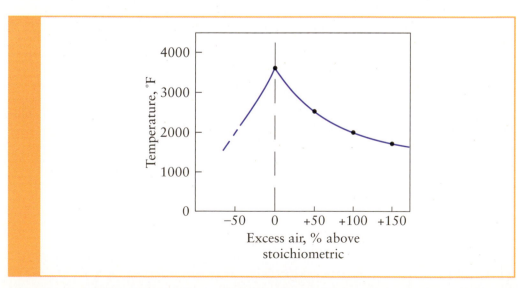

Figure 7-12 Excess underfire air and temperature relationship in MSW combustion.

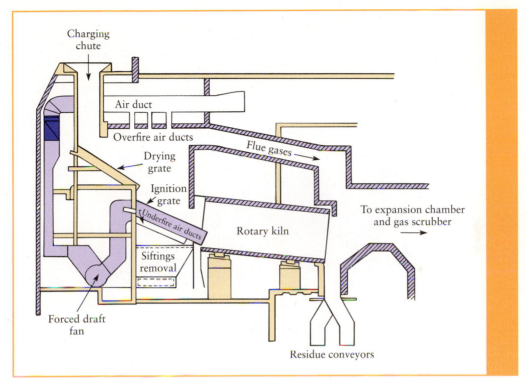

Figure 7-13 Rotary kiln.

A modification of the combustion chamber is the *rotary kiln*, shown in Figure 7-13. In this unit, the refuse travels down an ignition grate by gravity and into a rotating kiln where the combustion occurs. Rotary kilns provide the most turbulence of any grate system and thereby enhance the rate and completion of combustion.

The furnace walls of modern combustors are lined with metal tubing through which water is circulated. This *water wall* is then part of the boiler or heat recovery system.

The water tubes protect the combustion chamber housing by transferring the heat into the water. Figure 7-14 shows how the water wall is placed in the furnace. Figure 7-14 shows a close-up photograph of a water wall in an MSW combustor.

The steam-generating system has been a source of problems with solid waste fuels—in particular the generation of superheated steam, which is needed when generating electricity. The superheated steam, between 600 and 950°F (315 and 510°C), allows the use of multistage high-efficiency turbine generators. The exhaust gas from the combustion unit is still very hot when it encounters the superheater tubes, causing rapid corrosion of the tubes. These corrosion problems have required the retrofitting of refractory materials on the tubes, which decreases the efficiency of heat transfer. One solution has been to move the superheater tubes to the back of the boiler.

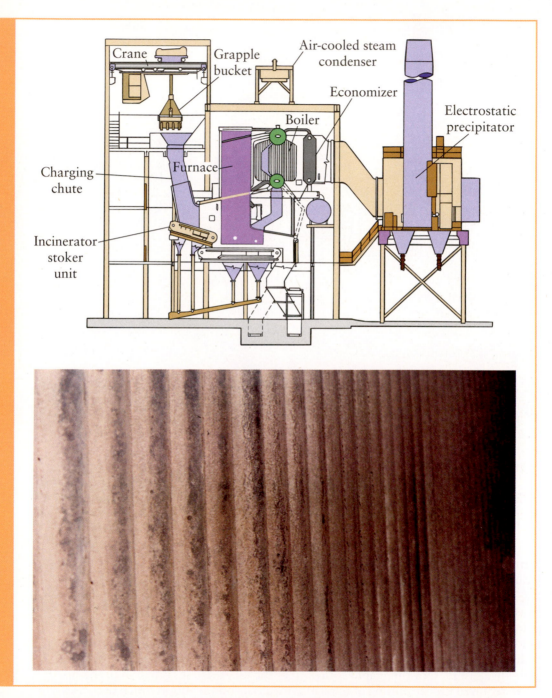

Figure 7-14 Water-wall tubes lining the furnace of an MSW combustor. (Courtesy William A. Worrell)

Table 7-7 Steam Production as Related to Quality of MSW as a Fuel

	As-Received Heat Value, Btu/lb				
Refuse	6500	6000	5000	4000	3000
Percent moisture	15	18	25	32	39
Percent noncombustible	14	16	20	24	28
Percent combustible	71	66	55	44	33
Steam generated, tons/ton refuse	4.3	3.9	3.2	2.3	1.5

Source: [1]

The efficiency of steam production also depends on the quality of the fuel. As shown in Table 7-7, the production of steam drops dramatically with increases in moisture in the fuel as in well as the fraction of noncombustibles.

7-3-2 Modular Starved-Air Combustors

A third type of combustion unit is the *modular starved-air combustor*, shown in Figure 7-15. These units are characterized by a two-stage combustion system, with the first stage being operated in a starved-air mode, producing a large quantity of suspended carbon that is then burned using a fossil fuel in the second stage. These units are batch fed using a double-ram system, as shown in Figure 7-16, and range in size from 15 to 100 tons per day. Typically, such units do not incorporate heat recovery systems.

Modular combustors are useful in cases where the waste has to be combusted but the quantity is insufficient to warrant the construction of a large refractory-lined or water-wall combustor. Modular units are also flexible in that more units can be purchased as the need arises. For example, if the average production of solid waste is 100 tons per day, three 50-ton/day units can be purchased with one remaining idle as a standby and for scheduled maintenance. One of the most widely used applications of these units is in the destruction of some hazardous materials, such as biohazards waste from hospitals.

7-3-3 Pyrolysis and Gasification

Pyrolysis is destructive distillation or combustion in the absence of oxygen. The products of pyrolysis include a solid, a liquid, and a gas. In true pyrolysis, heat is added to the complex organic feed. For example, if the feed material is pure cellulose, gaseous, liquid, and solid products are formed[47]:

Gases: $CH_4(g)$, $H_2(g)$, $CO(g)$, $H_2O(g)$, $C_2H_4(g)$ (ethylene), etc.

Liquids: Levoglucosan ($C_6H_{10}O_5$, b.p. 384°C), Formic acid (CH_2O_2, b.p.100.8°C), Glycolaldehyde ($C_2H_4O_2$, b.p. 131.3°C), Naphtalene ($C_{10}H_8$, b.p. 218°C), tars, etc.

Solids: Char

A modification of pyrolysis is *gasification*, in which a limited quantity of oxygen is introduced as pure oxygen, air, or as water, and the resulting oxidation produces

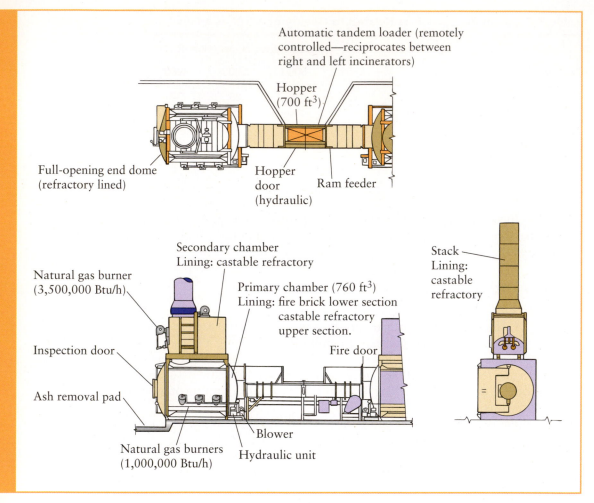

Figure 7-15 Modular combustor.

enough heat to make the system self-sustaining. Thus, the gasification reaction can be exo- or endothermic, depending on the amount of heat and oxygen added.

The process of pyrolysis or gasification can be manipulated in order to achieve a desired end product. Four general modes of operation (15) can be identified: See Table 7-8.

The choices of these variables determine the products obtained from the pyrolysis system, as shown in Figure 7-17. At very high temperatures, the product is mostly gas, while at low temperatures, mostly solid product results.[12]

Pyrolysis (and gasification) has a lot to recommend it theoretically.[13] The process is environmentally excellent, producing little pollution, and it results in the production of various useful fuels. It would seem therefore that pyrolyzing MSW would be an ideal application. Unfortunately, pyrolysis has had a sorry history with MSW. Large facilities were constructed in the 1970s to produce both liquid fuel (oil from garbage!)[14] and solid fuels, and all of these facilities failed because

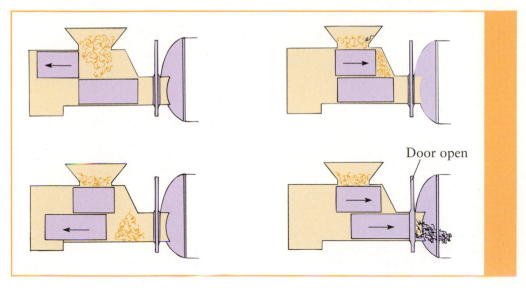

Figure 7-16 Ram feed system for a modular incinerator.

of operating problems. The pyrolysis of predictable and homogeneous fuels (such as sugarcane bagasse) would seem reasonable and logical, and these units are successful. The heterogeneous and unpredictable nature of MSW has resulted in numerous failures.

In recent decades, Japan has started to use gasification technology to combust MSW. In Nagoya, a full-size gasification plant opened in 2009. This plant was built to support the existing three conventional waste-to-energy plants and processes both MSW and ash from these existing waste-to-energy plants. By melting the ash from the conventional waste-to-energy plants, the gasification plant can convert the ash into a useful product that does not have to be landfilled. Unfortunately, this gasification technology that significantly reduces ash generation and has very low air emissions also costs significantly more than a conventional waste-to-energy plant and produces less energy. This challenge (see Figure 7-18) is of general

Table 7-8 **Four General Modes of Operation for Manipulation of Pyrolysis**

Slow pyrolysis	proceeds at a very slow rate of temperature increase, generally less than 1°C/sec, and the final temperature range is between 500 and 750°C.
Intermediate pyrolysis	takes place at a more rapid temperature rise, of 5 to 100°C/sec, and reaches temperatures of between 750 and 1000°C.
Rapid pyrolysis	occurs when the temperature rise is fast, between 500 and 106°C/sec. The temperatures reached with this process are over 1000°C.
Flash pyrolysis	takes place when the temperature rise is essentially instantaneous, of over 106°C/sec. The temperatures attained in this process exceed 1200°C.

Source: [15]

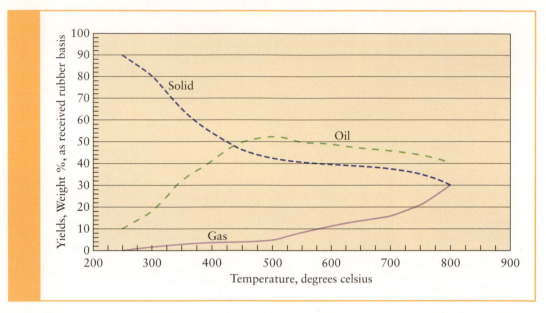

Figure 7-17 Effect of temperature in the formation of pyrolysis products. Source: *Final Report. Environmental Factors of Waste Tire Pyrolysis, Gasification, and Liquefaction.* California Integrated Waste Management Board, July 1995, table 4-1, p. 4-4. © 1995 by the California Department of Resources Recycling and Recovery (CalRecycle).

nature. To obtain high quality materials which are inert for disposal or which can be recovered in new products, energy is required which will be lost for the energy generation; that is, for the distribution of power to the grid. However, in order to judge if the treatment is justified, an overall assessment is needed that considers the additional benefits.

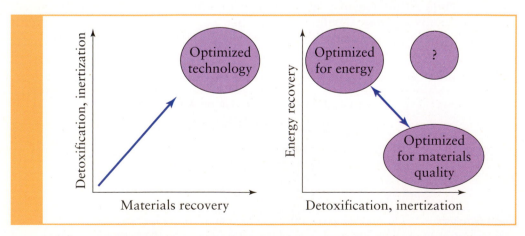

Figure 7-18 Detoxification and materials recovery go hand in hand (left), whereas these measures compete with energy generation (right). To achieve simultaneously high energy and materials recovery is therefore very challenging.

Several gasification plants have been proposed in the United States. The proponents hope that the reduction in air emission will convince the regulatory agencies, the elected officials, and, most importantly, the public, that this technology is an improvement over the conventional waste-to-energy technology and is therefore viable.

7-3-4 Mass Burn versus RDF

Combustion systems are characterized as either *mass burn* units or *refuse-derived fuel* (RDF) units. A mass burn unit has no pre-processing of the solid waste prior to being fed into the combustion unit. The solid waste is loaded into a feed chute, which leads to the grates: the solid waste in the chute provides an air lock, which allows the amount of combustion air to be controlled. From the chute, the solid waste moves directly onto the combustion grate.[16]

In an RDF system, the solid waste is processed prior to combustion to remove noncombustible items and to reduce the size of the combustible fraction, thus producing a more uniform fuel at a higher heat value. The RDF is fed through a rotary feeder and injected into the combustion unit above the grate. Some combustion takes place above the grate, with the remaining combustion occurring on the grate. This is referred to as *semi-suspension firing*. If no grate exists and all combustion occurs in the air, this would be called *suspension firing*. Suspension firing is common when burning pulverized coal but is not generally applicable to RDF.

An initial advantage of an RDF plant is that the heat value of the fuel is more uniform, and thus, the amount of excess air required for combustion is reduced. The amount of combustion air used is important, because if there is insufficient oxygen in the combustion chamber, a reducing atmosphere is created, which leads to corrosion problems. For RDF systems, the excess air (above stoichiometric requirements) is about 50%, while in mass burn plants (because of the large variation in fuel value between items) about 100% excess air is needed. However, with modern furnace design and flue gas recirculation, the excess air requirements in mass burn plants have been significantly reduced to levels of RDF plants or below. When pre-processing the solid waste, some of the potential problem items—such as batteries that contain mercury—can be removed.[17] In addition, by removing noncombustible material prior to burning, less ash is produced.

While there appear to be several theoretical advantages of RDF over mass burn plants, they have had their share of operating problems. Processing solid waste is not easy, and RDF plants have encountered corrosion and erosion problems. Whether there is an advantage to incurring the additional cost related to the production of RDF over mass burn is still unclear; however, many more mass burn plants have been built than RDF plants.

If the RDF is produced from mixed paper that has been sorted as part of the recycling effort, the fuel is a valuable resource. The characteristics of shredded mixed paper as a fuel are shown in Table 7-9. The data in this table are the average values obtained from 12 different communities.[18]

The heat value of mixed paper is surprisingly high, considering the high ash content. Overall, the HHV of mixed paper is about 7200 Btu/lb (16,700 kJ/kg).[18]

Table 7-9 **Characteristics of Mixed Paper as a Refuse-Derived Fuel**

	Component Average as Percent of Total Weight
Moisture	7.3
Ash	9.9
Volatile	82.8
Fixed carbon	7.1
Carbon	41.1
Hydrogen	6.2
Nitrogen	0.1
Sulfur	0.1
Chloride	0.1

	Average Concentration as Parts per Million (mg/kg)
Arsenic	0.48
Barium	46.21
Beryllium	0.70
Cadmium	0.55
Cobalt	6.78
Chromium	6.48
Copper	18.12
Manganese	27.34
Mercury	0.05
Molybdenum	7.42
Nickel	6.92
Lead	7.51
Antimony	3.88
Selenium	0.06
Tin	7.92
Zinc	149.21

Source: [18]

The amount of heat value contributed by various components of mixed paper varies, as shown previously in Table 7-4 and in Appendix D. In addition to mixed papers, some other secondary fuels can become valuable sources of energy, as shown in Table 7-10.

The American Society for Testing and Materials (ASTM) has developed designations of refuse-derived fuels, as shown in Table 7-11. RDF-1 is mixed refuse without any processing, while RDF-2 is refuse that has been shredded. This homogenizes the fuel and makes it much easier to handle in combustion. RDF-3 is shredded refuse from which most of the inorganic materials have been removed. Such a fuel would be produced in a typical materials recovery facility (MRF), which might process mostly pre-separated solid waste. Further shredding into a *fluff* (RDF-4) (Figure 7-19) or pelletized into dog-food-sized pellets (RDF-5) further improves the usefulness of the fuel (Figure 7-20).[20] The last categories, RDF- 6 and 7, have been tried on a pilot basis but have not been found to be successful at full-scale plants.

Table 7-10 Combustion Properties of Various Fuels

Fuel	Percent Moisture	Heat Value, Btu/lb	Percent Ash
Bark and sawdust	55	3160	2.0
Cardboard boxes	5	7180	3.6
Carpet edges	1	11,600	22.1
Peat	35	5800	5.0
Tires	1	16,200	1.8
Wood chips	38	5230	0.2

Note: To convert from Btu/lb to kJ/kg, multiply by 2.32.
Source: [19]

Table 7-11 ASTM Refuse-Derived Fuel Designations

Name	Description
RDF-1	Unprocessed MSW (the mass burn option).
RDF-2	MSW shredded but no separation of materials.
RDF-3	Organic fraction of shredded MSW. This is usually produced in a materials recovery facility (MRF) or from source-separated organics such as newsprint.
RDF-4	Organic waste produced by a MRF that has been further shredded into a fine form (almost powder), which is also called *fluff*.
RDF-5	Organic waste produced by a MRF that has been densified by a pelletizer or a similar device. These pellets often can be fired with coal in existing furnaces.
RDF-6	Organic fraction of the waste that has been further processed into a liquid fuel, such as oil.
RDF-7	Organic waste processed into a gaseous fuel.

Figure 7-19 Fluff, or finely shredded RDF. (Courtesy P. Aarne Vesilind)

Figure 7-20 Pellets used as RDF. (Courtesy NCRR)

One advantage of RDF over mass burn is that the fuel can be stored in bulk storage containers, as shown in Figure 7-21, and burned as needed—unlike the raw MSW in mass burn facilities, where the storage time is limited.

7-4   UNDESIRABLE EFFECTS OF COMBUSTION

The combustion of fuel such as RDF can have undesirable side effects. In this discussion, we cover the waste heat generated by a power plant, the ash requiring disposal, the materials that can escape in the stack and cause air pollution, and the production of carbon dioxide.

7-4-1 Waste Heat

MSW is a low-grade fuel that can be used for the production of steam. The boiler generates this steam, which is used for driving turbines, but the exhaust stream

Figure 7-21 Bulk storage unit. (Courtesy P. Aarne Vesilind)

from the turbine has little industrial use (unless it is located sufficiently close to buildings that it can be used for heating).

Often the exhaust steam is condensed back to water using either a water cooler or air-cooled condenser or heat exchanger. Usually the boiler water is reused in the boiler, because it is too expensive to be used only once. Small amounts—less than 10%—are discharged, and fresh water is added to the recycled water to keep dissolved solids low.

If a water-cooled condenser is used and the water used to cool the steam is discharged into a watercourse, the result can be serious deleterious ecological effects in streams, rivers, and estuaries. To prevent this from occurring, heat discharges are governed by environmental regulations. Typically, the limit on heat discharges is that the temperature of the receiving water cannot be raised by more than 1°C.

The calculation of the temperature of a receiving body of water is straightforward. Heat energy is easy to analyze by energy balances, since the quantity of heat energy in a material is simply its mass times its absolute temperature. This is true if the heat capacity is independent of temperature. In particular, if phase changes do not occur (as in the conversion of water to steam), an energy balance for heat energy in terms of the quantity of heat is

$$\begin{bmatrix} \text{heat} \\ \text{energy} \end{bmatrix} = \begin{bmatrix} \text{mass of} \\ \text{energy} \end{bmatrix} \times \begin{bmatrix} \text{absolute temperature} \\ \text{of the material} \end{bmatrix}$$

Energy flows are analogous to mass flows. When two heat-energy flows are combined, for example, the temperature of the resulting flow at equilibrium is calculated using the energy balance:

$$0 = (\text{heat energy IN}) - (\text{heat energy OUT})$$

or, stated in other words, assuming two input and one output stream,

$$0 = (T_1 Q_1 + T_2 Q_2) - (T_3 Q_3)$$

or

$$T_3 = \frac{T_1 Q_1 + T_2 Q_2}{Q_3}$$

where

T = absolute temperature
Q = flow, mass/unit time (or volume constant density)
1 and 2 = input streams
3 = output stream

The mass/volume balance is

$$Q_3 = Q_1 + Q_2$$

Although strict thermodynamics requires that the temperature be expressed in *absolute* terms, in the conversion from Celsius (C) to Kelvin (K), the 273 simply cancels out, and T can be conveniently expressed in degrees Celsius. Recall that $0°C = 273°K$.

EXAMPLE 7-9

A coal-fired power plant discharges 3 m³/sec of cooling water at 80°C into a river, which has a flow of 15 m³/sec and a temperature of 20°C. What will be the temperature in the river immediately below the discharge?

SOLUTION

The confluence of the river and cooling water can be thought of as a black box, and

$$T_3 = \frac{T_1 Q_1 + T_2 Q_2}{Q_3}$$

$$T_3 = \frac{[(80 + 273)(3)] + [(20 + 273)(15)]}{(3 + 15)} = 303°K$$

or 303 − 273 = 30°C. Note that the use of absolute temperatures is not necessary since the 273 cancels out (the right-hand side of the equation is really 30 + 273).

Since most states restrict thermal discharges to a less than 1°C rise above the ambient stream temperature, the heat in the cooling water must be dissipated into the atmosphere before the cooling water is discharged. Various means are used for dissipating this energy, including large shallow ponds and cooling towers. A cutaway drawing of a typical cooling tower is shown in Figure 7-22. Cooling towers represent a substantial additional cost to the generation of electricity.

Even with this expense, watercourses immediately below cooling water discharges are often significantly warmer than normal. This results in the absence of ice during hard winters and the growth of immense fish. Stories about the size of fish caught in artificially warmed streams and lakes abound, and these places become not only favorite fishing sites for people but winter roosting places for birds. Wild animals similarly use the unfrozen water during winter when other surface waters are frozen. The heat clearly changes the aquatic and terrestrial ecosystem, and some would claim that this change is for the better. Everyone seems to benefit from having the warm water. Yet changes in aquatic ecosystems are often unpredictable and potentially disastrous. Heat can increase the chances of various types of disease in fish, and heat will certainly restrict the types of fish that can exist in the warm water. Many cold-water fish (such as trout) cannot spawn in warmer water; they will die out, and their place will be taken by fish that can survive (such as catfish and carp). It is unclear what the net environmental effect is due to governmental restrictions on thermal discharges—such as "no more than 1°C rise in temperature."

Given the potential disadvantages of water-cooled condensers, the obvious question is why not use an air-cooled condenser. In an air-cooled condenser, air is used to cool the exhaust steam and condense it back to water. Air-cooled condensers are more expensive and less efficient than water-cooled condensers.

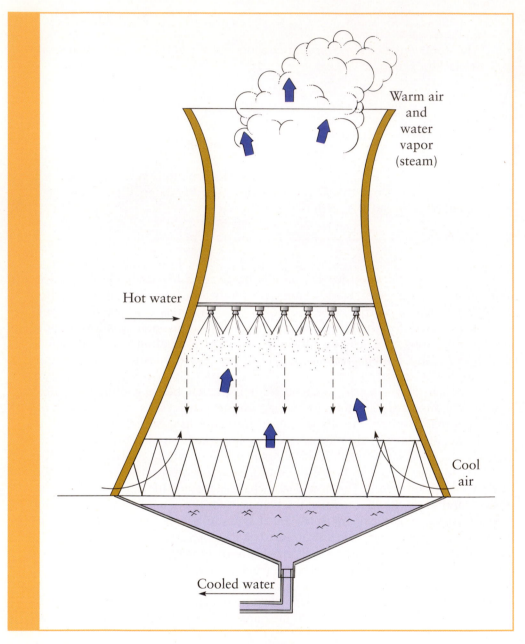

Warm air
and
water
vapor
(steam)

Hot water

Cool
air

Cooled water

Figure 7-22 Cooling towers typically used in fossil-fuel and solid waste combustion facilities.

7-4-2 Air Pollutants

In the United States, much has been done to minimize the impact of waste-to-energy plants on air pollution. The Clean Air Act required all large and small plants to be retrofitted to meet Maximum Achievable Control Technology (MACT) by 2000 and 2005, respectively. As shown in Table 7-12, this retrofit significantly reduced

Table 7-12 **Emissions from Large and Small Municipal Waste Combustion Units**

Pollutant	1990 Emissions (tpy)	2005 Emissions (tpy)	Percent Reduction
CDD/CDF, TEQ basis*	4400	15	99+
Mercury	57	2.3	96
Cadmium	9.6	0.4	96
Lead	170	5.5	97
Particulate matter	18,600	780	96
HCl	57,400	3,200	94
SO_2	38,300	4,600	88
NO_x	64,900	49,500	24

*Dioxin/furan emissions are in units of grams per year toxic equivalent quantity (TEQ), using 1989 NATO toxicity factors; all other pollutant emissions are in units of tons per year. Source: [33]

air emissions. The EPA recognized this air-emission reduction by concluding that WTE plants produce electricity with less environmental impact than almost any other source of electricity.[34] In Switzerland, where over 50% of waste is burned, combustion of waste is not a significant source of air pollution (see Figure 7-23).

The air pollutants of concern in municipal waste combustion can be classified as *gases* or *particulates*. Another classification is *primary pollutants* and *secondary pollutants*. Primary pollutants are products of the combustion process that can be shown to be harmful in the form they are emitted. Secondary pollutants are those that are formed in the atmosphere as a direct result of the emission of primary pollutants.

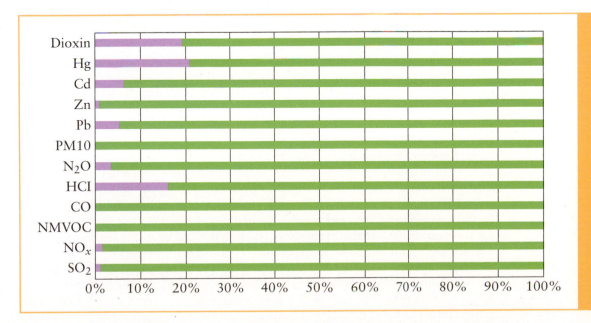

Figure 7-23 Share of all emissions from combustion in Switzerland (red bars) in comparison to the total emissions in Switzerland (100%). Source [51].

Because the combustion of MSW is a complex process, many undesirable reactions can occur that produce undesirable gases.

A particularly difficult problem is the sulfur oxides produced during combustion. In the presence of high temperature, the sulfur in the fuel can combine with oxygen to produce sulfur dioxide.

$$S + O_2 \rightarrow SO_2$$

Sulfur dioxide is itself a primary pollutant and can cause respiratory distress and even property damage. But sulfur emissions also can be a secondary pollutant when sulfur dioxide (in the presence of water vapor and oxygen) can oxidize further to sulfur trioxide:

$$O_2 + 2SO_2 \rightarrow 2SO_3$$

Sulfur trioxide can then dissolve in water to form sulfuric acid,

$$SO_3 + H_2O \rightarrow H_2SO_4$$

The sulfuric acid, in combination with the hydrochloric acid produced in the reaction of chlorine compounds and the nitric acid produced from nitrogen oxides, produces *acid rain*. Normal, uncontaminated rain has a pH of about 5.6, but acid rain can be as low as pH 2 or even lower.

The effect of acid rain production from the burning of coal has been devastating. Hundreds of lakes in North America and Scandinavia have become so acidic that they no longer can support fish life. In a study of Norwegian lakes, more than 70% of the lakes having a pH of less than 4.5 contained no fish, while nearly all lakes with a pH of 5.5 and above contained fish. The low pH not only affects fish directly, but contributes to the release of potentially toxic metals such as aluminum, thus magnifying the problem. In North America, acid rain has already wiped out all fish and many plants in 50% of the high mountain lakes in the Adirondacks. The pH in many of these lakes has reached such levels of acidity as to replace the trout and native plants with acid-tolerant mats of algae.

The deposition of atmospheric acid on freshwater aquatic systems prompted the EPA to suggest a limit of from 10 to 20 kg SO_4^{-2} per hectare per year. If "Newton's law of air pollution" is used (what goes up must come down), it is easy to see that the amount of sulfuric and nitric oxides emitted from power plants, municipal combustors, cars, and many other sources is vastly greater than this limit. For example, just for the state of Ohio alone, the total annual emissions are 2.4×10^6 metric tons of SO_2 per year. If all of this were converted to SO_4^{-2} and deposited on the state of Ohio, the total would be 360 kg per hectare per year.

But not all of this sulfur falls on the folks in Ohio, and much of it is exported by the atmosphere to places far away. Similar calculations for the sulfur emissions for the northeastern United States indicate that the rate of sulfur emissions is 4 to 5 times greater than the rate of deposition.

Another secondary pollutant of concern in municipal waste combustion is the formation of *photochemical smog*. The well-known and much-discussed Los Angeles smog is a case of secondary pollutant formation. Table 7-13 lists (in simplified form) some of the key reactions in the formation of photochemical smog.

The reaction sequence illustrates how nitrogen oxides formed in the combustion of fuels (such as gasoline and municipal waste) and emitted to the atmosphere

Table 7-13 Simplified Reaction Scheme for Photochemical Smog

NO_2	+ Light	→	NO + O
O	+ O_2	→	O_3
O_3	+ NO	→	$NO_2 + O_2$
O	+ (HC)x	→	HCO°
HCO°	+ O_2	→	$HCO_3°$
$HCO_3°$	+ HC	→	Aldehydes, ketones, etc.
$HCO_3°$	+ NO	→	$HCO_2° + NO_2$
$HCO_3°$	+ O_2	→	$O_3 + HCO_2°$
$HCO_x°$	+ NO_2	→	Peroxyacetyl nitrates

are acted upon by sunlight to yield ozone (O_3), which is a compound not emitted as such from sources and hence is considered a *secondary pollutant*. Ozone reacts with hydrocarbons to form a series of compounds that includes aldehydes, organic acids, and epoxy compounds. The atmosphere can be viewed as a huge reaction vessel, wherein new compounds are being formed while others are being destroyed.

The formation of photochemical smog is a dynamic process that begins with the production of nitrogen oxides from automobiles, industrial facilities, and MSW combustion. As the nitrogen oxides react with sunlight, O_3 and other oxidants are produced. The hydrocarbon level similarly increases at the beginning of the day and then drops off in the evening.

The escape of *heavy metals* with the emission gases has been another concern with combustion of MSW. Lead, cadmium, and mercury have been the most studied and represent the metals of most likely health concern which are presently regulated under the Clean Air Act. Mercury is especially difficult to control, because it volatilizes so readily and escapes with the gaseous emissions. Difficulties in the control of mercury are exacerbated by the difficulties in controlling its inclusion in the waste stream. However, the best practice for controlling mercury emissions is dry scrubbing, activated carbon injection, and fabric filters, which result in a greater than 95% reduction in mercury emissions![54] The effect of efficient scrubbers is also shown in Figure 7-24 which is showing the emissions of different MSW incinerators.

Another concern with solid waste combustion is the formation of *global warming gases*. The earth acts as a reflector to the sun's rays, receiving the radiation from the sun, reflecting some of it into space (called *albedo*), and adsorbing the rest—only to reradiate this into space as heat. In effect, the earth acts as a wave converter, receiving the high-energy, high-frequency radiation from the sun and converting most of it into low-energy, low-frequency heat to be radiated back into space. In this manner, the earth maintains a balance of temperature, so that

$$\begin{bmatrix} \text{energy from the sun} \\ \text{IN} \end{bmatrix} = \begin{bmatrix} \text{energy radiated back to space} \\ \text{OUT} \end{bmatrix}$$

Unfortunately, some gases—such as methane (CH_4) and carbon dioxide (CO_2)—adsorb radiation at wavelengths approximately the same as the heat radiation trying to find its way back to space. Because the radiation is adsorbed in the atmosphere by these gases, the temperature of the atmosphere increases, heating the earth. The system works exactly like a greenhouse in that light energy

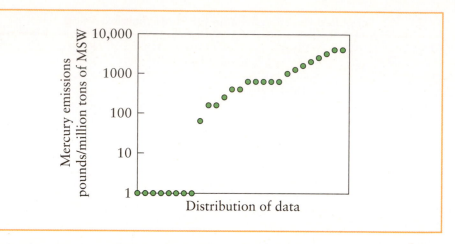

Figure 7-24 Mercury emissions during combustion of MSW. They strongly depend on the implemented air pollution control (APC) system, the operation conditions, and the amount of mercury in the waste. Strong variations can occur over a longer period. The data given in the Figure are ordered with increasing concentrations and therefore show no underlaying physical-chemical relationship or trend. Source: [25]

(short-wave, high-frequency radiation) passes through the greenhouse glass, but the long-wavelength, low-frequency heat radiation is prevented from escaping. The gases that adsorb the heat energy radiation are properly referred to as *greenhouse gases*, since they in effect cause the earth to heat up just like a greenhouse.

Not all gases have the same effect in global warming. Methane, for example, is much more potent as a greenhouse gas than carbon dioxide. The so-called greenhouse warming potential (GWP) of methane depends on the time scale. A common time scale considers 100 years. The Intergovernmental Panel on Climate Change (IPCC) has adopted the GWP value for methane in the past several times, from 21 in 1996 to 34 in 2013.[53] It therefore can be argued that methane is 34 times more effective as greenhouse gas than CO_2, and therefore the production of CO_2 in solid waste combustion is far better for the global environment than the production of methane in landfills. The requirement of flaring the gases in modern landfills is directly related to the reduction in the production of global warming gases. According to the EPA, nearly one ton of CO_2 equivalent emissions is avoided for every ton of MSW combusted by a waste-to-energy plant.

The combustion of renewable fuels such as paper and wood does not contribute to the increase in global carbon dioxide and thus does not increase the threat of global warming. The combustion of a wood fiber results in the production of carbon dioxide and water, but this is exactly what occurs when that same fiber decays naturally. Combustion simply increases the rate of decay. Thus the only deleterious gases produced by a MSW combustor come from plastics and any other products made from fossil fuels.

Control of Particulates

The Clean Air Act emission standards allow states to set strict emissions limits for various sources. Municipal waste combustors are regulated under Subpart E. This regulation limits the emission of particulates to 24 mg/dry standard m^3, corrected to 7% oxygen. The oxygen is specified because the concentration of the emissions can be reduced by the simple expedient of sucking in more air to be emitted through the stack. Because normal air has about 20% oxygen, the requirement to maintain less than 7% oxygen (by volume) allows a uniform standard among various plants. In addition, the federal regulations limit the emissions to less than 10% opacity.

The simplest devices for controlling particulates are *settling chambers*, which consist of nothing more than wide places in the exhaust flue where larger particles can settle out, usually with a baffle to slow the emission stream. Obviously, only very large particulates (>100 μm) can be efficiently removed in settling chambers, and these are used (if ever) only as pretreatment in modern combustion units. Economizers that transfer heat from the exhaust gas to combustion air are examples of a settling chamber.

Possibly the most popular, economical, and effective means of controlling particulates from many industrial sources is the *cyclone*, similar to the cyclones used for removing shredded particles following air classification (see Chapter 5). Figure 7-25 shows a simple single-stage cyclone and a bank of high-efficiency cyclones. The dirty air coming to the cyclone is blasted into a conical cylinder off the centerline. This creates a violent swirl within the cone, and the heavy solids migrate to the wall of the cylinder, where they slow down due to friction, slide down the cone, and finally exit at the bottom. The clean air is in the middle of the cylinder and exits out the top. Cyclones are not sufficiently effective for removing small particles and thus need to be backed up by other particulate-removal devices.

Bag (or *fabric*) *filters* used for controlling particulates (Figure 7-26) operate like the common vacuum cleaner. Fabric bags are used to collect the dust, which must be periodically shaken out of the bags. The fabric will remove nearly all particulates, including submicron sizes. Bag filters are widely used in many industrial applications, including MSW combustion. The basic mechanism of dust removal in fabric filters is thought to be similar to the action of sand filters in water quality management. The dust particles adhere to the fabric due to entrapment and surface forces. They are brought into contact by impingement and/or Brownian diffusion. Since fabric filters commonly have an air-space-to-fiber ratio of 1:1, the removal mechanisms cannot be simple sieving.

The *scrubber* (Figure 7-27) is another method for removing large particulates. More efficient scrubbers promote the contact between air and water by violent action in a narrow throat section into which the water or a chemical slurry is introduced. Generally, the more violent the encounter (hence the smaller the gas bubbles or water droplets), the more effective the scrubbing. Scrubbers are used in MSW combustion mostly for the removal of gaseous pollutants, but they also help in the removal of particulates. *Wet scrubbers*, which use water sprays, are efficient devices but have three major drawbacks:

1. They produce a visible plume, albeit only water vapor. The lay public seldom differentiates between a water vapor plume and any other visible plume, and hence, public relations often dictate no visible plume.
2. The waste is now in liquid form, and some manner of water treatment is necessary.
3. The ash is wet, and recovery of metals is difficult.

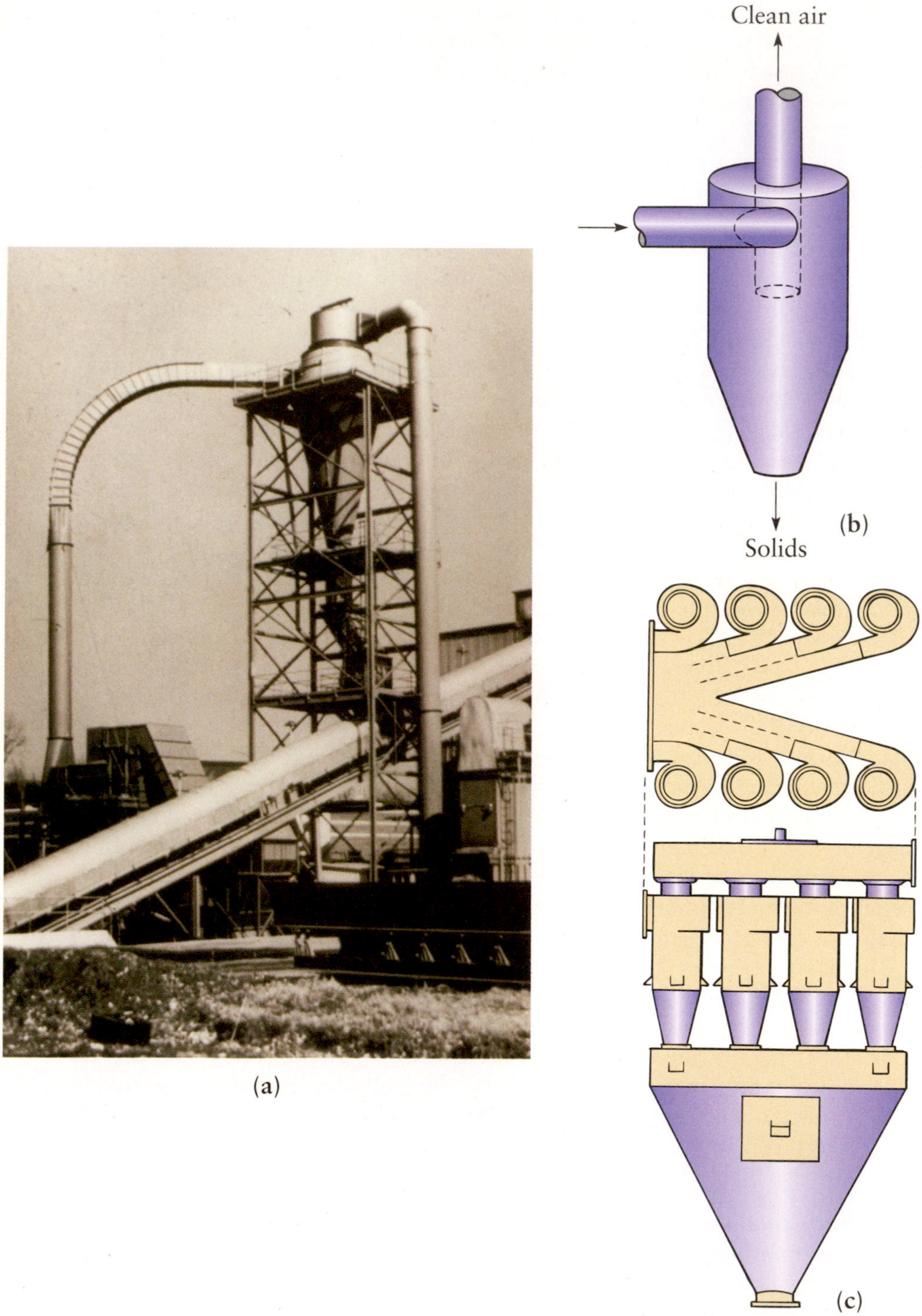

Figure 7-25 Cyclones. (a) Simple cyclone, (b) simple cyclone, (c) bank of high-efficiency cyclones. (Courtesy P. Aarne Vesilind)

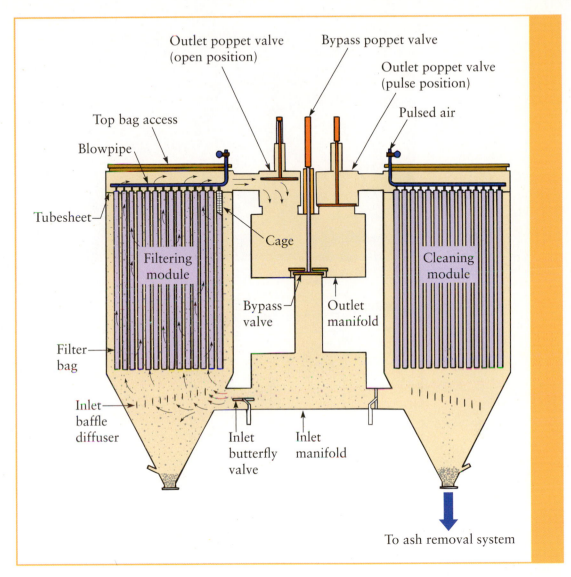

Figure 7-26 Bag filters. (Courtesy Bundy Environmental Technology, Inc.)

Dry scrubbers inject a chemical slurry such as lime. This type of scrubber does not produce a visible plume, and the waste is a powder—not a liquid.

Electrostatic precipitators (Figure 7-28) are widely used in power plants. The particulate matter is removed by first being charged by electrons jumping from one electrode to the other. The negatively charged particles then migrate to the positively charged collecting electrode. The type of electrostatic precipitator shown in Figure 7-28 consists of a negatively charged wire hanging down the middle of positively charged plates. The particulates collect on the plates and must be removed by banging with a hammer. Electrostatic precipitators have no moving parts, require only electricity to operate, and are extremely effective in removing submicron particulates.

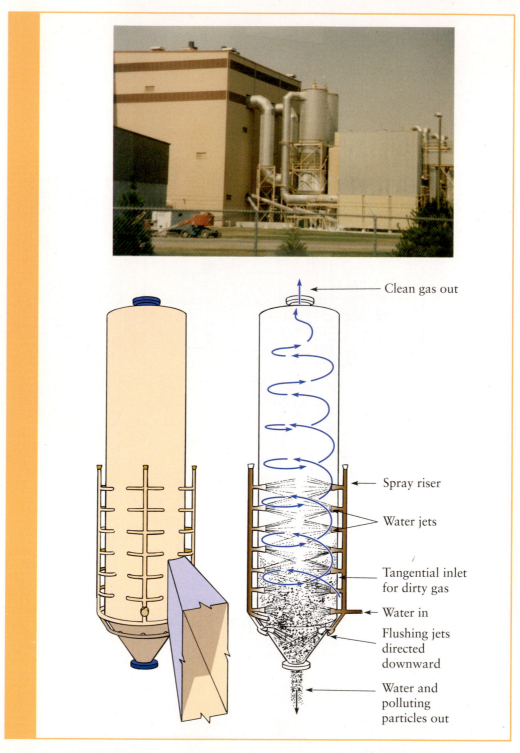

Clean gas out

Spray riser

Water jets

Tangential inlet
for dirty gas

Water in

Flushing jets
directed
downward

Water and
polluting
particles out

Figure 7-27 Scrubbers used for air pollution control. (Courtesy William A. Worrell)

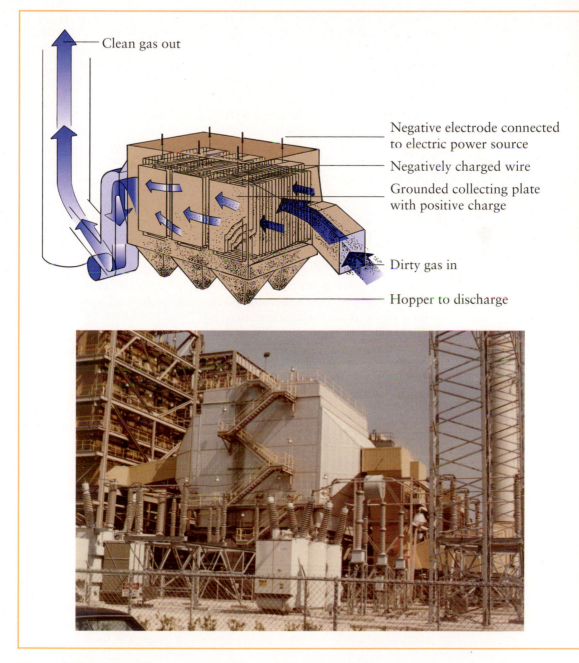

Figure 7-28 Electrostatic precipitator. (Courtesy P. Aarne Vesilind)

Control of Gaseous Pollutants

The control of gases involves the removal of the pollutant from the gaseous emissions, a chemical change in the pollutant, or a change in the process producing the pollutant. In MSW combustion, where the fuel composition is seldom under control, the gaseous removal processes must be robust and effective.

Wet scrubbers, which can be used for partial particulate removal, can also remove gaseous pollutants by simply dissolving them in the water. Alternatively, a chemical may be injected into the scrubber water, which then reacts with the pollutants. This is the basis for most SO_2 removal techniques, as discussed next. Because of the moisture carryover, wet scrubbers are usually placed after the ESP or baghouse.

Dry scrubbers are very effective in controlling sulfur dioxides. A lime slurry is injected into this unit, but the liquid evaporates due to the high temperature of the exhaust gas. The lime slurry also reduces hydrogen chloride in the exhaust gases.

In simple terms, the reaction in the dry scrubber is

$$Ca(OH)_2 + heat \rightarrow CaO + H_2O$$
$$SO_2 + CaO \rightarrow CaSO_3$$

or if limestone is used,

$$SO_2 + CaCO_3 \rightarrow CaSO_4 + CO_2$$

Waste-to-energy combustors also are a source of nitrogen oxides that can lead to the formation of photochemical smog and contribute to acid rain. Nitrogen oxides, designated by the general expression NO_x, are produced in two ways. *Thermal NO_x* results from the reaction of excess oxygen (from air) with nitrogen (from air) at high temperatures. *Fuel NO_x* are produced when the nitrogen in the fuel produces the oxide during combustion. Generally, the thermal NO_x represents only about 25% of the total nitrogen oxide production.[26]

Unlike sulfur, nitrogen cannot be readily removed from fossil fuels. The reduction of nitrogen oxide is accomplished by improved combustion control and using a $DeNO_x$ system that converts nitrogen oxides to nonpolluting nitrogen gas, N_2, and water vapor.

Covanta, the largest waste-to-energy provider in the United States, has developed a combustion control system for existing plants that can reduce nitrogen oxide emissions to 50% below EPA standards and another system for new plants that can reduce nitrogen oxide to 70% below EPA standards. These systems greatly reduce the amount of excess oxygen in the furnace needed for complete combustion, and for new plants, they also provide a furnace air-recirculation system.

In addition to combustion control, one of two types of $DeNO_x$ systems is used. The *selective non-catalytic reduction* (SNCR) process involves the injection of ammonia and steam into the furnace. This is an attractive retrofit option to existing plants because of the low capital cost to retrofit the equipment.

In a *selective catalytic reduction* (SCR) process, a catalyst is used to achieve higher reductions than are possible with a SNCR process. The catalysts are sensitive to particulate contamination and are typically located after the scrubber and baghouse. However, for the catalyst to function properly, the flue gas needs to be 450°F (230°C) and thus must be reheated prior to the catalyst. Therefore, some processes have the SCR system before the wet scrubber and baghouse. So that the catalyst can work properly, an electrostatic filter needs to be implemented before the SCR step. While many United States plants use the SNCR process, no plant uses the SCR process. However, the SCR process is being used

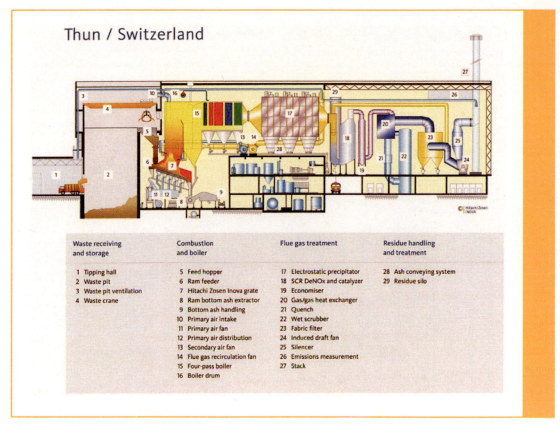

Thun / Switzerland

Waste receiving and storage	Combustion and boiler	Flue gas treatment	Residue handling and treatment
1 Tipping hall	5 Feed hopper	17 Electrostatic precipitator	28 Ash conveying system
2 Waste pit	6 Ram feeder	18 SCR DeNOx and catalyzer	29 Residue silo
3 Waste pit ventilation	7 Hitachi Zosen Inova grate	19 Economiser	
4 Waste crane	8 Ram bottom ash extractor	20 Gas/gas heat exchanger	
	9 Bottom ash handling	21 Quench	
	10 Primary air intake	22 Wet scrubber	
	11 Primary air fan	23 Fabric filter	
	12 Primary air distribution	24 Induced draft fan	
	13 Secondary air fan	25 Silencer	
	14 Flue gas recirculation fan	26 Emissions measurement	
	15 Four-pass boiler	27 Stack	
	16 Boiler drum		

Figure 7-29 Modern WTE Plant, Thun, Switzerland (Courtesy Hitachi, Zosen Inova)

in plants in both Europe and Japan. While the SCR process achieves a higher nitrogen-oxide reduction compared to the SNCR process, it is more expensive to install and operate.

The control of air pollutants from MSW combustion plants has continued to evolve. A typical scheme used in European facilities, shown in Figure 7-29, consists of:

1. A boiler operating at high temperatures ($>1200°C$), destroying dioxins that may have been produced during the combustion of the bulk materials at lower temperatures or entered the burning chamber with the solid waste.
2. An electrostatic precipitator for particle removal.
3. A SCR Denox system with a catalyzer in order to reduce NO_x.
4. An economizer and a gas-gas heat exchanger to increase efficiency.
5. A quench with water.
6. A wet scrubber for removing acid gases.
7. A fabric filter to capture fine particles.

The cleaned gases are then emitted from a stack.

Table 7-14 lists typical gas emissions values reported by a WTE plant in Switzerland.[50]

Table 7-14 Gas Emissions in 2014 of an WTE Plant which Started Its Operation in 2006. The discharge gas flow corresponds to 52,900 Nm³/h (1 Nm³ equals 1 normal cubic meter reported at 0°C and 1013 mbar)

Line 1	Limit given by law		Requirements of the plant	Guaranteed values	Measured values August 19th, 2014	Reduction according to law limit
Volumetric flow rate	Nm³/h				52,900	
Dust	mg/Nm³	10	10	5	0.6	−94.0%
HCl	mg/Nm³	20	10	5	1.4	−93.0%
HF	mg/Nm³	2	1	0.5	0.12	−94.0%
SO_2	mg/Nm³	50	50	20	2.3	−95.4%
NO_2	mg/Nm³	80	200	50	39	−51.3%
NH_3	mg/Nm³	5	aucune	5	1.1	−78.0%
Hg	mg/Nm³	0.1	0.05	0.05	0.0086	−91.4%
Cd	mg/Nm³	0.1	0.05	0.05	< 0.0005	−99.5%
Pb + Zn	mg/Nm³	1	0.5	0.5	0.061	−93.9%
CO	mg/Nm³	50	50	25	7	−86.0%
Dioxins and furans	ngTE/Nm³	0.1	0.1	0.1	0.0058	−94.2%

*Swiss Ordinance on Air Pollution Control

*TE = Toxicity Equivalents

7-4-3 Dioxin

Of particular concern in waste combustion is the production of *dioxin*. Dioxin emission concentrations have been shown and discussed in previous chapters. Dioxin has become a big issue since the chemical accident in Seveso (Italy) occurred in 1976, and the following chapter is devoted to dioxin in combustion. Dioxin is actually a combination of many members of a family of organic compounds called *polychlorinated dibenzodioxins* (PCDDs). Members of this family are characterized by a triple ring structure of two benzene rings connected by a pair of oxygen atoms (Figure 7-30). A related family of organic chemicals are the *polychlorinated diben-zofurans* (PCDFs), which have a similar structure except that the two benzene rings are connected by only one oxygen. Since any of the carbon sites are able to attach either a hydrogen or a chlorine atom, the number of possibilities is great. The sites that are used for the attachment of chlorine atoms are identified by number, and this signature identifies the specific form of PCDD or PCDF. For example, 2,3,7,8-tetra-chloro-dibenzo-*p*-dioxin (or 2,3,7,8-TCDD in shorthand) has four chlorine atoms at the four outside corners, as shown in Figure 7-31. This form of dioxin is especially toxic to laboratory animals and is often identified as a primary constituent of contaminated pesticides and emissions from waste-to-energy plants.[27]

All of the PCDD and PCDF compounds have been found to be extremely toxic to animals, with the LD_{50} for guinea pigs being about 1 µg/kg body weight.

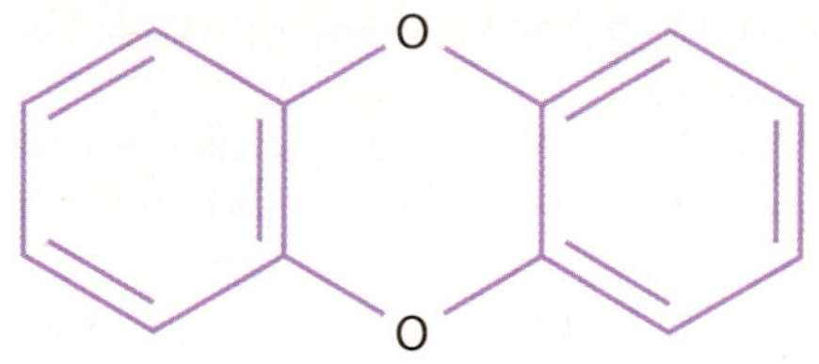

Figure 7-30 A dioxin molecule.

Neither PCDD nor PCDF compounds have found any commercial use and are not manufactured. They do occur, however, as contaminants in other organic chemicals.[28]

Various forms of dioxins have been found in pesticides (such as Agent Orange, widely used during the Vietnam War) and in various chlorinated organic chemicals such as chlorophenols. Curiously, recent evidence has not borne out the same level of toxicity to humans, and it seems likely that dioxins are actually less harmful than they might appear from laboratory studies. The chemical spill in Seveso was expected to result in a public health disaster based on extrapolations from animal experiments, but thus far, this has not materialized. One estimate of the dose suffered by the children in Sevesco was 3,000,000 pg of dioxin/kg of body weight—compared to the EPA risk-specific level of 0.006 pg/kg body weight or the Canadian tolerable daily intake of 10 pg/kg body weight. Although the children suffered from chloracne (a temporary skin condition), none have seemed to have the more serious cancers predicted. Nevertheless, dioxins are able (at very low concentrations) to disrupt normal metabolic processes, and this has caused the EPA to continue to place severe limitations on the emission of dioxins from WTE plants.[29]

Dioxins emitted from waste-to-energy facilities come from two sources: dioxins that are in the waste and are not combusted in the furnace and *de novo* dioxins that are created during combustion.[30] Tests on waste-to-energy plants that have included slug loads of chlorinated plastics (thought to be the main precursors

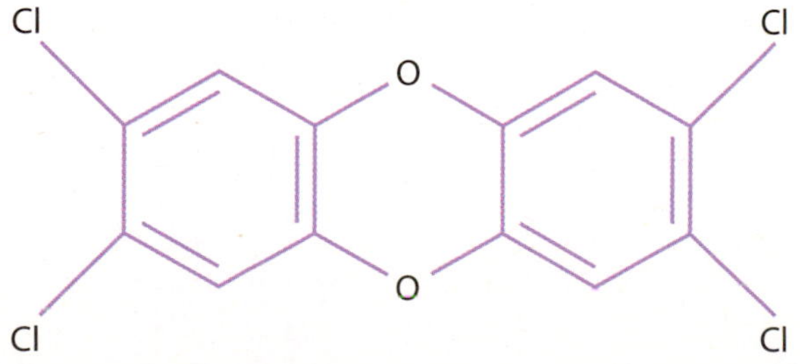

Figure 7-31 A particularly toxic form of dioxin: 2,3,7,8-TCDD.

in dioxin formation) have produced negative results.[31] Nevertheless, the EPA estimates that a significant part of environmental dioxins (of all forms, including difurans) comes, from combustion.

There is little doubt that waste-to-energy plants emit trace amounts of dioxins, but nobody knows for sure how the dioxins originate. Some dioxin is in the waste, and this may escape in the bottom ash, the fly ash, or the off-gases. On the other hand, it is also likely that *any* combustion process that has even trace amounts of chlorine produces dioxins, and these are simply an end product of the combustion process. Another theory is that the high temperatures in the combustion unit destroy the dioxins but that new dioxins form as the exhaust gas is cooled. The presence of trace quantities of dioxins in emissions from wood stoves and fireplaces seems to confirm this view. Whatever the mechanism, it seems clear that it is not a simple chemical equation and that many parallel reactions are occurring to produce the various forms of dioxins and difurans.[29]

It might be well to remember here that the two sources of risk, WTE plants versus fireplaces, are clearly unequal. The effect of the latter on human health is greater than the effect of WTE plant emissions. But the fireplace is a *voluntary risk*, whereas the WTE plant is an *involuntary risk*. People are willing to accept voluntary risks 1000 times higher than involuntary risks, and they are therefore able to vehemently oppose WTE plants while enjoying a romantic fire in the fireplace.[32]

7-4-4 Ash

Power plants produce both *bottom ash* and *fly ash*. Bottom ash is recovered from the combustion chamber and consists of the inorganic material as well as some unburned organics (>3%), while fly ash is the particulates removed from the flue gas. Considering both types of ash together, MSW combustion typically achieves 75 to 80% reduction in material by weight (and 90 to 95% reduction by volume). Thus, about 25% of the original mass is ash with a high density of about 710 to 1070 kg/m³ (1200 to 1800 lb/yd³). The materials in typical MSW ash are shown in Table 7-15.

The major problem with ash from MSW combustion is the presence of heavy metals. Table 7-16 shows a representative array of some heavy metals found in combined fly ash and bottom ash from a MSW waste-to-energy unit.

Table 7-15 Materials Found in Typical MSW Ash

Material	Percent by Weight
Metals	16.1
Combustibles	4.0
Ferrous metal	18.3
Nonferrous metal	2.7
Glass	26.2
Ceramics	8.3
Mineral, ash, other	24.1

Source: [21]

Table 7-16 **Total Metal in Combined Ash**

Metal	Ash by Weight, mg/kg
Aluminum	17,800
Calcium	33,600
Sodium	3,800
Iron	20,400
Lead	3,100
Cadmium	35
Zinc	4,100
Manganese	500
Mercury	Less than 3

Source: Modified from [22]

Ash from MSW combustion comes perilously close to being classified as a hazardous waste by the EPA. The test usually used is an extraction procedure where the ash is shaken with a solvent and the amount of metal extracted is measured. If this concentration exceeds 100 times the drinking water standard, the waste is classified as hazardous and requires special and very expensive disposal. If fly ash is measured by itself, its constituents often do not pass the test, and it is classified as hazardous. Combined with the bottom ash, however, the mixture most often meets the requirements for a nonhazardous waste. However, some legislation also considers the toxic potential of the waste, such as by defining limit concentrations for toxic and potentially hazardous compounds. Under this approach, the fly ash would be tested separately even if it was combined with the bottom ash.

Treatment of the emissions with lime or limestone usually produces an ash in the alkaline range, and this helps in maintaining the metals in a less soluble hydroxide form. Fly ashes are potentially more problematic than bottom ashes. In some cases a stabilization is performed using an alkaline material such as cement. Also, mixing fly ashes with wood ashes has led to good results. Ash disposal is either in special or in regular municipal solid waste landfills. If the ash is compacted, the density increases to as high as 1950 kg/m³ (3300 lb/yd³). At this density the ash is highly impermeable, with a permeability as low as 1×10^{-9} cm/sec.

As more and more ash is being produced and landfill space becomes too valuable to use it for the disposal of such ash, alternative uses are being sought. In Europe, where a large fraction of MSW is incinerated and where landfill space is very expensive, ash processing has been quite successful.[24] Some of the uses of ash include

• Road base material

• Structural fill

• Gravel drainage ditches

• Capping strip mines

• Mixing with cement to make building blocks

In addition, ash from MSW combustion contains metals, such as steel, stainless steel, copper, and aluminum, and these can be reclaimed in a mechanical separation process. This process is increasingly feasible for both bottom and fly ashes. For example, concentrations of zinc in fly ashes are of the order of magnitude of zinc concentrations in some ores, and copper can be recovered from bottom ashes. The recovery of the metals has the secondary benefit of reducing the toxicity of the ashes and thus facilitating their disposal.

7-5   FINAL THOUGHTS

Earle Phelps was the first to recognize that most environmental regulatory decisions are made using what he called the *principle of expediency*. A sanitary engineer known for his work with stream sanitation and the development of the Streeter-Phelps oxygen sag curve equation, Phelps described expediency as "the attempt to reduce the numerical measure of probable harm, or the logical measure of existing hazard, to the lowest level that is practicable and feasible within the limitations of financial resources and engineering skill." He recognized that "the optimal or ideal condition is seldom obtainable in practice, and that it is wasteful and therefore inexpedient to require a nearer approach to it than is readily obtainable under current engineering practices and at justifiable costs." Most importantly for today's standard setters, who often find it difficult to defend their decisions, he advised that "the principle of expediency is the logical basis for administrative standards and should be frankly stated in their defense."

Phelps saw nothing wrong with the use of standards as a kind of speed limit on pollution affecting human health. He also understood the laws of diminishing returns and a lag time for technical feasibility. Yet he always pushed toward reducing environmental hazards to the lowest expedient levels. Just as utilitarianism is an ethical model that can resolve moral dilemmas, Phelps's expediency principle can be used to resolve the moral dilemmas of setting environmental regulations. The regulator must balance two primary moral values—do not deprive liberty, and do no harm. Setting strict regulations would result in unwarranted reduction in liberty, while the absence of adequate regulations can damage public health. By using the principle of expediency, the regulator can establish the proper balance and resolve a moral dilemma.

All ethical models, if they are to be useful, need adequate information. Using the utilitarian ethical model, for example, a just decision is possible only if the amount of happiness and unhappiness that results from decisions can be calculated. Similarly, the regulator must have scientific evidence on pollutant quantities, concentrations, vectors, and health effects to make a just environmental decision. Unfortunately, there always will be gaps between what we know and what we would like to know about environmental hazards, and the absence of adequate scientific knowledge makes the regulatory decision difficult.

A classic case is the setting of the dioxin standard for MSW combustion. The best we can do at this time, in the absence of adequate information, is to set the standard as low as we can without completely shutting down all waste burning combustors. It is not expedient to set the standard too high and allow the facilities not to worry about dioxin emissions. This chemical is exceedingly toxic, and we should be concerned. As better information becomes available, we should refine our standard. But in the meantime, we set the standard as low as is expedient.

Finally, we know that combustion remains problematic due to the release of toxic heavy metals such as mercury. With improved air pollution control equipment, these metals are removed from the emission gases and captured with the fly and bottom ashes. The volatile metal salts that are transferred to the fly ashes can be easily leached out when these ashes are exposed to the weather. Thus the ash is treated before landfilling. However, the long-term behavior of landfills containing stabilized fly ashes is not well known, and there is the potential that the weathering processes may lead to a release of the toxic metals to the environment over the long term. Today's expedient decision could have a significant environmental impact in the future.

References

1. Wilson, D. L. 1972. "Prediction of Heat of Combustion of Solid Wastes from Ultimate Analyses." *Environmental Science and Technology* 13 (June).

2. Brunner, C. 1994. "Waste-to-energy." In *Handbook of Solid Waste Management*, F. Keith (ed.). New York: McGraw-Hill.

3. Test Method E1037–84, Standard Test Method for Measuring Particle Size Distribution of RDF-5, ASTM, Philadelphia, Pa., 1996.

4. Neissen, W. R. 1977. "Properties of Waste Materials." In *Handbook of Solid Waste Management*, D. G. Wilson (ed.). New York: Van Nostrand Reinhold.

5. Ali Khan, M. Z. A. and Z. H. Abu-Ghararah. 1991. "New Approach for Estimating Energy Content of Municipal Solid Waste." *Journal of the Environmental Engineering Division* ASCE 117, no. 3:376–380.

6. Liu, J. and R. D. Paode. 1996. "Modeling the Energy Content of MSW Using Multiple Regression Analysis." *Journal of the Air & Waste Management Association* 46, no. 7:650–656.

7. Brunner, C. and S. Schwartz. 1983. *Energy and Resource Recovery from Wastes*. Park Ridge, N.J.: Noyes.

8. Erdincler, A. U. and P. A. Vesilind. 1993. "Energy Recovery from Mixed Paper." *Waste Management and Research* 11:507–513.

9. Wilson, D. G. 1977. *Handbook of Solid Waste Engineering*. New York: Van Nostrand Reinhold.

10. Neissen, W. R. 1995. *Combustion and Incineration Processes: Applications in Environmental Engineering*, 2nd edition. New York: Marcel Decker.

11. Vesilind, P. A., W. P. Martello, and B. Gullett. 1981. "Calorimetry of Refuse Derived Fuels." *Conservation and Recycling* 4, no. 2:89–97.

12. Wen, C. Y. and E. S. Stanley. 1979. *Coal Conversion Technology*. Reading, Mass.: Addison-Wesley.

13. Levy, S. J. 1974. "Pyrolysis of Municipal Solid Waste." *Waste Age* (October).

14. *Pyrolysis*. 1973. Washington, D.C.: National Center for Resource Recovery.

15. Neissen, W. R. 1996. "Evaluating Gasification and Advanced Thermal Technologies for Processing MSW." *Solid Waste Technologies*, no. 4:28–36.

16. International Solid Waste Association. 1997. *Energy from Waste: State of the Art Report.* Copenhagen: ISWA.

17. Boley, G. L. 1991. "Refuse-Derived Fuel (RDF)—Quality Requirements for Firing in Utility, Industrial, or Dedicated Boilers." *Proceedings* International Joint Power Generation Conference, San Diego.

18. Kersletter, J. D. and J. K. Lyons. 1991. "Mixed Waste Paper as a Fuel." *Waste Age* (November): 41–43.

19. Philipson Industries. 1999. Private communication. Mississauga, Ontario, Canada.

20. Eimers, J. L. and P. A. Vesilind. 1984. "Physical Properties of Densified Refuse Derived Fuel." *Journal of Testing and Evaluation* 12, no. 4:238–240.

21. Chesner, W. H., R. J. Collins, and T. Fung. 1994. "Assessment of the Potential Stability of Southwest Brooklyn Incinerator Residue in Asphaltic Concrete Mixes." Quoted in Hasselriis, F. "Ash disposal." In *Handbook of Solid Waste Management*, R. Keith (ed.). New York: McGraw-Hill.

22. Forrester, K. 1994. "State-of-the-art in Refuse-to-energy Facility Ash Residue Characteristics." Quoted in Hasselriis (35).

23. Sawell, S. E., T. R. Bridle, and T. W. Constable. 1988. "Heavy Metal Leachability from Solid Waste Incinerator Ashes." *Waste Management and Research* 6:227–238.

24. "Ash Use on the Rise in United States." 1999. *World Wastes*, pp. 16–18.

25. Hasselriis, F. 1994. Data published in *Waste Age* (November), p. 96.

26. Harrison, K. W., R. D. Dumas, S. R. Nishtala, and M. A. Barlaz. 2000. "A Life Cycle Inventory Model of Municipal Solid Waste Combustion." *Journal of the Air and Waste Management Association* 50:993–1003.

27. Chang, N-B. 1994. "The Impact of PCDD/PCDF Emissions on the Engineering Design of Municipal Incinerators." Reported in Hasselriis (31).

28. Chang, N-B and S-H Huang. 1995. "Statistical Modeling for the Precision and Control of PCDDs and PCDFs Emissions from Municipal Solid Waste Incinerators." *Waste Management and Research* 13:379–400.

29. McAdams, C. L. and J. T. Aquino. 1994. "Dioxin: Impact on Solid Waste Industry Uncertain." *Waste Age* (November): 103–106.

30. Johnke, B. and E. Stelzner. 1992. "Results of the German Dioxin Measurement Programme at MSW Incinerators." *Waste Management and Research* 10:345–355.

31. Hasselriis, F. 1987. "Optimization of Combustion Conditions to Minimize Dioxin Emissions." *Waste Management and Research* 5:311–325.

32. Elliott, S. J. 1998. "A Comparative Analysis of Public Concern over Solid Waste Incinerators." *Waste Management and Research* 16, no. 4:351–364.

33. USEPA Memorandum, Emissions from Large and Small MWC Units at MACT Compliance, August 10, 2007.

34. Letter from USEPA to Integrated Waste Services Association, February 14, 2003.

35. FOEN, 1986. "Leitbild für die Schweizerische Abfallwirtschaft (= Guidelines for Swiss Waste Management)", *Schriftenreihe Umweltschutz*, Federal Office for the Environment (FOEN), Berne, Switzerland.

36. Ludwig, Chr., Hellweg, S. and Stucki, S., 2003. *Municipal Solid Waste Management.* New York: Springer.

37. Fahrni, H.-P., 2010. "Von der wilden Deponie zu den

Verbrennungsrückständen," in: *K. Schenk (Ed.), KVA-Rückstände in der Schweiz. Der Rohstoff mit Mehrwert*, Swiss Federal Office of the Environment (FOEN).

38. IWB, 2006. "Aus Energie wird Abfall, KVA Basel," IWB, Margarethenstrasse 40002 Basel, http://www.iwb.ch.

39. Ministry of the Environment 2014. "History and Current State of Waste Management in Japan," Compiled by the Japan Environmental Sanitation Center.

40. EPA, 2014. United States Environmental Protection Agency, Municipal Solid Waste Generation, Recycling and Disposal in the United States: Facts and Figures for 2012. Report EPA-530-F-14-001.

41. Ministry of the Environment 2013. *Solid Waste Management and Recycling Technology of Japan—Toward a Sustainable Society*, Compiled by the Japan Environmental Sanitation Center.

42. Analyst Information 2014. Waste to Energy 2014/2015, *Technologies, plants, projects, players and backgrounds in the global thermal waste treatment business*, 7th edition 2014, via Research and Markets http://www.researchandmarkets.com.

43. Boie, W., 1953. "Fuel technology calculations," *Energietechnik* 3:309–16.

44. Lemann, M. 1997. Fundamentals of Waste Technology, published by C. Herrmann Consulting, Kilchberg, Switzerland.

45. van Berlo, M.A.J. and Wandschneider, J., 2003. "High Efficiency Waste-to-Energy Concept," in *Municipal Solid Waste Management*, Eds.: Chr. Ludwig, S. Hellweg, S. Stucki, Springer.

46. Rognon, F., 2008. Effizientere Nutzung von fossilen Brennstoffen und Reduktion der CO2-Emissionen bei der Erzeugung von Raumwärme und Elektrizität in der Schweiz. Swiss Federal Office of Energy (SFOE).

47. Patwardhan, P. R., Satrio, J. A., Brown, R. C., and Shanks, B. H., 2009. "Product distribution from fast pyrolysis of glucose-based carbohydrates," *Anal. Appl. Pyrolysis*, 86:323–330

48. FOEN, 1990. Technical Ordinance on Waste (TOW, from December 10, 1990; Status on July 1, 2011), in German: Technische Verordnung über Abfälle (TVA; SR 814.600). Swiss Federal Office of the Environment (FOEN), Berne Switzerland.

49. Ludwig, C., Hellweg, S. and Stucki, S., 2003. *Municipal Solid Waste Management*. New York: Springer.

50. TRIDEL, 2014. http://www.tridel.ch, document: Comparatif_emissions _TRIDEL_2012_13_14.pdf.

51. Hügi M., Gerber P. et al., 2008. "Abfallwirtschaftsbericht 2008. Zahlen und Entwicklungen der schweizerischen Abfallwirtschaft 2005–2007". Umwelt-Zutsand Nr. 0830. Swiss Federal Office of the Environmemt (FOEN), Bern. 119 S.

52. Howarth, R. W., 2014, "A Bridge to Nowhere: Methane Emissions and the Greenhouse Gas Footprint of Natural Gas," *Energy Sci. & Eng.* 2(2):47–60.

53. Themelis, Nicholas, and Assaf-Anad, Nada, "To MACT or Not to MACT, Mercury Emissions from Waste to Energy and Coal-Fired Power Plants," 13th American Waste to Energy Conference, May 23–25, Orlando, Florida.

Abbreviations Used in This Chapter

APC = air pollution control
ASTM = American Society for Testing and Materials

EfW = energy from waste
ESP = electrostatic precipitator
GC = gas chromatograph

HHV = higher heating value, also high heat value

KEV = Kostenlose Einspeisevergütung; i.e., cost difference which is covered by subsidies to cover the difference between power generation costs and the market price

LHV = lower heating value, also termed low heat value, see also NCV

MSW = municipal solid waste

NB = *nota bene* = take notice

NCV = net calorific value; it is the same as LHV or also low or net energy value

PCDDs = polychlorinated dibenzodioxins

PCDFs = polychlorinated dibenzofurans

RDF = refuse-derived fuel

SCR = selective catalytic reduction

SNCR = selective non-catalytic reduction

TG = thermo-gravimeter

WFPP = waste-fired power plant

WTE = waste-to-energy

TE = toxicity equivalent

Problems

7-1. Describe the difference between pyrolysis and gasification.

7-2. The ideal equation for the combustion of cellulose is $C_6H_{10}O_5 + 6O_2 \rightarrow 6O_2 + 5H_2O$. What is a similar idealized equation for pyrolysis? What end products might you expect?

7-3. You serve as the president of XYZ Corporation, which is an energy-intensive manufacturing operation. You have four boilers to produce steam and electricity for your operations. One is a semi-suspension fired boiler with a capacity of 100 tons of coal per hour. The boiler can also burn supplemental fuel such as coconut or macadamia nut shells. The other three boilers are peaking units and burn fuel oil. The City of Podunk has approached you about the possibility of burning solid waste as a supplemental fuel. As the president of the XYZ Corporation, you need to develop answers to many political, social, and technical questions before you make a commitment to the city.

    a. What specific questions would you ask your own power plant engineer about the possibility of accepting Podunk's MSW?

    b. What specific additional data would you need from the city before you discuss this further?

7-4. A furnace dedicated to paper (assume pure cellulose, $C_6H_{10}O_5$) operates at 15 tons/hour, at 100% excess air. How much air is required?

7-5. Draw flow diagrams of unit operations for producing from raw MSW the following products:

    a. RDF-1

    b. RDF-2

    c. RDF-3

    d. RDF-4

No other materials are to be recovered, and the purity of the fuels is important.

7-6. Suppose your nonengineering roommate asked you to explain how a refuse waste-to-energy facility worked. In one page ($8\frac{1}{2} \times 11$) using words and sketches, help him/her understand how such facilities take in MSW and produce electric power.

7-7. Why are automobile tires troublesome in MSW combustion?

7-8. Two members of Greenpeace climb the stack of your waste-to-energy plant and chain themselves to it. You are the city public works director, and the waste-to-energy plant operator calls you in a panic. What actions do you take?

7-9. A *Citizens' Guide to the Care of the Environment* states the following: "Avoid buying milk in white plastic containers. They are virtually indestructible, except by burning, which produces a poisonous gas." Comment on this assertion.

7-10. Cellulose is to be burned in a waste-to-energy facility. The chemical equation for cellulose is $C_6H_{10}O_5$. The atomic weights of C, H, and O are 12, 1, and 16, respectively.

   a. Calculate the stoichiometric oxygen necessary for the combustion of cellulose.

   b. Calculate stoichiometric air.

7-11. Draw the organic chemical formula for a polychlorinated dibenzodioxin, and describe how dioxins are controlled in MSW combustion.

7-12. A sample of refuse has a moisture content of 20%. A 1.2 g sample is placed in a calorimeter, and a gross calorific value of 6200 Btu/lb is measured based on the temperature rise. Further analysis shows that 0.3 g of ash remains in the bomb calorimeter after combustion. Calculate the HHV in terms of (a) moisture-free and (b) moisture- and ash-free.

7-13. Using words and sketches, show how a bomb calorimeter is used to measure the heating value of a sample of RDF.

7-14. Explain the difference between the HHV and LHV. Why is the HHV always higher than the LHV?

7-15. What are the two primary reasons for using underfired air in a solid waste combustion system? Use a sketch to show how underfired air is used.

7-16. Based on Figure 7-12, what would be the temperature inside an RDF-fired unit and a mass burn unit?

7-17. Find the location of the waste-to-energy plant closest to your campus. Is it an RDF-fired unit or a mass burn unit?

CHAPTER 8

Landfills

Regardless of how much reduction, reuse, recycling, and energy recovery is achieved, some fraction of the MSW must be returned to the environment. This chapter discusses how residues that have no value can be best managed and disposed of. Although the majority of the text is devoted to the disposal of raw, untreated MSW, it is equally applicable to any fraction of MSW that remains after the recoverable materials have been removed.

Discounting outer space (although the U.S. Environmental Protection Agency (EPA) did a feasibility study of sending waste into space in the 1970s—it was not economical) and air (from whence it eventually comes back to the earth's surface), the only two locations for the ultimate disposal of wastes are (1) in the oceans and other large bodies of water or (2) on or in land. With rare exceptions, most solid waste may no longer be legally dumped into oceans. Most developed countries have enacted strong ocean-dumping legislation, and the once ubiquitous refuse-or sludge-loaded barges have all but disappeared. New York City no longer uses barges to transfer waste to Staten Island since the Fresh Kills Landfill has closed. The remaining ocean disposal problems appear to result from the discharge of refuse from ships and debris washing from the land into the ocean. The total amount of refuse deposited on the sea bottom in some of the more frequently traveled shipping lanes (such as the North Atlantic) and areas such as the North Pacific Gyro is impressive as well as depressing. Philosophically, therefore, since ocean disposal is merely a storage process and not a method of treatment, it makes little sense to use the oceans as dumping sites. Little is said in this chapter on ocean disposal, and only the option of land disposal (as our only remaining alternative) is discussed here.

A landfill is an engineered method for land disposal of solid or hazardous wastes in a manner that protects the environment. Within the landfill biological, chemical, and physical processes occur that promote the degradation of wastes and result in the production of **leachate** (polluted water emanating from the base of the landfill) and gases. In the United States, MSW landfills are regulated under Subtitle D of the Resource Conservation and Recovery Act (Public Law 94-580) passed in 1976. Specific landfill design and operational criteria were issued under 40 CFR Part 258 in 1991.

The European Union (EU) Council Directives on Landfilling of Waste have also identified the need to optimize final waste disposal methods and seek to ensure uniform high standards of landfill operation and regulation throughout the European Union. These standards require a strategy that, by 2016, limits the quantity of biodegradable wastes entering the landfill to 35% of the total amount by weight of biodegradable waste produced in 1995. But many European countries have moved more aggressively to end the practice of landfilling biodegradable waste. For example, Denmark has already reached the last reduction target by banning the landfilling of all waste suitable for incineration. German landfills may accept only municipal waste that has been incinerated or that has undergone mechanical-biological treatment (MBT). Consequently, most waste in Europe is incinerated or treated prior to landfilling. In the United States, some states (such as California) have adopted a goal of 75% reduction of solid waste being landfilled by the year 2020.

Some communities in the United States have taken a different approach by operating a landfill as a **bioreactor**. The bioreactor landfill provides control and process optimization, primarily through the addition of leachate or other liquid amendments, if necessary. Beyond that, bioreactor landfill operation may involve the addition of wastewater sludge and other amendments, temperature control, and nutrient supplementation. The bioreactor landfill attempts to control, monitor, and optimize the waste stabilization process rather than contain the wastes as prescribed by most regulations.

8-1 PLANNING, SITING, AND PERMITTING OF LANDFILLS

A solid waste engineer, when asked to take on the job of managing the solid waste system for a major city, had only one question: "Do the existing landfills have enough remaining capacity to last until I retire?"

No essential public facility, with the possible exception of an airport, is more difficult to plan, site, and permit than a landfill. In the last 25 years, the number of operating landfills has decreased from about 8000 to under 1700, primarily due to stringent requirements under the RCRA (40 CFR Part 258). However, landfill capacity is greater now than it was a decade ago due to the expansion of many sites to take advantage of economies of scale achieved in operating the large modern landfill.

Commonly, it now takes over 10 years to go through the process of opening a new landfill. During the 10-year process, many of the rules will change, including regulations, permits, and approval requirements. In addition, public opposition and even lawsuits during the process are more than likely. The difficulty of siting landfills makes disposal in outer space appear to be a more attractive alternative! Clearly, one quality that is needed when siting a landfill is perseverance.

8-1-1 Planning

In planning for solid waste disposal, communities must look many years into the future. A 10-year time frame is considered short-term planning. Thirty years seems to be an appropriate time frame, because after 30 years, it becomes difficult to anticipate solid waste generation and new disposal technology.

The first step in planning for a new landfill is to establish the requirements for the landfill site. The site must provide sufficient landfill capacity for the selected design period and support any ancillary solid waste functions, such as leachate treatment, landfill gas management, and special waste services (i.e., tires, bulky items, household hazardous wastes). Some sites also house facilities for handling recyclable materials (materials recovery facilities) and composting. To determine landfill capacity, the disposal requirements for the community or communities must be estimated. A landfill that is too small will not have an adequate service life and will not justify the expense of building it. On the other hand, a landfill that is too large may eliminate many potential sites and will result in high up-front capital costs that would preclude the community from constructing other needed public facilities.

Records from the past several years provide a historical guide of disposal quantities. In some cases, this information can be very accurate if scales were used at existing landfills. In other cases, the information is suspect because the quantities of solid waste delivered to the landfills were only estimated. Finally, by using the historical population, a per capita disposal rate can be calculated. Since 1990, U.S. per capita solid waste generation rates have remained at about 4.5 pounds per person per day. On the other hand, per capita disposal has decreased from 3.12 pounds per person per day to 2.43 pounds per person per day. Finally, the annual amount of waste landfilled each year has remained fairly constant since 1990, ranging from 134 to 142 million tons per year.

Table 8-1 Bulk Densities of Some Uncompacted Refuse Components

Material	g/cm³	lb/ft³
Light ferrous (cans)	0.100	6.36
Aluminum	0.038	2.36
Glass	0.295	18.45
Miscellaneous paper	0.061	3.81
Newspaper	0.099	6.19
Plastics	0.037	2.37
Corrugated cardboard	0.030	1.87
Food waste	0.368	23.04
Yard waste	0.071	4.45
Rubber	0.238	14.9

Source: [1]

If possible, in-place density (the density once the refuse has been compacted in the ground) should be estimated. If an existing landfill is available, density can be easily determined by routinely conducting aerial surveys of the landfill and then calculating the volumes. This method includes the volume of cover material in the calculation. If dirt is used as daily and final cover, 20 to 50% of the volume of the landfill may be cover material. An in-place density in the range of 1200 lb/yd³ (700 kg/m³) is typical.

When some materials are recovered from solid waste, the compaction characteristics may change, and it may be necessary to estimate the compaction of the waste by individual refuse components. For example, if yard waste is diverted from the landfill, the landfill will have a more uniform compaction. Bulk densities of the components can be used in such calculations, even though the actual compaction may be quite different. Table 8-1 lists some bulk densities that can be used for this purpose, and Example 8-1 illustrates the procedure. A more complete listing of bulk densities is found in Appendix B.

Volume, mass, and density calculations for mixed materials can be simplified by considering a container that holds a mixture of materials, each of which has its own bulk density. Knowing the volume of each material, the mass is calculated for each contributing material, added, and then divided by the total volume. In equation form,

$$\frac{(\rho_A \times V_A) + (\rho_B \times V_B)}{V_A + V_B} = \rho_{(A+B)}$$

where

ρ_A = bulk density of material A
ρ_B = bulk density of material B
V_A = volume of material A
V_B = volume of material B

When there are more than two different materials, this equation is extended.

If the two materials at different densities are expressed in terms of their weight fraction, then the equation for calculating the overall bulk density is

$$\frac{M_A + M_B}{\left[\dfrac{M_A}{\rho_A}\right] + \left[\dfrac{M_B}{\rho_B}\right]} = \rho_{(A+B)}$$

M_A = mass of material A
M_B = mass of material B
ρ_A = bulk density of material A
ρ_B = bulk density of material B

If more than two materials are involved, this equation is extended.

The volume reduction achieved in refuse baling or landfill compaction is an important design and operational variable. If the original volume of a sample of solid waste is denoted by V_o, and the final volume, after compaction, is V_c, then the calculation of the volume reduction is

$$\frac{V_c}{V_o} = F$$

where

F = fraction remaining of initial volume as a result of compaction
V_o = initial volume
V_c = compacted volume

Because the mass is constant (the same sample is compacted so there is no gain or loss of mass) and volume = mass/density, volume reduction also can be calculated as

$$\frac{\rho_o}{\rho_c} = F$$

where

ρ_o = initial bulk density
ρ_c = compacted bulk density
F = fraction remaining of initial volume as a result of compaction

EXAMPLE 8-1

For illustrative purposes only, assume that refuse has the following components and bulk densities.

Component	Percentage (by weight)	Uncompacted Bulk Density (lb/ft³)
Miscellaneous paper	50	3.81
Garden waste	25	4.45
Glass	25	18.45

Assume that the compaction in the landfill is 1200 lb/yd³ (44.4 lb/ft³). Estimate the fraction remaining of the initial volume achieved during compaction of the waste. Estimate the overall uncompacted bulk density if the miscellaneous paper is removed.

SOLUTION

The overall bulk density prior to compaction is

$$\frac{50 + 25 + 25}{\dfrac{50}{3.81} + \dfrac{25}{4.45} + \dfrac{25}{18.45}} = 4.98 \text{ lb/ft}^3$$

The fraction remaining of the initial volume which has been achieved during compaction is

$$\frac{4.98}{44.4} = 0.11$$

So the required landfill volume is approximately 11% of the volume required without compaction. If the mixed paper is removed, the uncompacted density is

$$\frac{25 + 25}{\dfrac{25}{4.45} + \dfrac{25}{18.45}} = 7.18 \text{ lb/ft}^3$$

Over the life of a landfill, the volume of waste may change due to the following:

- **New Regulations.** Many states have adopted waste diversion/recycling goals. If fully implemented, diversion rates of 70% and beyond may be achieved.

- **Competing Facilities.** Other landfills may exist or be planned that would receive some of the planned waste, or other existing landfills may close, resulting in the importation of solid waste.

- **Different Cover Options.** Dirt used as daily and final cover at a landfill may consume 20 to 50% of the available landfill volume, depending on the size of the landfill. The new landfill may rely on foam, tarps, or mulched green waste for cover and thus significantly increase the volume available for the solid waste.

- **Nonresidential Waste Changes.** A per capita generation rate takes into account all nonresidential waste currently going into the landfill. In some cases, the non-residential waste can account for more than 50% of the landfill volume needed. Some examples are large military facilities, agriculture operation, manufacturing facilities, and cities with a large percentage of communities and tourists. Either the closing or opening of these facilities can significantly affect future waste projections.

EXAMPLE 8-2

Calculate the required 20-year landfill capacity for a community with the population projection, per capita waste generation rate, and diversion rate shown in the following table. Note that the waste generation is expected to increase at approximately 3% per year through year 5 and then remain constant. Note also that the community is expected to increase its rate of waste diversion to 35% in year 4 through an aggressive recycling and yard waste composting program. Assume a soil daily cover is used that accounts for 25% of the landfill volume.

		Per Capita Generation			
Year	Population (000)	Rate, lb/cap/ day	Diversion, Fraction	Waste to Landfill, tons	Waste to Landfill, yd³
1	105.4	5.6	0.25	8.08E+04	1.35E+05
2	108.6	5.8	0.28	8.28E+04	1.38E+05
3	109.8	6.0	0.30	8.42E+04	1.40E+05
4	112.2	6.2	0.35	8.25E+04	1.38E+05
5	115.2	6.4	0.35	8.75E+04	1.46E+05
6	117.7	6.4	0.35	8.94E+04	1.49E+05
7	121.1	6.4	0.35	9.19E+04	1.53E+05
8	124.7	6.4	0.35	9.47E+04	1.58E+05
9	128.4	6.4	0.35	9.75E+04	1.62E+05
10	133.4	6.4	0.35	1.01E+05	1.69E+05
11	139.1	6.4	0.35	1.06E+05	1.76E+05
12	144.5	6.4	0.35	1.10E+05	1.83E+05
13	150.7	6.4	0.35	1.14E+05	1.91E+05
14	155.6	6.4	0.35	1.18E+05	1.97E+05
15	163.1	6.4	0.35	1.24E+05	2.06E+05
16	169.4	6.4	0.35	1.29E+05	2.14E+05
17	175.3	6.4	0.35	1.33E+05	2.22E+05
18	181.4	6.4	0.35	1.38E+05	2.30E+05
19	187.7	6.4	0.35	1.43E+05	2.38E+05
20	194.3	6.4	0.35	1.48E+05	2.46E+05
Total				2.15E+06	3.59E+06

SOLUTION

A typical calculation for one year's volume is

$$\frac{(\text{population}) \times (\text{per capita generation rate}) \times (1 - \text{diversion}) \times (365 \, \text{days/yr})}{1200 \, \text{lb/yd}^3}$$

Total landfill waste volume is 3.59×10^6 yd³. To account for the volume requirement for the cover soil,

$$0.25(T) + 3.59 \times 10^6 = T$$
$$T = 4.79 \times 10^6 \, \text{yd}^3$$

8-1-2 Siting

Once the size of the landfill has been determined, it is necessary to find an appropriate site. While the concept is simple, the execution is far from easy. A button (NOPE—Not On Planet Earth) seen at one landfill siting public hearing can best describe the challenges faced when siting a landfill. This attitude is referred to as NIMBY: not in my back yard. Politicians are more likely to embrace the principles of NIMTO—not in my term of office.

Historically, for a local community or regional agency, an appropriate landfill site is within the geographic boundary of the agency. The relationships between municipalities are such that few would allow a neighboring community to site a landfill in its jurisdiction if the host community was not a participant. The reality has become that few local communities can site new landfills. The barriers that local planning and engineering staff face when trying to site a new landfill are numerous and include lack of political support, high development costs, and public disclosure laws.

For a private company, the geographic location is not as critical. Recently, private companies have been siting large remote landfills in rural areas. Host fees are then paid to the local community. For example, both Seattle, Washington, and Portland, Oregon ship waste to a private regional landfill in eastern Oregon. New York City also exports waste to other states.

Once the geographical boundary of the potential site has been determined, unsuitable locations should be identified. This process is a pass/fail test referred to as a *fatal flaw analysis*. Some fatal flaws are established in regulations such as Subtitle D of the Resource Conservation and Recovery Act. For example, no landfill can be sited near an active seismic fault or an airport. Other criteria are subjective and may be established by the local community, such as no landfill will be sited within one mile of a school. Because of local regulations and criteria, it is not possible to provide a complete list of potential fatal flaws, but the following should be considered as fatal flaws:

- The site is too small.
- The site is on a flood plain.
- The site includes wetlands.
- A seismic zone is within 200 ft of the site.
- An endangered species habitat is on the site.
- The site is too close to an airport (not within 5000 ft for propeller aircraft or 10,000 ft if turbine engine aircraft).
- The site is in an area with high population density.
- The site includes sacred lands.
- The site includes a groundwater recharge area.
- Unsuitable soil conditions (e.g., peat bogs) exist on the site.

All of the areas exhibiting one or more of the above fatal flaws should be highlighted on the search map. If no unhighlighted areas remain within the geographic boundary, then it will not be possible to site a new landfill. However, if areas do remain, the siting process goes to the next step, which involves developing a ranking system. In addition to the fatal flaws, there are subjective ranking criteria, which are also used. In many cases, the public is asked to participate in developing the ranking criteria. For example, a site that has good access might receive five points, while a site with poor access might receive only one point. Other qualitative features might include relative population density, land use designation, groundwater quality, visual and noise impacts, site topography, site ownership, soil conditions, and proximity to the centroid of solid waste generation.

Once the criteria are developed, they are applied to any remaining areas that do not have fatal flaws. Potential sites are developed and ranked. The top several sites are then designated for more detailed investigation that includes on-site analysis of habitat, groundwater, and soil conditions plus physical surveys and initial environmental assessment. Once this information is available, additional public hearings are held, and the elected officials select a site or sites for the preliminary engineering design, permitting application, and detailed environmental impact report. Typically, the public is opposed to landfills, even though they may be far distances from their point of interest (such as their home), but their opposition is most intense when the landfill is to be constructed close to their home. Much of this opposition can be attributed to such undesirable developments as additional traffic, noise, odor, and litter. Most important is the belief that the siting of a landfill will reduce property values and decrease residents' quality of life.

There is little question that siting a landfill near residential areas will reduce property values. In one study, a typical town was configured with various neighborhoods having different property values, and a hypothetical landfill was placed in this town. Real estate assessors were asked to estimate the effect of this new landfill on the property values of existing homes. As expected, as the value of the property increased, so did the percentage drop in the property value.[2] Most dramatic was the far-reaching effect of the landfill. For the more expensive homes, the effect on property values exceeded a three-mile radius from the proposed landfill site. These findings were confirmed in a study using actual home values.[3] Figure 8-1 shows that the largest impact on private property values is on the most expensive homes that are closest to the landfills.[4]

Because there is no compensation for the loss in the value of homes, is it any wonder that there is such strong opposition to the siting of landfills? Should society pay property owners some fraction of their loss as a fair compensation to the siting of undesirable public facilities close to their property? However, if these owners are compensated, what about the owners near other locally undesirable land uses (LULUs), such as an airport or a wastewater treatment plant?

8-1-3 Permitting

On October 9, 1991, the EPA issued 40 CFR Part 258, regulations pertaining to landfills under Subtitle D of the Resource Conservation and Recovery Act; the regulations were implemented two years later. States were required to incorporate the federal standards into their regulations. In addition, the states had the flexibility of adding more stringent requirements. Subtitle D addressed such issues as location, design requirements, operating conditions, groundwater monitoring, landfill closure and post-closure, and financial assurance.

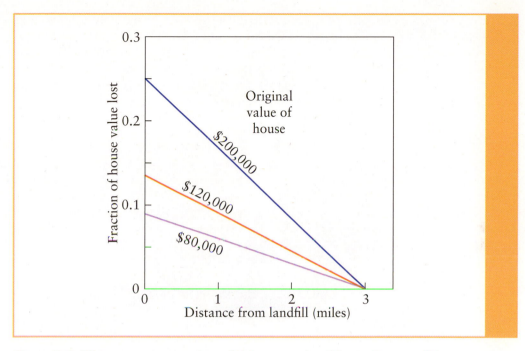

Figure 8-1 Effect on property values of siting a new landfill at a distance of three miles. Source: Vesilind, P. A., and E. I. Pas. 1998. "Discussion of A. C. Nelson, J. Genereux, and M. M. Genereux, 'A Price Effect of Landfills on Different House Value Strata.'" *Journal of Urban Planning and Development*, ASCE, 123, n. 3: 59–68. With permission from ASCE.

After extensive hearings across the country, the EPA decided to use a combination of *performance standards* and *design standards*. A design standard specifies a specific design—for example, a requirement for a composite liner. On the other hand, the requirement for no off-site migration of gas is a performance standard. How this standard is met is up to the applicant. In addition to Subtitle D requirements, landfills are usually subject to permitting for land use conformance, air emissions, groundwater and surface water discharge, operations, extraction for cover material, and closure.

8-2 LANDFILL PROCESSES

8-2-1 Biological Degradation

Refuse is approximately 75 to 80% organic matter composed mainly of proteins, lipids, carbohydrates (cellulose and hemicellulose), and lignins. Approximately two-thirds of this material is biodegradable, while one-third is recalcitrant (Figure 8-2). The biodegradable portion can be further divided into a readily biodegradable fraction (food and garden wastes) and a moderately biodegradable fraction (paper, textiles, and wood). Figure 8-3 summarizes the predominant biodegradation pathways for the decomposition of major organic classes in solid waste.

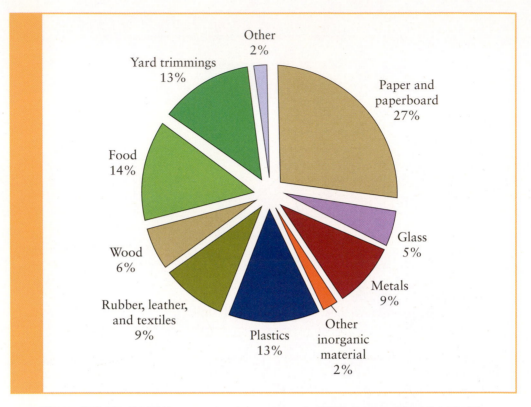

Figure 8-2 Total MSW generation by category in 2012. Source: Municipal Solid Waste Generation, Recycling, and Disposal in the United States: Detailed Tables and Figures for 2012, U.S. Environmental Protection Agency Office of Resource Conservation and Recovery.

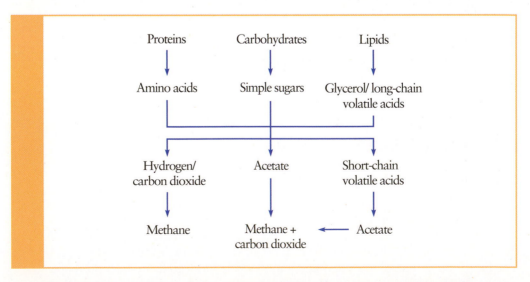

Figure 8-3 Predominant decomposition pathways for common organic waste constituents.

Table 8-2 Important Microbial Groups Promoting Anaerobic Waste Degradation

Microbial Group	Substrate
Amylolytic bacteria	Starches
Proteolytic bacteria	Proteins
Cellulolytic bacteria	Cellulose
Hemicellulolytic bacteria	Hemicellulose
Hydrogen-oxidizing methanogenic bacteria	Hydrogen
Acetoclastic methanogenic bacteria	Acetic acid
Sulfate-reducing bacteria	Sulfate

Source: This article was published in *Sanitary Landfilling: Process, Technology and Environmental Impact*, Sleats, R., C. Harries, I. Viney, and J. F. Rees, "Activities and Distribution of Key Microbial Groups in Landfills," Copyright Academic Press 1989.

The landfill ecosystem is quite diverse due to the heterogeneous nature of waste and landfill operating characteristics. The diversity of the ecosystem promotes stability; however, the system is strongly influenced by environmental conditions (such as temperature, pH, the presence of toxins, moisture content, and the oxidation-reduction potential). The landfill environment tends to be rich in electron donors, primarily organic matter. The dominant electron receptors are carbon dioxide and sulfate. Seven key physiological microbial groups that participate in rate-limiting stabilization steps of fermentation and methanogenesis are listed in Table 8-2.

A number of landfill investigation studies[6] have suggested that the stabilization of waste proceeds in five sequential and distinct phases. During these phases, the rate and characteristics of leachate produced and gas generated from a landfill are not only dissimilar, but also reflect the microbially mediated processes taking place inside the landfill. The phases experienced by degrading wastes are described next. Leachate characteristics during the waste degradation phases are summarized in Table 8-3.

Table 8-3 Landfill Constituent Concentration Ranges as a Function of the Degree of Landfill Stabilization

Parameter	Transition	Acid Formation	Methane Fermentation	Maturation
Chemical oxygen demand, mg/l	480–18,000	1500–71,000	580–9760	31–900
Total volatile acids, mg/l as acetic acid	100–3000	3000–18,800	250–4000	0
Ammonia, mg/l-N	120–125	2–1030	6–430	6–430
pH	6.7	4.7–7.7	6.3–8.8	7.1–8.8
Conductivity, μS/cm	2450–3310	1600–17,100	2900–7700	1400–4500

Source: [6]

Phase I—Initial Adjustment Phase

This phase is associated with initial placement of solid waste and accumulation of moisture within landfills. An acclimation period (or initial lag time) is observed until sufficient moisture develops and supports an active microbial community. Preliminary changes in environmental components occur in order to create favorable conditions for biochemical decomposition.

Phase II—Transition Phase

In the transition phase, the field capacity is often exceeded, and a transformation from an aerobic to an anaerobic environment occurs, as evidenced by the depletion of oxygen trapped within the landfill media. A trend toward reducing conditions is established in accordance with shifting of electron acceptors from oxygen to nitrates and sulfates and the displacement of oxygen by carbon dioxide. By the end of this phase, measurable concentrations of chemical oxygen demand (COD) and volatile organic acids (VOAs) can be detected in the leachate.

Phase III—Acid Formation Phase

The continuous hydrolysis (solubilization) of solid waste, followed by (or concomitant with) the microbial conversion of biodegradable organic content, results in the production of intermediate volatile organic acids at high concentrations throughout this phase. A decrease in pH values is often observed, accompanied by metal species mobilization. Viable biomass growth associated with the acid formers (acidogenic bacteria), and rapid consumption of substrate and nutrients are the predominant features of this phase.

Phase IV—Methane Fermentation Phase

During Phase IV, intermediate acids are consumed by methane-forming consortia (methanogenic bacteria) and converted into methane and carbon dioxide. Sulfate and nitrate are reduced to sulfides and ammonia, respectively. The pH value is elevated, being controlled by the bicarbonate buffering system and, consequently, supports the growth of methanogenic bacteria. Heavy metals are removed from the leachate by complexation and precipitation.

Phase V—Maturation Phase

During the final state of landfill stabilization, nutrients and available substrate become limiting, and the biological activity shifts to relative dormancy. Gas production dramatically drops, and leachate strength stays steady at much lower concentrations. Reappearance of oxygen and oxidized species may be observed slowly. However, the slow degradation of resistant organic fractions may continue with the production of humic-like substances.

   The progress toward final stabilization of landfill solid waste is subject to the physical, chemical, and biological factors within the landfill environment; the age and characteristics of landfilled waste; the operational and management controls applied; and the site-specific external conditions.

8-2-2 Leachate Production

Leachate Quantity

The total quantity produced can be estimated either by using empirical data or a water balance technique that sets up a mass balance among precipitation, evapotranspiration, surface runoff, and soil moisture storage.[7,8,9] Landfills operated in the Northeast have been designed using leachate generation rates of 1200 to 1500 gallons per acre per day (gpad) (11,200 to 14,000 liters/hectare/day) during active phases, 500 gpad (4,700 L/ha/day) following temporary closure, and 100 gpad (930 L/ha/day) following final closure. A 1993 study found landfills in arid regions generate only 1 to 7 gpad (9 to 60 L/ha/day).[10] Clearly, climatic conditions significantly affect leachate generation rates.

A water balance uses site-specific data to track water volumes, as shown schematically in Figure 8-4. Prior to closure, which generally involves use of an impermeable cap, the water balance can be described. Some fraction of precipitation (dependent on runoff characteristics and soil type and conditions) will percolate through the cover soil, and a fraction of this water is returned to the atmosphere

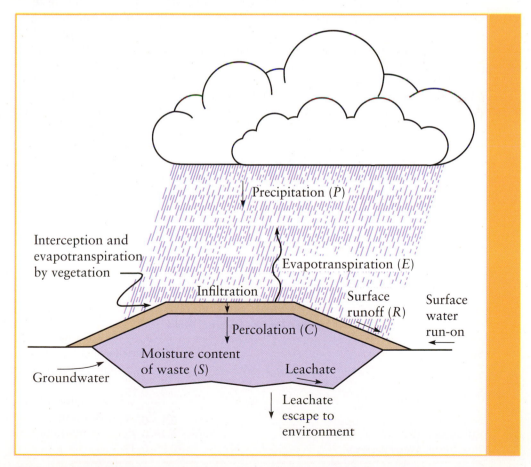

Figure 8-4 Schematic of components of water balance within a landfill.

through evapotranspiration. If the percolation exceeds evapotranspiration for a sufficiently long time, the amount of water the soil is able to hold (its *field capacity*) will be exceeded. The field capacity is the maximum moisture the soil (or any other material, such as refuse) can retain without a continuous downward percolation due to gravity.

Without plants, the soil on top of a landfill eventually reaches its field capacity so that any excess moisture will displace the moisture in the soil. With plant cover, the plants may extract water from the soil and release it by evapotranspiration, thus drying the soil to below field capacity. In some communities with very wet rainy seasons, temporary plastic liners are placed on the unfinished side slopes of active landfills to minimize water infiltration.

All layers of a mixture of soil and refuse in the compacted landfill also have a field capacity (or the ability to retain moisture). If the field capacity of this mixture is exceeded, the liquid (leachate) will drop to the next lowest soil/refuse layer. A soil/refuse mixture that does not attain field capacity discharges essentially no water to the deeper layers.

In the finished landfill, if the field capacity of the soil covering waste is exceeded, the water percolates through the soil and into the buried solid waste. If, in turn, the field capacity of the waste is exceeded, leachate flows into the leachate collection system. The water balance method is a means of calculating if and by how much field capacities are exceeded and, thus, is a way of calculating the production of leachate by landfills.

Some rough estimates can be used to develop the necessary calculations. Surface runoff coefficients, for example, can be estimated for different soils and slopes, as shown in Table 8-4. Precipitation data are available through the Weather Bureau. Evapotranspiration rates are also available or can be calculated using the method of Thornthwaite,[11] which takes into account the fact that evapotranspiration is reduced as the soil moisture drops. In other words, plants use less water in dry weather.

The *lysimeter* can also be used to estimate evapotranspiration.[13] In most cases, however, yearly average figures are sufficiently accurate for design. Figure 8-5 shows average yearly evapotranspiration in the United States.

The field capacities of various soils and wastes are listed in Table 8-5. The field capacity of compacted waste has been estimated as 20 to 35% by volume (about 30%), which translates to 300 mm of water/m of MSW. This value must be corrected for moisture already contained in the refuse, which is about 15%. A net field capacity of 150 mm/m of refuse (1.8 in/foot) is therefore not unreasonable.[7]

Table 8-4 Runoff Coefficients for Grass-Covered Soils

Surface	Runoff Coefficient
Sandy soil, flat to 2% slope	0.05–0.10
Sandy soil, 2% to 7% slope	0.01–0.15
Sandy soil, over 7% slope	0.15–0.20
Heavy soil, flat to 2% slope	0.13–0.17
Heavy soil, 2% to 7% slope	0.18–0.22
Heavy soil, over 7% slope	0.25–0.35

Source: [12]

Table 8-5 **Field Capacities of Various Porous Materials**

Material	Field Capacity, as mm water/m of soil
Fine sand	120
Sandy loam	200
Silty loam	300
Clay loam	375
Clay	450
Solid waste	200–350

Source: [7]

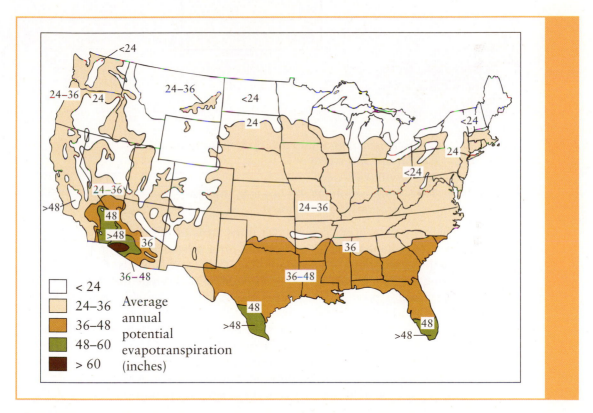

Figure 8-5 Average annual potential evapotranspiration in the continental United States. The potential evapotranspiration will occur if the soil is completely saturated; hence these figures are highest probable values, in inches of water. The actual evapotranspiration will always be lower. Source: [12]

The actual calculations of leachate production involve a one-dimensional analysis of water movement through soil and the compacted refuse, as shown in Figure 8-6. The figure also defines the symbols used in the calculation, which are based on the following mass balance equation:

$$C = P(1 - R) - S - E$$

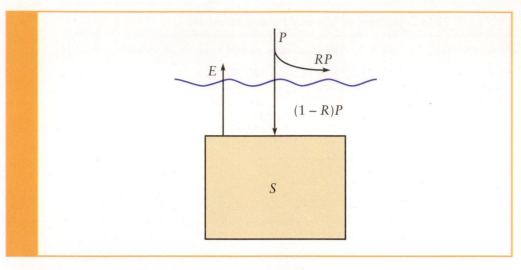

Figure 8-6 Mass balance of moisture in a landfill.

where

C = total percolation into the top soil layer, mm/yr
P = precipitation, mm/yr
R = runoff coefficient
S = storage within the soil or waste, mm/yr
E = evapotranspiration, mm/yr

The net percolation calculated for three areas of the United States is shown in Table 8-6. Note that the net percolation for a Los Angeles landfill is zero. This explains the absence of leachate in Southern California landfills.

Using the data from Table 8-6, it is possible to estimate the number of years before leachate is produced, since the soil/waste mixture will absorb the percolated water until its field capacity is reached. In Los Angeles, there seldom will be leachate generated by a landfill because of the lack of percolation. In Orlando, for a landfill 7.5 m deep, the first leachate should appear in 15 years. In Cincinnati, for a landfill depth of 20 m, the first leachate should appear in 11 years. These calculations are based on the assumption that moisture moves as a wetting front. Often there are channels and cracks in the waste that permit preferential flow. In

Table 8-6 Percolation in Three Landfills

	Precipitation (mm/yr) P	Runoff Coefficient R	Evapotranspiration (mm/yr) E	Percolation (mm/yr) C
Cincinnati	1025	0.15	658	213
Orlando	1342	0.07	1173	70
Los Angeles	378	0.12	334	0

Source: [7]

addition, some leachate is generated as a result of precipitation running over the landfill face. Therefore, leachate may be (and frequently is) produced long before these calculations suggest.

EXAMPLE 8-3

Estimate the percolation of water through a landfill 10 m deep, with a 1 m cover of sandy loam soil. Assume that this landfill is in southern Ohio, and that

P = 1025 mm/yr

R = 0.15

E = 660 mm/yr

Soil field capacity, F_s = 200 mm/m

Refuse field capacity, F_r = 300 mm/m, as packed

Assume further that the soil is at field capacity when applied, and that the incoming refuse has a moisture content of 150 mm/m and therefore has a net absorptive capacity of 150 mm/m. Percolation through the soil cover is

$$C = P(1 - R) - S - E = 1025(1 - 0.15) - 0 - 660 = 211 \text{ mm/y}$$

SOLUTION

The moisture front will move

$$\frac{211 \text{ mm/y}}{150 \text{ mm/m}} = 1.4 \text{ m/y}$$

or it will take

$$\frac{10 \text{ m}}{1.4 \text{ m/y}} = 7.1 \text{ y}$$

to produce a leachate that will be collected at a rate of (211 mm × area of landfill) per year.

Once a final cap is placed on the landfill, the water balance must include a consideration of the permeability of the cap and the reduced ability of the water to percolate into the waste. Obviously, leachate volume will be reduced following landfill closure.

The model for leachate generation most frequently used today, the Hydrologic Evaluation of Landfill Performance (HELP), was developed by the U.S. Army Corps of Engineers.[14] This model requires detailed on-site morphology and extensive hydrologic data to perform the water balance. The HELP computer program, which is currently in its third version (3.07), is a quasi-two-dimensional hydrologic model of water movement across, into, through, and out of a landfill.

Site-specific information is needed for precipitation, evapotranspiration, temperature, wind speed, infiltration rates, and watershed parameters (such as area, imperviousness, slope, and depression storage). The model accepts weather, soil, and design data and uses solution techniques that account for the effects of surface storage, snowmelt, runoff, infiltration, evapotranspiration, vegetative growth, soil moisture storage, lateral subsurface drainage, leachate recirculation, unsaturated vertical drainage, and leakage through soil, geomembrane, or composite liners. A variety of landfill systems can be modeled, including various combinations of vegetation, cover soils, waste cells, lateral drain layers, low-permeability barrier soils, and synthetic geomembrane liners. The HELP model is most useful for long-term prediction of leachate quantity and comparison of various design alternatives; however, it is not suitable for prediction of daily leachate production.

Leachate Quality

Within a landfill, a complex sequence of physically, chemically, and biologically mediated events occurs. As a consequence of these processes, refuse is degraded or transformed. As water percolates through the landfill, contaminants are leached from the solid waste. Mechanisms of contaminant removal include leaching of inherently soluble materials, leaching of soluble biodegradation products of complex organic molecules, leaching of soluble products of chemical reactions, and washout of fines and colloids. The characteristics of the leachate produced are highly variable, depending on the composition of the solid waste, precipitation rates, site hydrology, compaction, cover design, waste age, sampling procedures, interaction of leachate with the environment, and landfill design and operation. Some of the major factors that directly affect leachate composition include the degree of compaction and composition of the solid waste, climate, site hydrogeology, season, and age of the landfill. Table 8-7 shows the wide variation of leachate quality as determined by various researchers.

Table 8-7 Ranges of Various Parameters in Leachate

Parameter	Ehrig 1989	Qasim and Chiang 1994	Florida Landfills Grosh, 1996 (mean value)	National Database (mean value)
BOD (mg/l)	20–40,000	80–28,000	0.3–4660 (149)	0–100,000 (3761)
COD (mg/l)	500–60,000	400–40,000	7–9300 (912)	11–84,000 (3505)
Iron (mg/l)	3–2100	0.6–325	—	4–2200
Ammonia (mg/l)	30–3000	56–482	BDL–5020 (257)	0.01–2900 (276)
Chloride (mg/l)	100–5000	70–1330	BDL–5480 (732)	6.2–67,000 (3691)
Zinc (mg/l)	0.03–120	0.1–30	BDL–3.02 (0.158)	0.005–846 (0.23)
Total P (mg/l)	0.1–30	8–35	—	0.02–7 (3.2)
pH	4.5–9	5.2–6.4	3.93–9.6	6.7–8.2
Lead (mg/l)	0.008–1.020	0.5–1.0	(29.2 ± 114)	0.00–2.55 (0.13)
Cadmium (mg/l)	< 0.05–0.140	< 0.05	(7.52 ± 23.9)	0.0–0.564 (0.0235)

BDL = below detection limit
Source: Reinhart, D. R., and C. J. Grosh. 1997. "Analysis of Florida MSW Landfill Leachate Quality Data." *Report to the Florida Center for Solid and Hazardous Waste Management* (March).

The capacity for leachate polluting streams and lakes is expressed by the Biochemical Oxygen Demand (BOD), which estimates the potential for using oxygen. For the sake of comparison, the BOD of raw sewage is about 180 mg/l. Another means for expessing pollutional potential is COD, or Chemical Oxygen Demand. COD is always greater than BOD.

Leachate tends to contain a large variety of organic and inorganic compounds at relatively low concentration that can be of concern if groundwater and surface water contamination occurs. These compounds are often constituents of gasoline and fuel oils (aromatic hydrocarbons such as benzene, xylene, and toluene), plant degradation by-products (phenolic compounds), chlorinated solvents (such as used in dry cleaning), and pesticides. Inorganic compounds of concern are lead and cadmium, which come from batteries, plastics, packaging, electronic appliances, and light bulbs.

The quality of leachate directly affects viable leachate treatment alternatives. Because leachate quality varies from site to site and over time, neither biological treatment nor physical/chemical treatment processes separately are able to achieve high treatment efficiencies. A combination of both types of treatment is one of the more effective process trains for the treatment of leachate. Physical/chemical processes are needed for the pretreatment of young leachate to make it amenable to biological treatment and to hydrolyze some refractory organic compounds found in leachate from older landfills. Biological treatment is primarily used to stabilize degradable organic matter found in young and middle-aged leachates.

8-2-3 Gas Production

Gas Quantity

Landfill operators, energy recovery project owners, and energy users need to be able to project the volume of gas produced and recovered over time from a landfill. Recovery and energy equipment sizing, project economics, and potential energy uses depend on the peak and cumulative landfill-gas yield. The composition of the gas (percent methane, moisture content) is also important to energy producers and users. Proper landfill management can enhance both yield and quality of gas.

Mathematical and computer models for predicting gas yields are based on population, per capita generation, waste composition and moisture content, percent actually landfilled, and expected methane or landfill-gas yield per unit dry weight of biodegradable waste.[16] Mathematical models also can be used to model extraction systems, including layout, equipment selection, operation optimization, failure simulation, and problem determination and location within existing extraction systems.[17] Four parameters must be known if gas production is to be estimated with any accuracy: gas yield per unit weight of waste, the lag time prior to gas production, the shape of the lifetime gas production curve, and the duration of gas production.

In theory, the biological decomposition of one ton of MSW produces 15,600 ft³ (442 m³) of landfill gas containing 55% methane (CH_4) and a heat value of 530 Btu/ft³ (19,730 kJ/m³). Since only part of the waste converts to CH_4 due to moisture limitation, inaccessible waste (plastic bags), and nonbiodegradable fractions, the actual average methane yield is closer to 3900 ft³/ton (100 m³/tonne) of MSW. Theoretical landfill gas production potential for the

United States is estimated at 1.4 trillion ft³/y (33 × 10¹² m³/y). However, a more conservative estimate would put the figure at roughly 500 billion ft³/y (14 billion m³/y) with an oil equivalent energy potential of 5.6 million tons/y.[18]

Gas production data collected at landfills across the United States show that there is significant variation in the data due to differences in environmental conditions and landfill management. The methane generation usually is between 1 to 2 ft³/lb (0.06 to 0.12 m³/kg) of waste on a dry basis—over a period of 10 to 40 years. The recoverable fraction for energy use is between 50 and 90%.[19]

Once the expected yield is determined, a model is selected to describe the pattern of gas production over time. As mentioned above, many models have been proposed, ranging from single valued, to linear increase/linear decline, to exponential decline. The EPA has published a model called LandGEM based on the following equation:

$$Q_T = \sum_{i=1}^{n} 2kL_oM_ie^{-kt_i}$$

where

Q_T = total gas emission rate from a landfill, volume/time
n = total time periods of waste placement
k = landfill gas emission constant, time^{-1}
L_o = methane generation potential, volume/mass of waste
t_i = age of the ith section of waste, time
M_i = mass of wet waste, placed at time i

This model can be downloaded at http://www.epa.gov/ttn/catc/products .html#software.

EXAMPLE 8-4

The following example shows how this model is applied for a simple case. A landfill cell is open for three years, receiving 165,700 tonnes of waste per year (recall that 1 tonne = 1000 kg). Calculate the peak gas production if the landfill-gas emission constant is 0.0307 yr^{-1} and the methane generation potential is 140 m³/tonne.

SOLUTION

For the first year,

$$Q_t = 2 \,(0.0307)\,(140)\,(165,700)(e^{-0.0307(1)}) = 1,381,000 \text{ m}^3$$

For the second year, this waste produces less gas, but the next new layer produces more, and the two are added to yield the total gas production for the second year.

These results are plotted as Figure 8-7. The annual peak gas production is found from this figure as about 4,000,000 m³/yr.

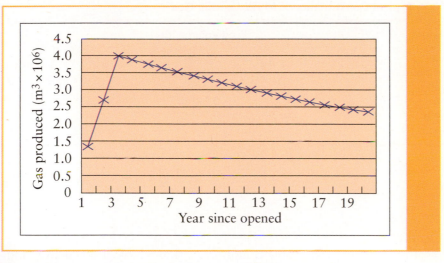

Figure 8-7 Landfill gas generated in a landfill. [See Example 8-4.]

The lag period prior to methane generation may range from a few weeks to a few years, depending on landfill conditions. The duration of gas production is also influenced by environmental conditions within the landfill. For the LandGEM model, no lag period is assumed, and the duration is controlled by the magnitude of the landfill-gas emission constant.

A practical guide to forecasting of yield—based on waste generation, the ability of wastes to decompose, and the efficiency of collecting gases from that decomposition—uses the following assumptions:[20]

- 80% of the MSW generated is landfilled.
- 50% of the organic material will actually decompose.
- 50% of the landfill gas generated is recoverable.
- 50% of the landfills are operating within a favorable pH range.

By using these percentages, the actual methane yield turns out to be one-tenth of the theoretical yield. Unfortunately, estimates and rules of thumb are of little use for sizing equipment, determining project life, and analyzing project economics for a specific landfill. Gas generation then may be more closely estimated by the use of models and/or by laboratory tests of waste samples, small- and large-scale pilot tests, and test cells incorporated in the landfill itself. Sources of error include the inability to select representative landfill samples or well sites, the inability to accurately determine the volume from which a well is actually extracting, and the variation in gas generation across the landfill and over time (seasonally). Of course, once a recovery system has been installed and is functioning at a steady state, reasonably precise recovery data can be collected to predict future results.

Table 8-8 **Typical Constituents of MSW Landfill Gas**

Component	% by Volume (dry)
Methane	45–60
Carbon dioxide	40–60
Nitrogen	2–5
Oxygen	0.1–1.0
Ammonia	0.1–1.0
Hydrogen	0–0.2

Source: Based on by Tchobanoglous, G., H. Theisen, and S. Vigil, *Integrated solid waste management: engineering principles and management issues,* McGraw-Hill, 1993.

Gas Quality

Because of the prevailing anaerobic conditions within a biologically active landfill, these sites produce large quantities of gas composed of methane, carbon dioxide, water, and various trace components such as ammonia, sulfide, and nonmethane volatile organic carbon compounds (VOCs). Tables 8-8 and 8-9 provide typical composition data for MSW landfill gas. Landfill gas is generally controlled by installing vertical or horizontal wells within the landfill. These wells are either vented to the atmosphere (if gas migration control is the primary intent of the system) or connected to a central blower system that pulls gas to a flare or treatment process.

The gas can pose an environmental threat, because methane is a potent greenhouse gas, and many of the VOCs are odorous and/or toxic. According to the U.S. EPA, methane is 23 times more potent as a greenhouse gas than carbon dioxide.

Over the past two centuries, methane concentrations in the atmosphere have more than doubled, largely due to human-related activities. Methane now accounts for 16% of global greenhouse gas emissions from human activities. Landfills are the second largest source of human-related methane emissions in the United States, accounting for approximately 23% of these emissions in 2007. Thus, it is important that landfill gas is not released to the atmosphere; rather, it should be collected and the methane converted to carbon dioxide and water.

Given the large impact that landfilling can have on the generation of greenhouse gas, the USEPA developed a computer model, the Waste Reduction Model (WARM), to measure greenhouse gas impacts of landfilling compared to recycling, composting, or source reduction.

However, the gas has a high energy content and can be captured and burned for power, steam, or heat generation. Treatment of landfill gas is required prior to its beneficial use. Treatment may be limited to condensation of water and some of the organic acids or may include removal of sulfide, particulates, heavy metals, VOCs, and carbon dioxide. Some of the more potential innovative uses of gas include power generation using fuel cells, vehicle fuel (compressed or liquid natural gas), and methanol production.

Table 8-9 Trace Organic Compounds Found in Florida Landfill Gas

Compound	Concentration in ppb	
	Average	Range
Benzene	240	BDL–470
Chlorobenzene	247	BDL–1978
Chloroethane	161	BDL–796
Cis-1,2-Dichloroethylene	76	BDL–379
Dichlorodifluoromethane	858	BDL–1760
1,1-Dichloroethane	241	BDL–1113
Dichlorotetrafluoroethane	229	BDL–544
Ethylbenzene	5559	2304–7835
M,P-Xylene	8872	3918–12675
Methylene chloride	94	BDL–749
o-Xylene	2872	1244–4378
p-Dichlorobenzene	372	BDL–699
Styrene	449	BDL–963
Tetrachloroethylene	131	BDL–561
Toluene	6313	2443–12215
Trichloroethylene	19	BDL–149
Trichlorofluoromethane	20	BDL–160
Vinyl chloride	333	BDL–1448
1,2,4-Trimethylbenzene	1328	244–2239
1,3,5-Trimethylbenzene	573	BDL–1140
1,1,1-Trichloroethane	268	BDL–1137
4-Ethyltoluene	1293	265–2646

BDL = below detection limit
Source: Reinhart, D. R., C. D. Cooper, and N. E. Ruiz. 1994. *Estimation of Landfill Gas Emissions at the Orange County Landfill.* Orlando: Civil and Environmental Engineering Department, University of Central Florida.

8-3 LANDFILL DESIGN

The landfill design and construction must include elements that permit control of leachate and gas. The major design components of a landfill, as shown in Figure 8-8, include the liner, the leachate collection and management system, gas management facilities, stormwater management, and the final cap.

8-3-1 Liners

The liner system is required to prevent migration of leachate from the landfill and to facilitate removal of leachate. It generally consists of multiple layers of natural material and/or geomembranes selected for their low permeability. Soil liners usually are constructed of natural clays or clayey soils. If natural clay materials are not readily available, commercial clays (bentonite) can be mixed with sands to produce a suitable liner material. Geomembranes are impermeable (unless perforated) thin sheets made from synthetic resins, such as polyethylene, polyvinyl chloride, or other polymers. High-density polyethylene (HDPE) tends to be used in MSW landfill liners most commonly, because it is resistant to most chemicals found in landfill leachates.

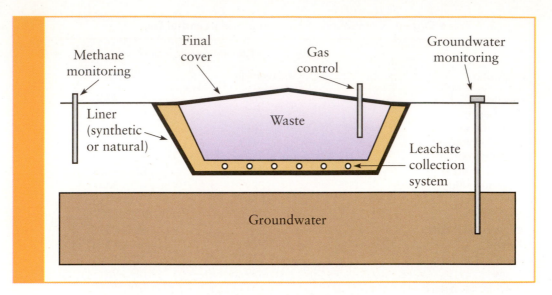

Figure 8-8 Design components in a Subtitle D landfill.

Landfills may be designed with single, composite, or double liners, depending on the applicable local, state, or federal regulations (see Figure 8-9). A single liner is constructed of clay or a geomembrane. A composite liner, which is the minimum liner required by RCRA Subtitle D, consists of two layers: The bottom is a clay material and the top layer is a geomembrane. The two layers of a composite liner are in intimate contact to minimize leakage. A double liner may be either two single liners or two composite liners (or even one of each). Figure 8-10 shows a synthetic liner on a side slope, ready for the earth cover. The clay layer already has been installed under this synthetic liner.

Each liner is provided with a leachate collection system. The collection system separating the two liners is a leak detection system—a series of pipes placed between the liners to collect and monitor any leachate that leaks through the top liner. Recently, the geosynthetic clay liner has been introduced for use as the top component in the double liner system. This liner is composed of a thin clay layer (usually sodium bentonite) supported by geotextiles (a geosynthetic filter) or geomembranes. The geosynthetic clay liner is easily placed in the field and uses up less volume, allowing for more volume to be used for waste deposition.

Clearly, the more layers that are included, the more protective the liner system will be. The costs, however, increase dramatically. A composite liner can cost as much as $250,000 per acre. Because the liner is so critical to groundwater protection, an exhaustive quality control/quality assurance program is required during liner installation.

8-3-2 Leachate Collection, Treatment, and Disposal

Leachate is directed to low points at the bottom of the landfill through the use of an efficient drainage layer composed of sand, gravel, or a geosynthetic material.

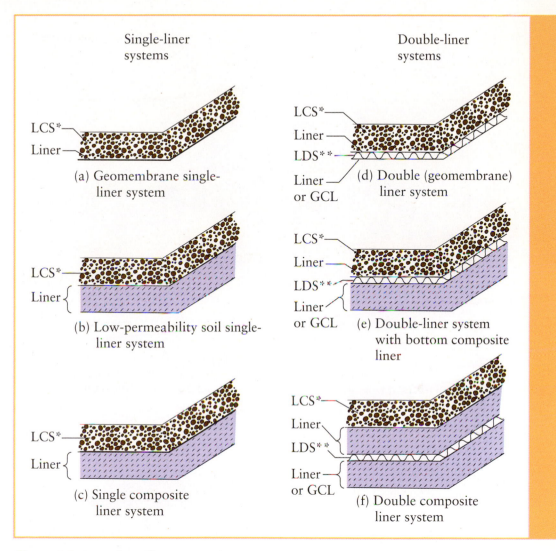

Figure 8-9 Examples of liner systems in municipal solid waste landfills. LCS = leachate collection system, GCL = geosynthetic clay liner, and LDS = leachate detection system.

Perforated pipes are placed at low points to collect leachate and are sloped to allow the moisture to move out of the landfill.

Leachate Collection and Storage

The primary purpose of lining a landfill cell is to minimize the potential for groundwater contamination. The liner serves as a barrier between the buried waste and the groundwater and forms a catch basin for leachate produced by the landfill. The leachate that is collected within the cell must be removed from above the liner as quickly as possible, since the RCRA Subtitle D regulations restrict the head of leachate (free liquid depth) on a liner system to 30 cm. Leachate is typically removed by two means: gravity flow or pumping. The various

Figure 8-10 Synthetic liner on slope, ready for earth cover. (Courtesy William A. Worrell)

components of a leachate collection system for an MSW landfill typically include the following:

- Protective and drainage layers
- Perforated collection lateral and header pipes
- Pump station sump
- Leachate pumps
- Pump controls
- Pump station appurtenances
- Force main or gravity sewer line

Table 8-10 provides general guidelines for leachate collection system components based on a survey of landfill design engineers.[23]

Leachate removed from the landfill cell(s) is temporarily stored on site until it can be treated, recirculated, or transported off site for final treatment and disposal. Storage of leachate is also important for equalization of flow quantities and constituent quality to protect downstream treatment facilities. The typical leachate storage alternatives are surface impoundments and tanks.

Leachate Collection System Design Equations and Techniques

Because of federal regulations[24] that restrict leachate head to 12 in (30 cm) on top of the liner, much attention has been devoted to predicting this value. The drainage length, drainage slope, permeability of the drainage materials, and the leachate impingement rate control this depth on the liner.

Table 8-10 Design Guidance for Leachate Collection System Components

Parameter	Range	Median
Leachate loading rate (gpd/ac)	600–1000	750
Maximum leachate head (in)	9–12	11
Pipe spacing (ft)	60–400	180
Collection pipe dia. (in)	6–8	8
Collection pipe material	PVC or HDPE	HDPE
Pipe slope (%)	0.5–2	1
Drainage slope (%)	0.2–2	1

Source: Reinhart, D. R., and T. Townsend. 1998. "Assessment of Leachate Collection System Clogging at Florida Municipal Solid Waste Landfills." *Report to the Florida Center for Solid and Hazardous Waste Management* (April).

Darcy's law (in conjunction with the law of continuity) can be used to develop an equation to predict the leachate depth on the liner based on anticipated infiltration rates, drainage material permeability, distance from the drain pipe, and slope of the collection system.[25,26]

For $R < 1/4$,

$$Y_{max} = (R - Rs + R^2S^2)^{1/2}\left[\frac{(1 - A - 2R)(1 + A - 2RS)}{(1 + A - 2R)(1 - A - 2RS)}\right]^{1/2.4}$$

For $R = 1/4$,

$$Y_{max} = (R - Rs + R^2S^2)^{1/2}\exp\left[\frac{1}{B}\tan^{-1}\left(\frac{2RS - 1}{B}\right) - \frac{1}{B}\tan^{-1}\left(\frac{2R - 1}{B}\right)\right]$$

For $R > 1/4$,

$$Y_{max} = \frac{R(1 - 2RS)}{1 - 2R}\exp\left[\frac{2R(S - 1)}{(1 - 2RS)(1 - 2R)}\right]$$

where

$R = q/(K \sin^2 \alpha)$, unitless
$A = (1 - 4R)^2$, unitless
$B = (4R - 1)^2$, unitless
$S = \tan \alpha$, slope of liner, unitless
$Y_{max} =$ maximum head on liner, ft
$L =$ horizontal drainage distance, ft
$\alpha =$ inclination of liner from horizontal, degrees
$q =$ vertical inflow (infiltration) per unit of horizontal area, ft/day
$K =$ hydraulic conductivity of the drainage layer, ft/day

These equations are obviously cumbersome to use, and a more conservative but far easier to use equation has been proposed:[27]

$$Y_{max} = \frac{P}{2}\left(\frac{q}{K}\right)\left[\frac{K\tan^2\alpha}{q} + 1 - \frac{K\tan\alpha}{q}\left(\tan^2\alpha + \frac{q}{K}\right)^{1/2}\right]$$

where

Y_{max} = maximum saturated depth over the liner, ft
P = distance between collection pipes, ft
q = vertical inflow (infiltration), defined in this equation as from a 25-year, 24-hour storm, ft/day

This equation can be used to calculate the maximum allowable pipe spacing based on the maximum allowable design head, anticipated leachate impingement rate, slope of the liner, and permeability of the drainage materials. The previous equation suggests that, holding all other parameters constant, the closer together the pipes are placed (at greater construction cost), the lower the head will be. A reduced head on the liner results in a lower hydraulic driving force through the liner, and the consequence of a puncture in the liner is likewise reduced.

EXAMPLE 8-5

Determine the spacing between pipes in a leachate collection system using granular drainage material and the following properties. Assume that in the most conservative design all stormwater from a 25-year, 24-hour storm enters the leachate collection system.

Design storm (25 years, 24 hours) = 8.2 in = 0.00024 cm/s
Hydraulic conductivity = 10^{-2} cm/s
Drainage slope = 2%
Maximum design depth on liner = 15.2 cm

SOLUTION

$$P = \frac{2Y_{max}}{\left(\dfrac{q}{K}\right)\left[\dfrac{K\tan^2\alpha}{q} + 1 - \dfrac{K\tan\alpha}{q}\left(\tan^2\alpha + \dfrac{q}{K}\right)^{1/2}\right]}$$

$$P = \frac{2(15.2)}{\left(\dfrac{0.00024}{0.01}\right)\left[\dfrac{0.01(0.02)^2}{0.00024} + 1 - \dfrac{0.01(0.02)}{0.00024}\left((0.02)^2 + \dfrac{0.00024}{0.01}\right)^{1/2}\right]} = 1428 \text{ cm}$$

Geosynthetic drainage materials (geonets) have been introduced to increase the efficiency of the leachate collection systems over such natural materials as sand or gravel. A geonet consists of a layer of ribs superimposed over each other, providing highly efficient in-plane flow capacity (or transmissivity). If a geonet is used between the liner and drainage gravel, the spacing between the collection pipes can be estimated from[28]

$$\theta = \frac{qp^2}{4Y_{max} + 2P\sin\alpha}$$

where θ = transmissivity of the geonet, ft²/day

Table 8-11 Summary of Leachate Treatment Options

Treatment Option	Removal Objective	Comments
Biological		Best used on "young" leachate
Activated sludge	BOD/COD	Flexible, shock resistant, proven, minimum SRT increases with increasing organic strength, > 90% BOD removal possible
Aerated lagoons	BOD/COD	Good application to small flows, > 90% BOD removal possible
Anaerobic	BOD/COD	Aerobic polishing necessary to achieve high-quality effluent
Powdered activated carbon/activated sludge	BOD/COD	> 95% COD removal, > 99% BOD removal
Physical/Chemical		Useful as polishing step or for treatment of "old" leachate
Coagulation/precipitation	Heavy metals	High removal of Fe, Zn; moderate removal of Cr, Cu, or Mn; little removal of Cd, Pb, or Ni
Chemical oxidation	COD	Raw leachate treatment requires high chemical dosages, better used as polishing step
Ion exchange	COD	10–70% COD removal, slight metal removal
Adsorption	BOD/COD	30–70% COD removal after biological or chemical treatment
Reverse osmosis	Total dissolved solids (TDSs)	90–96% TDS removal

Source: Reinhart, D. R., and C. J. Grosh. 1997. "Analysis of Florida MSW Landfill Leachate Quality Data." *Report to the Florida Center for Solid and Hazardous Waste Management* (March).

Leachate Treatment and Disposal

In wet areas, leachate treatment, use, and disposal represents one of the major expenses of landfill operations—not only during the active life of the landfill but also for a significant period of time after closure. Due to the cost implications, considerable attention must be given to selecting the most environmentally responsible, cost-effective alternative for leachate treatment and disposal. The optimal treatment solution may change over time as new technologies are developed, new regulations are promulgated, and/or leachate quality varies as a function of landfill age.

Leachate-treatment needs depend upon the final disposition of the leachate. Final disposal of leachate may be accomplished through co-disposal at a wastewater treatment plant or through direct discharge, both of which could be preceded by on-site treatment. Leachate treatment is often difficult because of high organic strength, irregular production rates and composition, variation in biodegradability, and low phosphorous content (if biological treatment is considered). Several authors have discussed leachate treatment options,[29-32] which are briefly summarized in Table 8-11. Generally, where on-site treatment and discharge is selected, several unit processes are required to address the range of contaminants present. For example, a leachate treatment facility at the Al Turi Landfill in Orange County, New York uses polymer coagulation, flocculation, and sedimentation, followed by anaerobic biological treatment, two-stage aerobic biological treatment, and filtration prior to discharge to the Wallkill River.[33] Pretreatment requirements may address only specific contaminants that may create problems at the wastewater treatment plant. High lime treatment has been practiced at the Alachua County, Florida, Southwest Landfill to ensure low heavy-metal loading on the receiving treatment facility.

The most economical alternative for leachate treatment is often to transport the wastewater off-site to a publicly owned treatment works (POTW) or commercial wastewater treatment facility. This option allows landfill owner/operators to focus on their primary solid waste management charge while letting wastewater experts handle the treatment of contaminated liquids. There are also economies of scale for larger treatment plants that cannot be realized for small plants designed to treat only the flow from a single landfill. Off-site treatment of leachate also alleviates some of the permitting, testing, monitoring, and reporting requirements for the landfill owner. Disposal to POTWs sometimes generates opposition from the plant owner, typically a municipal entity or a private operator running the plant on a contract basis for a municipal owner. While the acceptance of leachate certainly merits careful consideration, the waste stream should be evaluated as any other industrial wastewater stream. Leachate is generally more comprehensively characterized than other industrial waste streams. Modern Subtitle D landfills have rigorous waste acceptance criteria to keep hazardous, toxic, radioactive, and other noncompatible waste streams out of the landfill. Control of waste input has the effect of improving the quality of the leachate.

The Clean Water Act (CWA) mandates that sources directly discharging into surface waters must comply with effluent limitations listed in the National Pollution Discharge Elimination System (NPDES) permit(s). The CWA also requires the EPA to promulgate nationally applicable pretreatment standards that restrict pollutant discharges for those who discharge wastewater indirectly through sewers flowing to wastewater treatment plants. National pretreatment standards are established for those pollutants in wastewater from indirect discharges that may pass through or interfere with wastewater treatment plant operations. In addition, wastewater treatment plants are required to implement local treatment limits applicable to their industrial indirect discharges to satisfy any local requirements (40 CFR 403.5).

Emerging Technologies for the Treatment of Leachate

Because the treatment of leachate is both difficult and expensive, new techniques are being developed for managing this potential pollutant.[34]

Reverse osmosis has nothing to do with osmotic pressure or osmosis, but is the common name for forcing a fluid through a semipermeable membrane, thus separating soluble components on the basis of molecular size and shape. It is one of the treatment systems capable of removing dissolved solids. A high-pressure pump forces the leachate through a membrane, overcoming the natural osmotic pressure, and dividing the leachate into two parts: a water stream (permeate or product) and a concentrated (brine) stream. Molecules of water pass through the membrane, while contaminants are flushed along the surface of the membrane and exit as brine. Typical recovery rates (percentage of the feed that becomes permeate or clean water) range from 75 to 90%.

Direct osmosis concentration is a cold-temperature-membrane process that separates waste streams in a low-pressure environment. The system operates by placing a semipermeable membrane between the leachate and an osmotic agent, typically a salt brine with a concentration of approximately 10%. The semipermeable membrane allows the passage of water from the leachate (without outside pressure) into the salt brine but rejects the contaminants found in the leachate. As the process continues, the brine becomes dilute. This process is called osmosis, and it continues until the water concentrations on both sides of the membrane are equivalent.

Evaporation will actually dispose of the water component of water-based waste streams, such as leachate. This technology can reduce the total volume of leachate to be managed to less than 5% of the original volume. Other technologies (reverse osmosis, ultra/microfiltration, and conventional treatment systems) separate the waste stream into two components, but the water produced by these other treatment technologies must still be disposed of. Using landfill gas as the fuel in evaporation is a technology that effectively integrates the control of landfill gas and landfill leachate. Evaporative systems typically require an air permit for the flare and, usually, a modification to the landfill's solid waste permit to address the leachate management practices.

A *vapor compression distillation* process differs from an evaporation process in that a clean effluent is produced. Leachate is introduced into the VCD system through a recirculation loop. This loop constantly pumps leachate and concentrates at a high rate from the bottom of a steam disengagement vessel through a primary heat exchanger and back again. The leachate and concentrate enter the disengagement vessel through a tangential nozzle at a velocity sufficient to create a cyclonic separation of steam from the liquid. As the leachate and concentrate are rapidly recirculated, active boiling occurs on the inside of the primary heat exchanger and within the cyclonic pool formed inside the disengagement vessel.

The *mechanical vapor recompression* process uses the falling film principle in a vacuum. The core of the process is a polymeric evaporation surface (heat transfer element) on which water can boil at a temperature of 50 to 60°C (120 to 140°F). The process functions similarly to a heat pump. The raw leachate is pumped through two parallel heat exchangers and into the bottom of the evaporation vessel. From there, a circulation pump transfers a small volume of the leachate into the top of the vessel, where it is evenly distributed on the heat transfer element.

Several different types of *land treatment systems* are available for treating landfill leachates. However, these systems are usually sized for small leachate flows, since a significant amount of land could be required to handle larger flows. The three most common forms of land treatment systems found at MSW landfills are constructed wetlands, windrow composting, and growing poplar trees.

Leachate Recirculation

Landfills can be used as readily available biological reactors for the treatment of leachate. The landfill can be essentially transformed into an engineered reactor system by providing containment using liners and covers, sorting and discrete disposal of waste materials into dedicated cells, collection and recirculation of leachate, and management of gas. Such landfills are capable of accelerated biochemical conversion of wastes and effective treatment of leachate.

Most sanitary landfills are traditionally constructed so the leachate is collected and removed. The rate of stabilization in "dry" landfills may require many years, thereby extending the acid formation and methane fermentation phases of waste stabilization over long periods of time. Under these circumstances, decomposition of biodegradable fractions of solid waste will be impeded and incomplete, often preventing commercial recovery of methane gas and delaying closure and possible future reuse of the landfill site. Furthermore, during single-pass operations, the leachate must be drained, collected, and treated prior to final discharge.

In contrast, leachate recirculation may be used as a management alternative that requires the containment, collection, and recirculation of leachate back through the landfilled waste. This option offers more rapid development of active anaerobic microbial populations and increases reaction rates (and predictability) of these organisms. The time required for stabilization of the readily available organic constituents can be compressed to as little as two to three years rather than the usual 15- to 20-year period. This accelerated stabilization is enhanced by the routine and uniform exposure of microorganisms to constituents in the leachate, thereby providing the necessary contact time, nutrients, and substrates for efficient conversion and degradation. Hence, leachate recirculation essentially converts the landfill into a dynamic *anaerobic bioreactor* that accelerates the conversion of organic materials to intermediates and end products.

The advantages of leachate recirculation have been well documented. Application of leachate recirculation to full-scale landfills has occurred with increasing frequency in recent years. Leachate is returned to the landfill using a variety of techniques, including wetting of waste as it is placed, spraying of leachate over the landfill surface, and injecting of leachate into vertical columns or horizontal trenches installed within the landfill. It is important to design and operate other landfill components—such as gas management systems, leachate collection, and final and intermediate cover—so that they are compatible with bioreactor operation.

To optimize bioreactor operations, the waste moisture levels must be controlled by the rate of leachate recirculation, which is a function of waste hydraulic conductivity and the efficiency of the leachate introduction technique. To enhance the effect of leachate, the recirculation operations should be moved from one area to another, pumping at a relatively intense rate for a short period of time, then moving to another area. Many factors impact the rate of moisture input, such as the site specific capacity and the design of the recirculation system.

The quantity of liquid supplied is a function of such waste characteristics as moisture content and field capacity. In some cases, the infiltration of moisture resulting from rainfall is insufficient to meet the desired waste moisture content for optimal decomposition, and supplemental liquids (i.e., leachate from other areas, water, wastewater, or biosolids) may be required. Sufficient liquid supply must be ensured to support project goals. For example, the goal of moisture distribution might be to bring all waste to field capacity. The most efficient approach to reach field capacity is to increase moisture content through wetting of the waste at the working face and then uniformly to reach field capacity through liquid surface application or injection.

The addition of supplemental liquids increases the base flow of leachate from the landfill. This additional flow must be considered during design, especially following rain events when large amounts of leachate may be generated. Sufficient leachate storage must be provided to ensure that peak leachate generation events can be accommodated. While a properly designed and operated landfill will minimize extreme fluctuations of leachate generation with rainfall events, in wet climates, leachate generation will at times exceed the amount needed for recirculation. Other factors—such as construction, maintenance, and regulations—may also dictate that leachate not be recirculated from time to time. Therefore, it is very important to have contingency plans in place for off-site leachate management for times when leachate generation exceeds on-site storage capacity.

Leachate recirculation should be controlled to minimize outbreaks and to optimize the biological processes. Grading the cover to direct leachate movement away from side slopes, providing adequate distance between slopes and leachate injection, eliminating perforations in recirculation piping near slopes, and avoiding cover that has hydraulic conductivity significantly different from the waste can control seeps. In addition, it may be desirable to reduce initial compaction of waste to facilitate leachate movement through the waste. A routine monitoring program designed to detect early evidence of outbreaks should accompany the operation of any leachate recirculation system. Alternate design procedures—such as early capping of side slopes and installation of subsurface drains—also may be considered to minimize problems with side seepage.

The depth of leachate on the liner is a primary regulation in the United States to protect groundwater and is a major concern for regulators approving bioreactor permits. Control of head on the liner requires the ability to maintain a properly designed leachate collection system, monitor head on the liner, store or dispose of leachate outside of the landfill, and remove leachate at rates two to three times the rate of normal leachate generation. Several techniques are used to measure head on the liner, including sump measurements, piezometers, bubbler tubes, and pressure transducers. Measuring the head with currently available technology provides local information regarding leakage potential; however, for a more realistic evaluation, a more complete measurement may be required.

The construction, operation, and monitoring of leachate recirculation systems will affect daily landfill operations. If a leachate recirculation system is to be used, it should be viewed as an integral part of landfill operations. Installation of recirculation systems must be coordinated with waste placement and should be considered during planning of the fill sequence.

8-3-3 Landfill Gas Collection and Use

Gas generated within a landfill will move by pressure gradient, following paths of least resistance. Uncontrolled migrating gas can collect in sewers, sumps, and basements, leading to tragic consequences if explosions occur. To prevent gas migration, gas vents or wells must be provided. There are two basic systems for gas emissions control: passive collection and active extraction. *Passive collection systems* collect landfill gas using vent collectors and release the gas to the atmosphere without treatment or conveyance to a common point. Passive vents are often provided using natural convective forces within the landfill to direct gas to the atmosphere. In addition, passive vents might be placed at the property boundary of an old landfill that is producing minimal amounts of landfill gas. These vents would prevent the gas from moving off-site. Passive vents may reach only a few feet below the cap or may reach up to 75% of the landfill depth, designed in a similar manner to the active extraction well described next. Typical spacing for a passive vent is one per 9000 yd³ (7500 m³).

Active collection systems link collection wells with piping and extract the gas under vacuum created by a central blower. Active extraction wells may be vertical or horizontal wells, although vertical wells are more frequently employed. Vertical wells are installed in landfills using auger or rotary drills. A typical extraction well is shown in Figure 8-11. Wellheads provide a means of controlling the vacuum applied at the well as well as monitoring of gas flow rate, temperature, and gas

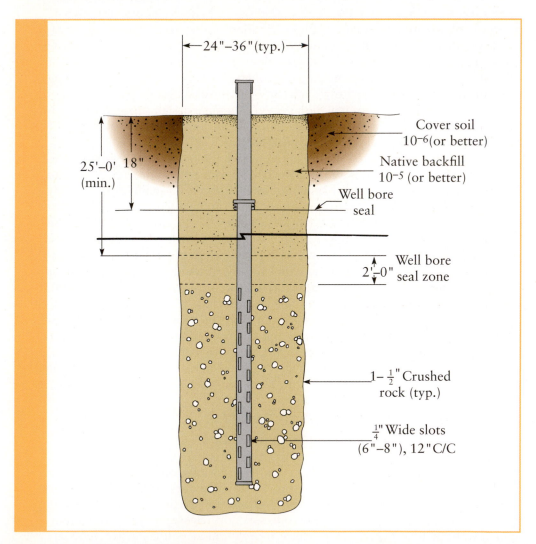

Figure 8-11 Typical vertical gas well.

Table 8-12 Guidelines for Construction of Vertical Gas Collection Systems

Parameter	Recommendation
Well depth	75% of depth or to water table, whichever comes first
Perforations	Bottom 1/3 to 2/3
	Minimum 25 ft below surface
Casing	3–8 in PVC or HDPE, telescoping well joint
Spacing (on center)	Interior collection system 200–500 ft
	Perimeter collection system 100–250 ft
Well density	One well per 1/2 to 2 acres
Minimum gas-collection piping slope	3%
Well bore diameter	12 to 36 in is standard (24, 30, 36 in most common)

Source: [35]

quality. Spacing of wells is a function of gas flow. Table 8-12 provides general guidance on well construction.

Landfill gas is extracted by central blowers that create negative pressure in the pipe network (Figure 8-12). These blowers are sized according to the volume of gas they must move. The greater the required flow, the greater must be the negative pressure that must be created, and therefore, the more energy that is required. The collection system should be designed to minimize head loss by providing sufficiently large pipes and by minimizing the number of valves and bends in the pipe. However, large pipes can be costly, and the design must balance the cost of the pipes and valves against the energy requirements of the blowers. The velocity of the flow through the piping system can be estimated using the continuity equation if compressibility is neglected.

$$Q = vA$$

where

Q = landfill-gas flow rate, ft^3/sec
v = landfill-gas velocity, ft/sec
A = cross-sectional interior area of the pipe, ft^2

The head loss through the pipe can be estimated using the Darcy-Weisbach equation, usually stated as

$$\Delta P = \frac{\rho f L v^2}{2gD}$$

where

ΔP = pressure drop, ft
f = Darcy-Weisbach friction factor, function of pipe roughness and diameter
L = length of pipe, ft
v = velocity, ft/sec
g = gravitational constant, ft/sec^2
D = diameter, ft
ρ = gas density, lb/ft^3

Figure 8-12 Landfill gas well and piping network. (Courtesy William A. Worrell)

The Darcy-Weisbach friction factor can be read from a Moody diagram, which is found in all fluid mechanics textbooks. The diagram is used by first calculating ϵ/D, where ϵ is the pipe roughness—a function of the pipe material. The rougher the pipe, the greater is the value of ϵ.

This equation gives the pressure drop in feet of water if all length units are in feet and the fluid is water. If the fluid is a gas, the equation can be modified (taking into account the density of the fluid) as

$$\Delta P = \frac{(0.0096)\rho f L v^2}{Dg}$$

where

ΔP = pressure drop when a gas is flowing in the pipe, ft of water

ρ = density of the gas, lb/ft³

The density of the gas is calculated using the universal gas law:

$$\rho = \frac{MP}{RT}$$

where

ρ = density of the gas, lb/ft³
M = molecular weight of the gas, lb/mole
P = pressure, lb/ft²
R = universal gas constant, lb-ft/mole-°R
T = absolute temperature, °R (°F + 460)

Typically, the pressure inside the collection pipes is close to atmospheric, about 30 inches of water (13.6 lb/in² or about 1958 lb/ft²). The molecular weight of the gas can be estimated as 28 lb/mole, with the universal gas constant then having units of lb-ft/°R.

EXAMPLE 8-6

Calculate the pressure drop in 800 ft of 4 in diameter PVC pipe carrying 500 ft³/min of landfill gas at 120°F. Assume the pressure in the pipeline is close to atmospheric (about 13.6 lb/in²) and the molecular weight of the landfill gas is 28 lb/mole. Assume the gas is incompressible.

The area of a 4 in diameter PVC Schedule 40 pipe (4.026 in ID) is 0.0884 ft².

SOLUTION

From the continuity equation:

$v = (500/60)/0.0884 = 94$ ft/sec

The roughness, ϵ, for PVC pipe is 0.000005 ft. Thus, $\epsilon/D = 0.000005/(4/12) = 0.000015$. From the modified Moody diagram, the friction factor is read as $f = 0.016$.

The gas density is estimated assuming the pressure at 1958 lb/ft², the molecular weight at 28 lb/mole, the universal gas constant at 1543 lb-ft/°R, and the absolute temperature at 120°F + 460. Thus,

$$\rho = \frac{(28)(1958)}{(1543)(120 + 460)} = 0.0613 \text{ lb/ft}^3$$

The pressure drop is then calculated as

$$\Delta P = \frac{(0.0096)(0.0613)(0.016)(800)(94)^2}{(4.02/12)(32.2)}$$

$$= 6.2 \text{ ft of water}$$

Repeating the calculations for 6 and 8 in pipes, the pressure drops are 0.8 ft and less than 0.1 ft, respectively, of water column.

Assuming head loss in the valves and fittings at 20% of the total pressure drop and adding an additional 10 in of water column to maintain negative pressure at the wellhead, the 4 in pipe requires a negative blower pressure at the suction side of the blower at a 84 in water column. The 6 in pipe requires about a 20 in water column, and the 8 in pipe requires less than a 12 in water column.

In designing the collection system it is important to recognize the purpose of the collection. A gas collection system can be designed and operated to achieve maximum gas collection so as to minimize the release of landfill gas or it can be designed and operated to collect landfill gas with the highest energy value. If the former is the objective, a high vacuum would be applied so that all gas is drawn to the collection wells. This could result in air being pulled through the cover and into the landfill. Since methane generation is an anaerobic process, pulling air with oxygen would have a negative impact on methane generation. Landfills with active collection systems routinely monitor oxygen levels at the well head as a way of adjusting the vacuum. If, on the other hand, the purpose is to produce the highest quality gas for subsequent use, then a low pressure system is required that will minimize the seepage of oxygen into the gas wells.

Technical Issues in Landfill Gas Use

Some of the technical issues regarding the use of landfill gas include gas composition, the effects of corrosives and particulates on equipment, potential energy losses, and gas extraction and cleanup. Energy users are concerned with the problems and solutions associated with the use of landfill gas as a fuel source. Much information can be obtained from equipment manufacturers as to the fundamentals and site-specific applications of gas extraction and cleanup.

The physical, chemical, and combustion characteristics of landfill gas can have significant impact on energy recovery equipment selection and operation. Trace organics (such gases as hydrogen sulfide and others) and particulates can cause corrosion and excessive wear. Carbon dioxide, nitrogen, and water vapor (with various inert materials) may reduce efficiency. Variability of gas composition and production rate over time exacerbates the problems associated with gas cleanup operations and energy applications.

The primary source of trace gases is from discarded volatile materials and their transformation by-products. Although most trace gases, primarily hydrocarbons, are harmless to energy use, halogenated hydrocarbons may cause problems upon combustion. Volatile and nonvolatile acidic hydrocarbons—such as the organic acids found in untreated landfill gas—are also highly corrosive. Trace constituents have been reported to cause corrosion, combustion chamber melting, and deposits on blades of turbine engines as well as internal combustion engines.[36]

Hydrogen sulfide and water vapor also can have corrosive effects. The use of landfill gas as a vehicle fuel requires removal of hydrogen sulfide and water vapor due to corrosion problems when they condense during gas compression and cooling. Hydrogen sulfide in concentrations as low as 100 ppm may lead to corrosion in piping, storage tanks, and engines.[37] Landfill gas with hydrogen sulfide and halide concentrations, chiefly chloride and fluoride, as low as 21 ppm and 132 ppm, respectively, requires pretreatment prior to application in fuel cells.[38] Stringent cleanup technology is also applied to remove the trace constituents from landfill gas when purifying it to pipeline quality natural gas.[36]

Landfills contain a large amount of soil and other particulate matter. Extraction systems can dislodge and take up these particulates into the gas stream. Deposits in engines and buildup in oil result in increased wear with a simultaneous decrease in lubricant capacity and life. Particulates can be removed by gas filtration or gas refrigeration. Dimethyl siloxane (a gaseous silicon compound) will combust to produce silica deposits in internal combustion engine cylinders and gas turbines, resulting in decreased combustion chamber volume, increased compression ratio, a tendency to detonate, and abrasion of valve stems and guides from hardened deposits.[19, 39]

The purpose of treatment systems is to remove particulates, condensates, and trace compounds and to upgrade landfill-gas quality for direct use or for energy/synthetic fuel conversion. Filtration and condensate removal are the more common cleanup approaches, whereas refrigeration and desiccation are used less frequently or as a means to enhance cleanup efficiency.[19]

Applications of Landfill Gas Use

Landfill gas can be *flared* (burned) on site, but this is not a beneficial application of this resource. Beneficial energy recovery systems include direct use, electricity generation, and conversion to chemicals or fuels. In 2010, the USEPA reported 519 operating projects generating 1597 MW and 306 mmscfd.

Boilers and other direct combustion applications are by far the cheapest and easiest options and represent about one-third of the current operating projects. Direct uses of landfill gas to replace or supplement coal, oil, propane, and natural gas have been successfully demonstrated. Applications include boiler firing, space heating, sludge drying, and leachate drying and incineration. In most cases, gas cleanup consists of little more than condensate removal. There is little risk for the end user in terms of gas quality, use, and continuity of supply. The payback period can be as little as a few months but is dependent on the negotiated price paid for the fuel replaced. The typical discounts of 10 to 20% are influenced by the pipeline distance and the gas quantity, quality, and variations permitted. The ideal situation is one where a user, located within a two-mile radius of the landfill, could accept all of the gas generated on a continuous basis.[40]

The potential for use as *vehicle fuel* exists if the gas is upgraded to natural gas quality, vehicles are modified to operate on some form of natural gas, and refueling stations are widely available and equipped for dispensing natural gas in the proper form. This technology is well established, and gas-powered, stationary internal combustion engines have been available for decades. Anaerobic digester gas is processed into fleet vehicle and engine generator fuels in several locations in Florida, including Plantation and Tampa.[42] In New Zealand, many vehicles can already run on upgraded landfill gas, and the number is increasing.[18] Worldwide, over 700,000 vehicles (many of them passenger cars) are fueled by natural gas.[39]

However, in most countries, the use of landfill gas as a vehicle fuel is limited to landfill or other municipal fleets with a limited range of operation. Until a greater number of vehicles (modified for processed landfill gas use) provide the demand necessary to support a more diverse infrastructure, the generation of landfill gas will outpace the fuel requirements of a landfill fleet.[18]

The benefits of using landfill gas as a vehicle fuel include improved air quality benefits that extend from reduced flare emissions to lower emissions from vehicles that burn landfill gas as an alternative to diesel. In 2010, the Altamont Landfill near San Francisco installed a world's largest landfill gas (LFG) to liquefied natural gas (LNG) plant. This plant is designed to produce up to 13,000 gallons of LNG per day or enough fuel for 300 garbage trucks.

Conversion of landfill gas to synthetic fuels and chemicals is also possible, if not economically feasible. The technologies include hydrocarbon production by the Fischer-Tropsch process and methanol synthesis by high-pressure chemical catalysis and partial biological oxidation. Gas-based chemical processes for synthesizing acetic acid and other compounds are also available. These technologies were developed for large-scale production of synfuels using coal-gas feedstocks. Production ventures were costly, and their products were expensive. The typical landfill can produce only 1 to 10% of the gas required for the size of plant considered for these technologies and processing techniques.[19]

Electrical power generation (internal combustion engines and gas turbines) is by far the most common landfill gas-to-energy application. The generation of electricity from LFG makes up about two-thirds of the currently operational projects in the United States. Electricity for on-site use or sale to the grid can be generated using a variety of different technologies, including internal combustion engines, turbines, microturbines, and fuel cells. The vast majority of projects use internal combustion (reciprocating) engines or turbines with microturbine technology being used at smaller landfills and in niche applications. Technologies such as Stirling and organic Rankine cycle engines and fuel cells are still in development.

Projects are set up according to the perceived electrical power generating capacity and the number of generating units. If landfill gas production is insufficient to support at least one MW of power generation, it is generally deemed economically unsuitable. Internal combustion engines are typically used at sites capable of producing less than three MW. Three to five engines are employed per project. Turbines, with higher horsepower and performance when operating at full load, are used at landfills where gas quantity can support greater than three MW, requiring only one or two turbine units for generating electricity.[40]

Electrical power generation (fuel cells) is another technology that is subject to unfavorable economics when using landfill gas. Fuel cells are basically electrochemical batteries utilizing molten carbonate or phosphoric acid, fueled by coal, petroleum, natural gas, or other such hydrocarbon feedstocks. Hydrogen from the converted fuel combines with oxygen to produce electricity. The advantages of fuel cells over other use options include higher energy efficiency, availability to smaller landfills, minimal by-product emissions, minimal labor and maintenance, and minimal noise impact.[40] Economic analysis reveals the benefits of fuel cells comparable to their environmental impacts. Locations with higher population densities generally have greater air emission problems. Likewise, commercial and industrial power rates are greater for these highly industrialized, highly urbanized areas.

Purification to pipeline-quality gas is also an attractive option for the use of landfill gas. There are major differences between landfill gas and pipeline-quality natural gas in composition and energy content. Landfill gas has a lower Btu content, combusts at a lower temperature, is more corrosive, and contains much greater concentrations of undesirable gases (CO_2, O_2, N_2) and harmful halocarbons than pipeline-quality natural gas. Diligent extraction and stringent cleanup are therefore necessary to render landfill gas virtually devoid of all components save methane. The required gas cleanup, an expensive and complex process for other use alternatives, includes nearly complete CO_2 removal. So expensive is the conversion of landfill gas to natural gas that only large landfills can attain the economies of scale necessary to support operations. Although other, less elaborate projects exist, fewer than 50 landfills in the United States produce gas for pipeline use.

Nontechnical Issues related to Landfill Gas Use

The beneficial use of landfill gas has safety, environmental, and energy implications. There also exist economic (as well as regulatory) issues for direct use, energy production, or co-generation. For any landfill gas project, these nontechnical issues must be identified, and their effects evaluated, quantified, and ultimately factored into an overall cost-benefit analysis.

Energy recovery for off-site use generally involves investor capital for defraying extraction, cleanup, and use project costs and generates revenue for landfill owners/operators. Generating revenue from landfill gas is easier said than done. Economic benefits from landfill-gas-to-energy projects exist in the form of federal (and in some cases, state) tax incentives and investment credits. However, the tendency of state and local governments to enact legislation and regulatory guidelines of greater stringency than those of federal standards may stifle the use of options economically available for the recovery and use of landfill gas. Despite all of these obstacles, a landfill-gas project must still have the potential to amass significant return on investment from operations (without the benefit of tax credits and other benefits) to attract investment. Many projects, however, are justified financially only through tax credits and favorable purchase price as required by regulation.

Consider two competing scenarios. Some states set goals for power utilities to generate a percentage of their electricity from green power. These utilities could build new solar and wind plants to generate electricity, or, if the utility has a gas-fired plant, they can switch to landfill gas. The utility would then be able to generate green electricity from its existing facility.

8-3-4 Geotechnical Aspects of Landfill Design

Landfill stability is an important aspect of design, particularly in light of the complex, multilayer construction of modern landfills. Failures can occur as slope failures during the construction of a landfill or after the landfill has been closed. The critical point of failure is usually the soil/geosynthetic and geosynthetic/geosynthetic interfacial surfaces as well as waste slopes. Landfill failures can have catastrophic results, including damage of leachate-and gas-collection systems, contamination of the surrounding environment, and even—in extreme cases—loss of life. The stability of landfills must be investigated under both static and seismic conditions, as required by RCRA Subtitle D.

Landfill stability is normally analyzed using readily available computer software. This analysis requires knowledge of properties of waste, materials used in the liners and caps (synthetic and natural), and foundation soils. These include such well-known geotechnical properties as unit weight, shear strength, shear moduli, internal friction angle, cohesion, and internal pore pressure, plus geosynthetic material properties such as tensile strength, surface roughness, flexibility, and surface wetness. Waste properties are often difficult to determine because of waste heterogeneity, changes in properties with time, and the difficulty in collecting samples necessary to evaluate these properties.

The operation of the landfill as a bioreactor can create unique geotechnical conditions. Increased moisture content leading to waste saturation can result in positive internal pore pressure and reduced angle of friction (angle at which slope failure occurs at an interface). The impact of waste decomposition on shear strength is also of critical concern. Many studies report finding mud-like conditions at the bottom of wet, deep landfills. Overaggressive leachate recirculation (compounded by the use of impermeable cover soil) has been cited as a contributing factor to the catastrophic slope failure of a landfill in Colombia, South America.

To minimize the probability of landfill failure, it is recommended that side slopes of completed and capped landfills be no greater than 1:3, with 1:4 being preferable. Shallow slopes mean reduced air space for waste disposal; however, the reduced risk of slope failure outweighs economic advantages of increased disposal volume. Use of textured geomembranes can also reduce slippage along interfaces. Finally, proper drainage and gas-pressure relief in the cap will reduce pore pressure and reduce the probability of failure. Given the uncertainties of design, significant factors of safety are highly recommended.

8-3-5 Stormwater Management

Many operating and design controls are available to minimize leachate production, including control of the size of the working face, placement of interim cover on the waste, and use of proper stormwater runoff and run-on controls. Control of stormwater run-on and runoff is also required by Subtitle D of RCRA (Section 258.26). Run-on control prevents the introduction of stormwater to the active area of the landfill, thus minimizing the production of leachate, erosion, and contamination of surface water. Reducing run-on also limits the production of runoff from the landfill surface.

Run-on can be prevented by diverting stormwater from active areas of the landfill. Any facility constructed to control run-on must be capable of handling peak volumes generated by a 24-hour, 25-year storm. Typical measures to control run-on include contouring the land surrounding the landfill cell or constructing ditches, dikes, or culverts to divert flow.

Runoff that is generated can be collected by swales (see Figure 8-13), ditches, berms, dikes, or culverts that direct contaminated runoff from active areas to storage and treatment facilities, and uncontaminated runoff from closed areas to detention facilities. Runoff from active areas must be collected and at least the volume generated from a 24-hour, 25-year storm must be controlled. Local regulations will dictate the design of uncontaminated stormwater management facilities. For example, in Florida, a detention pond must store the first inch of runoff for 14 days.

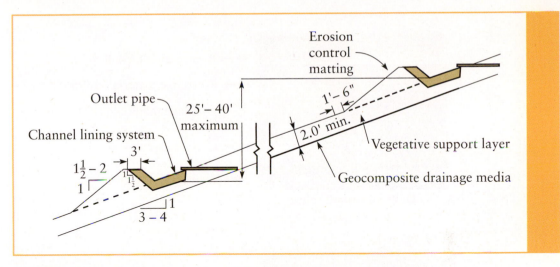

Figure 8-13 Side slope swale in a landfill final cover.

8-3-6 Landfill Cap

All Subtitle D landfills have to be capped, regardless of the potential for water intrusion. The purpose is to prevent the production of leachate that can contaminate groundwater. The effect of keeping water out of the landfill is to maintain dry conditions and hinder the process of biodegradation, making most landfills merely storage facilities.

Once the landfill reaches design height, a final cap is placed to minimize infiltration of rainwater, minimize dispersal of wastes, accommodate settling, and facilitate long-term maintenance of the landfill. The cap may consist (from top to bottom) of vegetation and supporting soil, a filler and drainage layer, a hydraulic barrier, foundation for the hydraulic barrier, and a gas-control layer. Figure 8-14 is a schematic of a recommended top slope cap. EPA regulations require that the top cap be less permeable than the bottom liner.

Slope stability and soil erosion are critical concerns for landfill caps. Typical side slopes are 1:3 to 1:4, and the interface friction between adjacent layers must resist seepage forces and may decrease the contact stresses between layers due to buildup of water and/or gas pressures. Consequently, on slide slopes, composite liner caps (geomembrane placed directly on top of a low permeability soil) are not recommended.

An alternative to the conventional landfill cap design—the *evapotranspiration cover*, also known as the *capillary barrier*—has been introduced. This cover places a thin (6 in) layer of silt over a 2.5 ft thick layer of uncompacted soil. The silt layer supports vegetation for transpiration, while the soil layer provides moisture storage. Because the uncompacted soil layer remains unsaturated and at lower suction pressure than the saturated silt layer, the flow tends to be driven upward by capillary forces, and percolation into the waste is prevented. Moisture is then removed by transpiration and evaporation. Studies have shown that the evapotranspiration cover is appropriate for most areas of the United States west of the Mississippi River.[43]

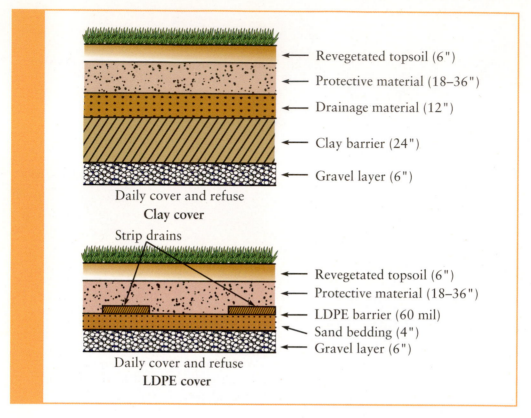

Figure 8-14 Typical caps used for closing landfills.

As landfills settle, caps can fail and allow stormwater to penetrate into landfills. Figure 8-15 shows three modes of cap failure. Although data are scarce, it seems reasonable to assume that the caps are at best short-range covers, and that within some years, they will become ineffective. Caps are expensive, approaching $200,000 per acre of landfill area, and it is unlikely that they contribute much to the protection of groundwater.

8-4   LANDFILL OPERATIONS

8-4-1 Landfill Equipment

The movement, placement, and compaction of waste and cover in a landfill require a variety of large machines, including tractors, loaders (track and wheel), and compactors. In addition, a variety of support equipment is needed, including motor graders, hydraulic excavators, water trucks, and service vehicles.

Originally, track-type bulldozers were used in landfills. Not only did they spread and compact the refuse, they could be used for placing and spreading cover material as well as preparing the site, building roads, hauling trees, and removing

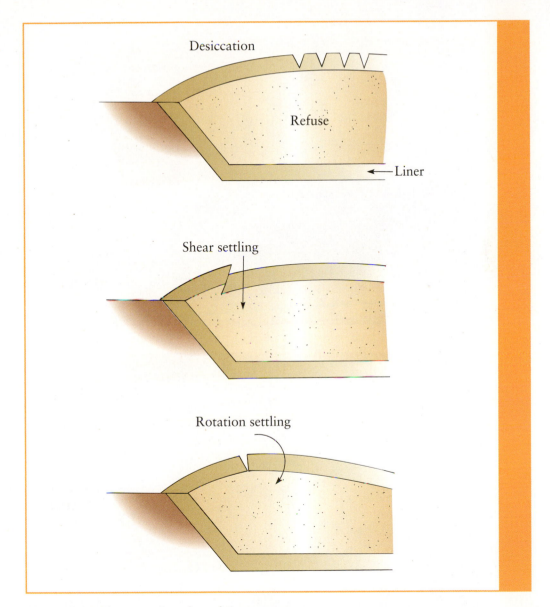

Figure 8-15 Three modes of cap failure.

stumps. They typically can achieve waste densities of 800 to 1000 lb/yd³ (475 to 590 kg/m³). See Figure 8-16a.

　　Specially built landfill compactors (Figure 8-16b) are now almost universally used. They are effective in spreading and compacting large quantities of waste, are quite heavy (exceeding 25 tons), and are equipped with knobbed steel wheels capable of tearing and compacting waste to densities of 1200 to 1600 lb/yd³ (710 to 950 kg/m³).

　　Daily dirt cover is excavated and placed on the operating face of a landfill using pans or scrapers. Designed for the sole purpose of scraping up dirt and

Figure 8-16 (a) A landfill dozer. (b) Landfill compactor. (Courtesy William A. Worrell)

hauling it to another location, these machines bring the cover dirt to the landfill and then run down the compacted slope, discharging the dirt evenly over the refuse.

The type and quantity of landfill vehicles are determined by the amount and type of waste handled, amount and type of soil cover, distance of moving waste and cover, weather requirements, compaction requirements, landfill configuration, budget, expected growth, and supplemental tasks anticipated.

8-4-2 Filling Sequences

At an active landfill, daily deliveries of waste are placed in *lifts* or layers on top of the liner and leachate collection system to depths of 20 m (65 ft) or greater. Typical waste placement methods used for landfill construction are shown in Figure 8-17. The area or mound type of landfill construction is commonly used in areas with a high groundwater table or subsurface conditions that prevent excavation. The excavation or trench technique provides waste placement in cells or trenches dug into the subsurface. The removed soil is often used as daily, intermediate, or final

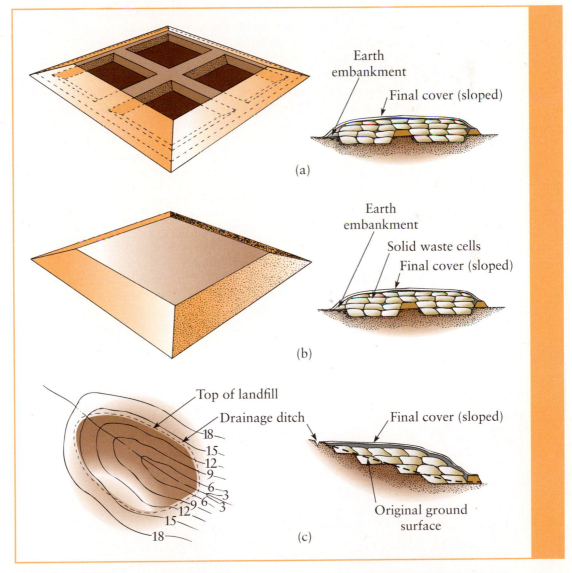

Figure 8-17 Commonly used landfilling methods: (a) excavated cell/trench, (b) area, and (c) canyon/depression. Source: [48]

cover. Where suitable, waste also can be placed against lined canyon or ravine side slopes. Slope stability and leachate and gas emission control are critical issues for this type of waste placement.

Because it is imperative that the leachate collection system be protected during landfill operations, waste in the first lift is selected to avoid heavy and sharp objects. This layer is often called the *operational layer*. The waste also must be placed in a manner to keep compactor wheels away from the leachate collection system. Filling then begins in subsequent lifts, generally in a corner and moving outward to form the next layer. The filling sequence is established at the time of landfill design and permitting. The working face must be large enough to accommodate several vehicles unloading simultaneously, typically 12 to 20 ft (4 to 6 m) per vehicle. Some landfill operations will, for safety reasons, provide a public dumping area in order to separate the public from the garbage trucks.

As waste is placed in the landfill, heavy equipment is used to compact the waste to maximize use of *airspace*. Airspace is the volume of space on a landfill site permitted for the disposal of waste. At first, the space is occupied by air, which is replaced by waste as the landfill fills. The degree of compaction expected is a function of several factors, including refuse layer thickness (see Figure 8-18), number of passes made over the waste (see Figure 8-19), slope (flatter slopes compact better by landfill compactors, steeper slopes (maximum 1:3) compact better by track-type tractors), and moisture content (wetter waste compacts more effectively than dry waste).

8-4-3 Daily Cover

Waste is covered at the end of each working day with soil (typically 6 inches) or alternative daily cover (such as textiles, geomembrane, carpet, foam, greenwaste, or other proprietary materials). Daily cover is required to control disease vectors and rodents; to minimize odor, litter, and air emissions; to reduce the risk of fire; and to minimize leachate production. The landfill sides are sloped to facilitate maintenance and to increase slope stability; generally, a maximum slope of 1:3 is maintained.

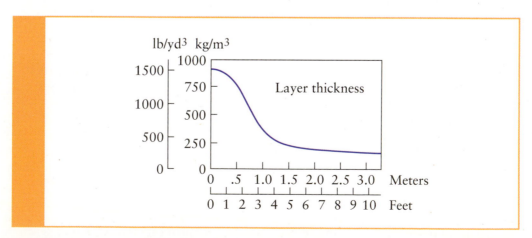

Figure 8-18 Waste density of a landfill as a function of layer thickness.

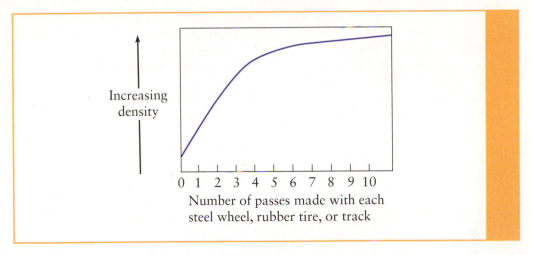

Figure 8-19 Waste density of a landfill as a function of the number of compactor passes.

8-4-4 Monitoring

Landfill monitoring is critical to the operation of a landfill. Most commonly, landfill operators monitor the following: leachate head on the liner, leakage through the landfill liner, groundwater quality, ambient air quality (to ensure compliance with the Clean Air Act), gas in the surrounding soil, leachate quality and quantity, landfill-gas quality and quantity, and stability of the final cover.

Control of head on the liner requires the ability to maintain a properly designed leachate collection system, monitor head on the liner, store or dispose of leachate outside of the landfill, and remove leachate at appropriate rates. Leakage through a single liner is often detected using a lysimeter, such as that shown in Figure 8-20. The locations and number of lysimeters are based on the landfill design; however, more than one should be provided to ensure redundancy. The lysimeter should be located below the crest of the landfill liner where the maximum head will be found at the point of maximum leakage potential.

Groundwater monitoring is generally accomplished through the construction and sampling of monitoring wells in the vicinity of the landfill. Optimum design would place a cluster of wells in each location to provide the means to evaluate groundwater quality at multiple depths. These well clusters should be placed up-gradient from the landfill to evaluate background groundwater quality, as well as immediately down-gradient from the landfill to determine the influence of the landfill on groundwater. Additional monitoring wells are also placed at the property boundary. These wells are sampled quarterly for a variety of organic and inorganic groundwater constituents as stipulated in RCRA Subtitle D regulations. If down-gradient constituents are found to have a statistically significant increase in concentration as compared with up-gradient concentrations, more complete monitoring must begin. If down-gradient levels continue to exceed drinking water standards, more extensive monitoring and a corrective action plan may be required.

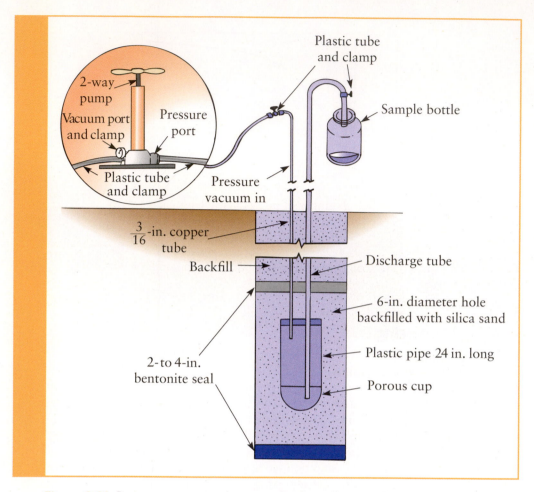

Figure 8-20 Porous cup suction lysimeter for the collection of liquid samples from the landfill. Source: Based on by Tchobanoglous, G., H. Theisen, and S. Vigil, *Integrated solid waste management: engineering principles and management issues*, McGraw-Hill, 1993.

The EPA permits a waiver of groundwater monitoring requirements if it can be demonstrated that there is no potential for the migration of hazardous constituents from the landfill to the uppermost aquifer during the active life of the landfill and the post-closure period. To accomplish this demonstration, computer models are generally used that simulate contaminant movement in the subsurface. The Multimedia Exposure Assessment Model (MULTIMED) was developed by the EPA specifically for this purpose.[44]

Landfill operators must ensure that the concentration of methane gas does not exceed 25% of the lower explosive limit for methane (5% concentration) in facilities, nor does it exceed the lower explosive limit at the landfill boundary. Gas

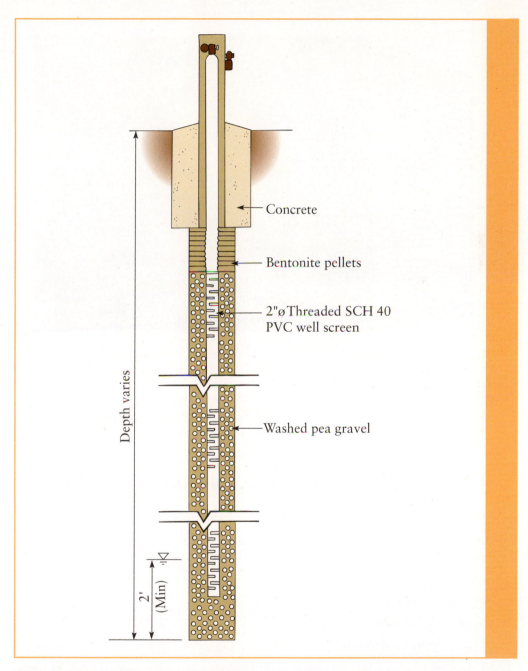

Concrete

Bentonite pellets

2"ø Threaded SCH 40
PVC well screen

Washed pea gravel

Depth varies

2'
(Min)

Figure 8-21 Landfill-gas monitoring probe.

monitoring probes, such as the one shown in Figure 8-21, are generally placed at the property boundaries and various other locations around the landfill property to allow the landfill owner to test for landfill gas. In addition, gas extraction wells, as shown in Figure 8-22, may be placed at the property boundaries and other locations to collect any landfill gas that is migrating toward the property boundaries.

Figure 8-22 Gas extraction well in place. (Courtesy William A. Worrell)

The hydrogeologic properties of the site, soil conditions, and the placement of landfill structures determine the locations of the probes.

8-5   POST-CLOSURE CARE AND USE OF OLD LANDFILLS

RCRA Subtitle D requires close attention to a landfill for 30 years following closure. The following are required during the 30-year post-closure period:

- Maintenance of the integrity and effectiveness of the final cover
- Operation of the leachate collection system
- Groundwater monitoring
- Gas-migration monitoring

A detailed post-closure care report must be prepared and approved prior to waste receipt at the landfill. The report must also include assurance that sufficient funds are available to meet the costs of closure, post-closure care, and corrective action for release of contaminants. These funds can be ensured using a variety of instruments such as surety bonds, cash deposits, irrevocable letters of credit, or private trust funds.

Once a landfill is closed, the land can be reused for other purposes. It is important to consider the final use of a landfill during planning and design phases to ensure that the landfill configuration is compatible with the selected final use. Final use alternatives include, but are not limited to, these:

- Golf courses
- Natural areas/open space
- Recreation parks
- Ski/toboggan slopes
- Parking lots
- Building construction

"Mount Trashmore" in the otherwise relatively flat state of Indiana is a popular ski location. The Harborside International Golf Complex near Chicago, Illinois—formerly the Chicago MSW landfill—hosts two world-class 18-hole golf courses and a 58-acre practice facility. A 50-acre landfill in urban Cambridge, Massachusetts, now provides a variety of sports and recreational facilities, including three soccer fields, three softball fields, two play areas, bocce courts, and jogging trails. The project increased the city's open space by 20%.

The reuse of a landfill site is complicated by leachate generation, gas emissions, and the potential for large differential settlement. Differential settlement can be particularly damaging to overlying structures and embankments, as well as any leachate- and gas-control facilities in place. Settlement is the result of added moisture, external loads such as final and daily cover, overlying wastes, and any added structures such as buildings and roads. Settlement typically occurs quite rapidly immediately after closure (first 1 to 12 months) as a result of the rapid dissipation of pore fluid and gas. This period of *primary settlement* is followed by slower settlement over the next 15 to 20 years when *secondary settlement* occurs as a result of waste creep and biodegradation. This long-term settlement may result in compaction of as much as 30% of the waste depth. The rate and magnitude of settlement are difficult to predict due to the heterogeneity of the waste and complex degradation processes. Poor foundations can result in unfortunate situations, such as the destruction of the motel shown in Figure 8-23.

Settlement must be considered when designing structures such as roads, parking lots, and structures supported on shallow foundations. Explosive gas can collect between the landfill surface and the supported floor of structures employing deep foundations built on top of closed landfills. To compensate for settlement, permanent structures may need to be equipped with costly foundations and articulated and telescoping pipe connections to maintain proper drainage and integrity of facilities.

8-6   LANDFILL MINING

If significant biodegradation occurs in a landfill, it might be possible to dig up old landfills, separate the nonbiodegradable fraction, and use the dirt and organic soil as a cover material for present landfills. This seems like a reasonable option for communities seeking new landfills, and at one time it appeared that *landfill mining* might be widely practiced. For example, a south Florida landfill mined an

Figure 8-23 An unfortunate motel built on a settling landfill. (Courtesy P. Aarne Vesilind)

old landfill and recovered some metals. However, with the continuing evolution of landfill regulations and control of landfill gas, it has become very difficult to mine landfills. Opening the cap of a landfill allows gas to escape to the atmosphere and permits rainwater to produce contaminated runoff. Finally, mined materials have marginal economic value because they are quite dirty and difficult to clean.

Landfill mining makes economic sense only if it can create new volume for continuing the life of an existing landfill and can be done at minimal environmental cost. Only shallow and wet landfills that have few vertical lifts are candidates for landfill mining, since they are most likely to have fully biodegraded. Landfill mining operations consist of excavating the buried refuse and screening it to separate out the useful new cover materials. Typically, the nonbiodegradable contents have little value and are re-landfilled. Successful operations significantly extend the time to closure for landfills and thus provide an economical alternative to the siting of new landfills.[45]

8-7 FINAL THOUGHTS

Landfills are engineering projects that require an unusual mixture of technical skills and public relations acumen, with the latter often outweighing the former. Little is mentioned here relative to the nontechnical problems associated with the planning, design, and operation of landfills, but this is not to imply that these aspects are insignificant.

A popular definition of solid waste is that "it is the stuff everyone wants picked up but no one wants put down." Studies on the psychology and sociology related to the siting of landfills are both fascinating and somewhat frightening. We know entirely too little about public reaction to and the impact of such projects, and the design engineer is often placed in a somewhat uncomfortable position of being the Solomon-like judge of public reaction and public good. Nowhere does the human nature aspect of the engineering profession become as important as in the design of the ultimate disposal of a community's residues, and too often, the engineer becomes "public enemy number one." Why is this?

Engineering is an old and honored profession. Some of the earliest engineers in the newly founded United States were our best minds and leaders. George Washington taught himself surveying and supervised the construction of roads, canals, and locks. Thomas Jefferson was an inveterate tinkerer, and some of his gadgets are clever even by modern mechanical engineering standards.

Engineers are indeed a special breed. A study performed some years ago evaluated the characteristics of students who entered college intending to study engineering. The study found that those students who continued in the engineering program after two years differed markedly from those who left engineering. Only certain persons choose to become engineers and undertake the rigors of the demanding college curriculum that allows entrance to the field. Engineering students and practicing engineers tend to have a very positive image of themselves and a lively *esprit de corps*. For them, engineering is *fun*.

Engineers also see themselves as performing a public service. The American Society of Civil Engineers motto describing the civil engineering profession as the "people-serving profession" neatly sums up engineers' perception of themselves. Engineers build civilizations. Engineers serve the public's needs.

But while engineers see themselves as successful problem solvers acting in the public interest, the public's perception of engineers is often somewhat different. Well-publicized engineering failures resulting in damage to human and environmental health have encouraged many people to perceive engineers as the creators, not solvers, of problems.

Part of the public attitude toward engineers might be explained by the view that engineers tend to conduct social experiments.[46] Engineers do not have all of the answers at hand when a problem arises and often cannot perform full-scale experiments to obtain such answers. For example, when designing a suspension bridge, engineers cannot construct a full-scale model of the bridge to test the design. Instead, they use the best knowledge and designs available and extrapolate, using sound judgment. Sometimes the extrapolation is faulty, and the experiment fails. Engineers learn from such failures and modify the design process in subsequent projects. What is important is that the experiments occur not in a private laboratory but on the public stage.

Engineering experiments affect humans in many ways. A road, for example, is both a technical and a social undertaking. The road might affect homes and businesses, create traffic and noise, or even divert funds from other projects. A new lawn mower is designed and manufactured, and it might cut grass effectively but at the expense of neighborhood quiet and perhaps even auditory damage to the user. A reservoir is constructed but at the cost to recreational whitewater rafting, despoliation of a bottomland virgin ecosystem, or the destruction of an ancestral homestead. A new biological weapon, designed to kill people, is developed with the knowledge that its use would devastate the global ecosystem. How is the engineer to deal with these conflicting public benefits?

The fundamental difference between how engineers see themselves and how the public sees them depends on how well engineers meet public concerns. Sometimes engineers are labeled the "tools of the establishment," the "despoilers of the environment," or the "diligent destroyers." With the public, it is often "us" against "them," and "them" is too often the engineers. Why are the engineers the villains?

The source of the problem is that engineers and the lay public hold different ethical attitudes. Because engineers tend to be utilitarians, they look at the overall and aggregate net benefits, thus diminishing the importance of harm to the individual. Because engineers are positivists, they tend to ignore or dismiss considerations for which reasons of a certain type cannot be given—that is, quantifiable or at least empirical data—thus ignoring intangibles. Engineers value people, of course, but they have a particular view of what is good for people and how this good is to be determined in a given case. Finally, engineers think of themselves as doing applied physical science, not applied social science. The physical science approach to engineering (ignoring the "people-serving profession" motto) allows engineers to think of their work as not being germane to the needs of society. This conflict of ethical outlooks is a root cause of much of the problem with engineers' interaction with the public.

References

1. O'Brien, J. 1977. Unpublished data. Durham, N.C.: Duke Environmental Center, Duke University.
2. Hershfield, S., P. A. Vesilind, and E. I. Pas. 1992. "Assessing the True Cost of Landfills." *Waste Management and Research* 10:471–484.
3. Nelson, A. C., J. Genereux, and M. M. Genereux. 1997. "Price Effects of Landfills on Different House Value Strata." *Journal of Urban Planning and Development*, ASCE, 123, no. 3:59–67.
4. Vesilind, P. A., and E. I. Pas. 1998. "Discussion of A. C. Nelson, J. Genereux, and M. M. Genereux, A Price Effect of Landfills on Different House Value Strata." *Journal of Urban Planning and Development*, ASCE, 123, no. 3:59–68.
5. Sleats, R., C. Harries, I. Viney, and J. F. Rees. 1989. "Activities and Distribution of Key Microbial Groups in Landfills." In *Sanitary Landfilling: Process, Technology and Environmental Impact.* London: Academic Press.
6. Pohland, F. G., and R. Englebrech. 1976. *Impact of Sanitary Landfills; An Overview of Environmental Factors and Control Alternatives.* New York: Report prepared for the American Paper Institute.
7. Fenn, D. G., K. J. Hanley, and T. V. DeGeare. 1975. *Use of the Water Balance Method for Predicting Leachate Generation from Solid Waste Disposal Sites.* EPA OSWMP, SW-168. Washington, D.C.

8. Remson, I. A., A. Fungaroli, and A. W. Lawrence. 1968. "Water Movement in an Unsaturated Sanitary Landfill." *Journal of the Sanitary Engineering Division*, ASCE, 94, no. SA2.

9. Reindl, J. 1977. "Managing Gas and Leachate Production on Landfills." *Solid Waste Management* (July): 30.

10. Othman, M. A., R. Bonaparte, B. A. Gross, and D. Warren. 1998. *Evaluation of Liquids Management Data for Double-Lined Landfills*. Draft Document Prepared for U.S. Environmental Protection Agency. Cincinnati, Ohio: National Risk Management Laboratory.

11. Thornthwaite, C. W., and J. R. Mather. 1957. *Instructions and Tables for Computing Potential Evapotranspiration and Water Balance*. Publ. Climatology, Drexel Institute of Technology, 10–185.

12. *Water Atlas*. 1998. Water Resources Information Service, USGS. Washington, D.C.

13. McGuinnes, J. L., and E. F. Bordne. 1972. *Comparisons of Lysimeter Derived Potential Evapotranspiration with Computed Values*. Agr. Res. Serv. Tech. Bull. 1452. Washington, D.C.: U.S. Dept. of Agriculture.

14. Schroeder, P. R., C. M. Lloyd, and P. A. Zappi. 1994. *The Hydrologic Evaluation of Landfill Performance (HELP) Model, User's Guide for Version 3*. EPA/600/R-94/168a.

15. Reinhart, D. R., and C. J. Grosh. 1997. "Analysis of Florida MSW Landfill Leachate Quality Data." Report to the Florida Center for Solid and Hazardous Waste Management (March).

16. Reinhart, D. R., and K. Lewis. 1994. "Beneficial Utilization of Landfill Gas." Report to the Florida Center for Solid and Hazardous Waste Management (September).

17. Young, A. 1989. "Mathematical Modeling of Landfill Gas Extraction." *Journal of Environmental Engineering* 115, no. 6:1073.

18. Nyns, E. J. 1992. *Landfill Gas: From Environment to Energy*, 1st ed. Luxembourg: Commission of the European Committee.

19. Augenstein, D., and J. Pacey. 1992. *Landfill Gas Energy Utilization: Technology Options and Case Studies*. EPA-600-R-92-116, EPA, Research Triangle Park.

20. Dessanti, D. J., and H. W. Peter. 1984. "Recovery of Methane from Landfill." *Proceedings* Symposium: Energy from Biomass and Wastes VIII:1053–1090.

21. Tchobanoglous, G., H. Theisen, and S. Vigil. 1993. *Integrated Solid Waste Management: Engineering Principles and Management Issues*, 3rd ed. New York: McGraw-Hill, Inc.

22. Reinhart, D. R., C. D. Cooper, and N. E. Ruiz. 1994. *Estimation of Landfill Gas Emissions at the Orange County Landfill*. Orlando: Civil and Environmental Engineering Department, University of Central Florida.

23. Reinhart, D. R., and T. Townsend. 1998. "Assessment of Leachate Collection System Clogging at Florida Municipal Solid Waste Landfills." Report to the Florida Center for Solid and Hazardous Waste Management (April).

24. *Criteria for Municipal Solid Waste Landfills*. 1988. Washington, D.C.: EPA. July.

25. McBean, E. A., F. G. Pohland, F. A. Rovers, and A. J. Crutcher. 1982. "Leachate Collection Design for Containment Landfills." *Journal of Environmental Engineering Division* ASCE 108, no. EE1:204–209.

26. McEnroe, B. M. 1993. "Maximum Saturated Depth over Landfill Liner." *Journal of Environmental Engineering* ASCE 119, no. 2:262–270.

27. Richardson, G., and A. Zhao. 2000. *Design of Lateral Drainage Systems for Landfills*. Baltimore: Tenax Corporation.

28. *Requirements for Hazardous Waste Landfill Design, Construction, and Closure.* 1989. EPA/625/4-89-022. Washington, D.C.: EPA.

29. Lema, J. M., R. Mendez, and R. Blazquez. 1988. "Characteristics of Landfill Leachates and Alternatives for Their Treatment: A Review." *Water, Air, and Soil Pollution* 40:223–250.

30. Chian, E. S. K., and F. B. DeWalle. 1977. *Evaluation of Leachate Treatment, Vol. 1, Characterization of Leachate.* EPA-600/2-77-186a. Cincinnati: EPA.

31. Cossu, R., R. Stegmann, G. Andreottola, and P. Cannas. 1989. "Biological Treatment." In *Sanitary Landfilling: Process, Technology and Environmental Impact,* ed. by T. H. Christensen, R. Cossu, and R. Stegmann, pp. 251–253.

32. Venkataramani, E. S., R. Ahlert, and P. Corbo. 1974. "Biological Treatment of Landfill Leachates." *CRC Critical Reviews in Environmental Control* 14, no. 4:333–376. Based on research conducted by W. C. Boyle and R. K. Ham. 1974. "Treatability of Leachate from Sanitary Landfills." *Journal Water Pollution Control Federation* 46, no. 5.

33. King, D., and A. Mureebe. 1992. "Leachate Management Successfully Implemented at Landfill." *Water Environment and Technology* 4, no. 9:18.

34. Harris, J. M., D. Purchewitz, and D. R. Reinhart. 1999. "Leachate Management for the New Millennium." *Proceedings 4th annual SWANA Landfill Management Symposium,* Denver.

35. Landtec. 1994. "Landfill Gas System Engineering Design: A Practical Approach." Courseware Landfill Gas System Engineering Design Seminar, Orlando.

36. Thorneloe, S. A. 1992. "Landfill Gas Utilization—Options, Benefits, and Barriers." *Proceedings* The Second United States Conference on Municipal Solid Waste Management, Arlington (June).

37. Wheless, E., S. Thalenberg, and M. Wong. "Making Landfill Gas into a Clean Vehicle Fuel." *Solid Waste Technologies* 7, no. 6:32.

38. Sandelli, G. J. 1992. *Demonstration of Fuel Cells to Recover Energy from Landfill Gas, Phase 1 Final Report: Conceptual Study.* EPA-600-R-92-007, EPA, Research Triangle Park.

39. Pacey, J. G., M. Doorn, and S. A. Thorneloe. "Landfill Gas Energy Utilization: Technical and Non-technical Considerations." *Proceedings 17th Annual Landfill Gas Symposium* GR-LG 0017, pp. 237–249. Long Beach, Calif.: Solid Waste Association of North America.

40. Thorneloe, S. A. 1992. "Landfill Gas Recovery/Utilization—Options and Economics." *Proceedings 16th Annual Conference on Energy From Biomass and Wastes* (March). Orlando, Fla.: Institute of Gas Technology.

41. Stahl, J., E. Wheless, and S. Thalenberg. "Compressed Landfill Gas as a Clean, Alternative Vehicle Fuel." *Proceedings 16th Annual Landfill Gas Symposium* GR-LG 0016, pp. 289–296. Louisville, Ky.: Solid Waste Association of North America.

42. Waner, J. A., C. A. Phillips, and N. A. Webster. 1994. "Methane Recovery Benefits Study Released." *Water, Environment & Technology* 4, no. 6:36.

43. Weand, B. L., and V. L. Hauser. 1997. "The Evapotranspiration Cover." *Environmental Protection* 8, no.1:40–42.

44. Sharp-Hansen, S., C. Travers, P. Hummel, and T. Allison. 1990. *A Subtitle D Landfill Application Manual for the Multimedia Exposure Assessment Model (MULTIMED).* Athens, Ga.: EPA.

45. Reinhart, D. R., and T. Townsend. 1997. *Landfill Bioreactor Design and Operation.* Boca Raton, Fla.: Lewis Publishers.

46. Martin, M. W., and R. Schinzinger. 1989. *Ethics in Engineering.* New York: McGraw-Hill.

47. Franklin Associates. 1999. *Characterization of Municipal Solid Waste in the United States: 1998 Update.* EPA 530, Washington, D.C.

48. Brunner, D. R., and D. J. Kelly. 1972. *Sanitary Landfill Design and Operation.* Washington, D.C.: EPA Office of Solid Waste Management.

Abbreviations Used in This Chapter

BDL = below detection limit
BOD = biochemical oxygen demand
COD = chemical oxygen demand
CWA = Clean Water Act
CO = carbon monoxide
CO_2 = carbon dioxide
EU = European Union
EPA = Environmental Protection Agency
GCL = geosynthetic clay liner
GPAD = gallons per acre per day
HDPE = high-density polyethylene
HELP = Hydrologic Evaluation of Landfill Performance (a computer model)
ID = inside diameter
LCS = leachate collection system
LDS = leachate detection system
LULU = locally undesirable land use
MBT = mechanical-biological treatment
MSW = municipal solid waste
MULTIMED = Multimedia Exposure Assessment Model

MVR = mechanical vapor recompression
MW = megawatt
NIMTO = Not In My Term of Office
NPDES = National Pollution Discharge Elimination System
NOPE = Not On Planet Earth
NO_x = nitrogen oxides
OA = osmotic agent
POTW = publicly owned treatment works
PVC = polyvinyl chloride
RCRA = Resources Conservation and Recovery Act
RO = reverse osmosis
TDS = total dissolved solids
USEPA = United States Environmental Protection Agency
VCD = vapor compression distillation
VOA = volatile organic acid
VOC = volatile organic compound

Problems

8-1. For your own community (assume appropriate quantities of refuse, densities, etc.), calculate how many years it would take to fill a football field 20 ft deep. Include the necessary earth cover.

8-2. Estimate the life of a landfill for a user population of 10,000. The available area for the area-type landfill is 10 acres. The water table is estimated at 20 ft below the ground surface. The pit must have a side slope of 1:3, and the final surface must have a slope of 1:4. Assume that the soil occupies 20% of the compacted volume.

8-3. Assume the following fraction, by weight, of material in a refuse:

Component	Percent, by weight
Food waste	10
Yard waste	20
Mixed paper	40
Corrugated cardboard	10
Glass	10
Ferrous (cans)	10

a. Estimate the in-place density of this refuse, using the bulk density values listed in Table 8-1.
b. If the following materials are recovered in a materials recovery operation, estimate the bulk density of the remaining refuse going to the landfill:

Component	Percent Recovered
Garbage	0
Yard waste	0
Mixed paper	50
Glass	60
Ferrous (cans)	95

c. If the life of the landfill for this community was originally calculated as 10 years, what is the expected life if the materials listed in part (b) are recovered?

8-4. You are the engineer hired by a community trying to site a modern (Subtitle D) landfill. At the public hearing, one of the neighbors of the proposed landfill asks you if her well water might become contaminated by the landfill leachate.
a. How would you answer her? Take your time and give a full answer.
b. The town hired you to site the landfill. Is the town your client, or do you have higher professional responsibility?

8-5. A landfill initially receives 1000 tons/day of waste. This receipt rate increases 10% per year. The landfill is closed at the end of the fourth year. Using the EPA exponential model, calculate the yearly gas production for 20 years after opening. What is the peak gas production in m^3/yr?

8-6. You are a municipal engineer who has just received a request from the city director of planning to comment on a proposed large multistory shopping center. You note that it is to be built on a site that you vaguely remember as having been a landfill many years ago. Nobody in the office remembers for sure when this was and what it was used for, when it closed, or how long it had been used. You do know that it has been closed for over 25 years. Write a letter to the planning director expressing your concerns and recommendations.

8-7. How does a modern (Subtitle D) landfill differ from the older existing landfills? List and explain all items that need to be considered in Subtitle D landfills that were not important in older landfills before the passage of RCRA legislation.

8-8. A community of 100,000 people decides to construct a landfill that is to have a volume of 2,000,000 cubic yards. How long can this landfill be expected to last? Use approximate numbers.

8-9. Determine the required spacing between pipes in a leachate collection system using a geonet and the following properties. Assume the most conservative design and that all stormwater from a 25-yr, 24-hr storm enters the leachate collection system.

Geonet transmissivity, $\theta = 10 \ cm^2/sec$
Design storm (25-year, 24-hour) = 8.2 in.
Drainage slope = 2%
Maximum design head on liner = 6 in.

8-10. Repeat Example 8-6 assuming that the highest flow rate in each collection leg is 750 cubic feet per minute at 115 °F.

8-11. A landfill is proposed for the area shown on the map in Figure 8-24. Why might this not be a good location? Where might be a better place? What

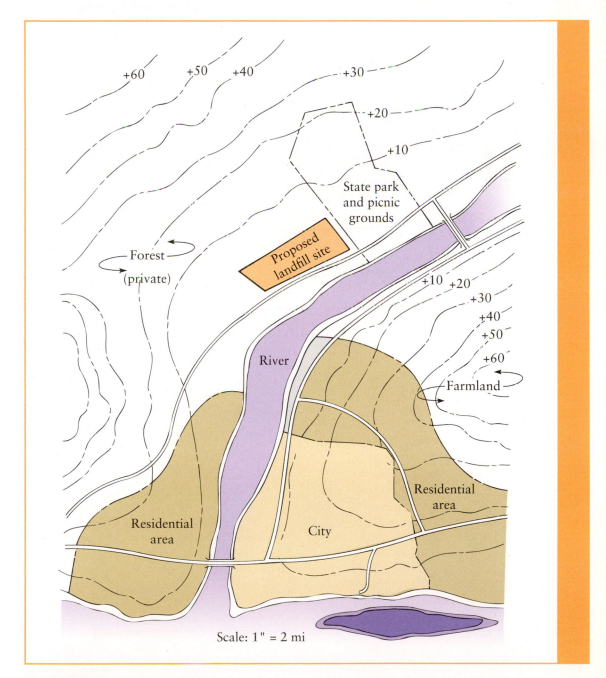

Figure 8-24 See Problem 8-11.

are the advantages and disadvantages of the new location? Some relevant facts: Groundwater at the proposed landfill site is 5 ft below surface; the soil is sand from −30 feet to +20 feet, and sandy loam above 20 feet.

8-12. A town of 100,000 people wants to construct a landfill on the site shown in Figure 8-25. Design such a landfill. Estimate the life of the landfill, and specify the equipment to be used,

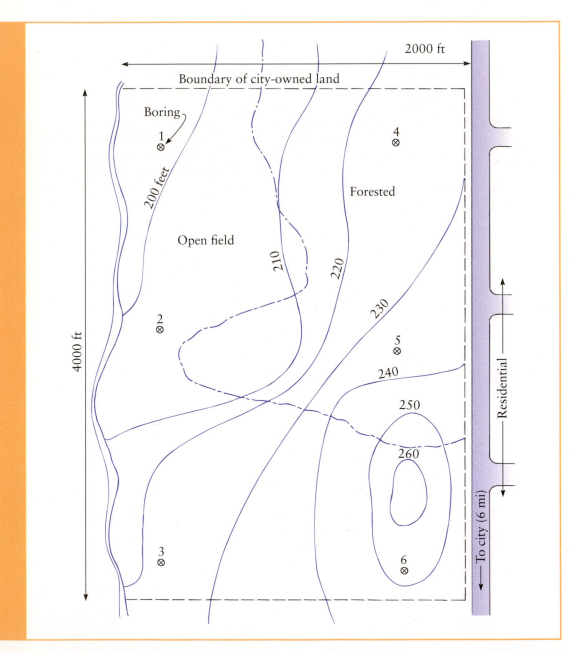

Figure 8-25 See Problem 8-12.

the method of operation, and other considerations necessary for acceptable operation. Draw in roads and any permanent installations on the map.

8-13. Describe, from a microbiological perspective, what goes on inside a sanitary landfill.

8-14. A town has hired a consulting firm to find a new landfill site. After it finds the ideal site, which is located 500 miles from the town, Megafill Inc. offers to transport and dispose of the waste for only $20 per ton more than the expected tipping fee at the new landfill.

a. What are some of the arguments in favor of, and against, the town signing up with Megafill?

b. To better understand the site, Megafill offers to fly you, the city's public works director, and the city council to the site in Megafill's corporate jet. Do you accept the offer? Why or why not?

8-15. What are the USEPA minimum design criteria for the two components that are in a composite landfill liner?

8-16. Which of the following are landfill design standards and which are landfill performance standards?

a. An operator must cover garbage with 6 in. of dirt each day.

b. There can be no off-site gas migration from a landfill.

c. No landfill odors can be detected at the boundary of the landfill.

d. A landfill must have gas collection wells every 40 ft.

CHAPTER 9

Toward Integrated Resources Management—Environmental, Political, and Economic Issues

While most of this book deals with the technical aspects of integrated solid waste management, many issues related to solid waste engineering are managerial, financial, regulatory, and even political. This chapter includes a number of current issues that influence solid waste engineering today and may have an increasing impact in the future.

9-1 LIFE CYCLE ANALYSIS AND MANAGEMENT

One means of understanding questions about material and product use and waste production is to conduct what has become known as a *life cycle assessment*. Such an assessment is a holistic approach to pollution prevention by analyzing the entire life of a product, process, or activity—encompassing raw materials, manufacturing, transportation, distribution, use, maintenance, recycling, and final disposal. In other words, assessing its *life cycle* should yield a complete picture of the environmental impact of a product.

The first step in a life cycle assessment is to gather data on the flow of a material through an identifiable society. Figure 9-1, for example, shows the flow of paper through Switzerland. Once the quantities of various components of such a flow are known, the environmental effect of each step in the production, manufacture, use, and recovery/disposal is estimated.

9-1-1 Life Cycle Analysis

Life cycle analyses are performed for several reasons, including the comparison of products for purchasing and the comparison of products by industry. In the former

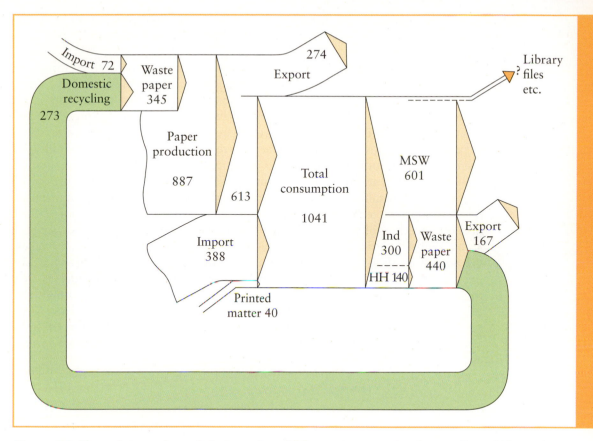

Figure 9-1 Flow of paper through Swiss society. All figures in tonnes/yr. Source: From "Alternative Methods for the Analysis of Municipal Solid Waste." Brunner, P. H., and W. R. Ernst, *Waste Management and Research* 4: 155, copyright © 1986 by SAGE Publications. Reprinted by Permission of SAGE.

case, the total environmental effect of glass returnable bottles, for example, could be compared to the environmental effect of non-recyclable plastic bottles. If all of the factors going into the manufacture, distribution, and disposal of both types of bottles are considered, one container might be shown to be clearly superior. In the case of comparing the products of an industry, we might determine if the use of phosphate builders in detergents is more detrimental than the use of substitutes, which have their own problems in treatment and disposal.

One problem with such studies is that they are often conducted by industry groups or individual corporations, and it should be no surprise that the results often promote their own product. For example, Procter & Gamble, the manufacturer of a popular brand of disposable baby diapers, found in a study conducted for P&G that cloth diapers consume three times more energy than the disposable kind. But a study by the National Association of Diaper Services found disposable diapers consume 70% more energy than cloth diapers. The difference was in the accounting procedure. If one uses the energy contained in the disposable diaper as recoverable in a waste-to-energy facility, then the disposable diaper is more energy efficient.[2]

Life cycle analyses also suffer from a dearth of data. Some of the information critical to the calculations is virtually impossible to obtain. For example, something as simple as the tonnage of solid waste collected in the United States is not readily calculable or measurable. And even if the data *were* there, the procedure suffers from the unavailability of a single accounting system. Is there an optimal level of pollution, or must all pollutants be removed 100% (a virtual impossibility)? If there are both air pollution and water pollution, how must these be compared?

This issue is schematically illustrated in Figure 9-2. For example, the cost to remove heavy metals from wastes increases with lower concentrations of metals to be recovered. The technology to reduce the metals in the waste gets more laborious as concentrations get lower. As a benefit, these metals may no longer pose an environmental risk but could be recycled. However, if no measures are taken, the potential risk that the metals pollute the environment in the future increases. In case of the landfill Kölliken, exorbitant costs were created in order to treat the wastes after they have been landfilled. The goal is to find the optimal level of treatment. But life cycle assessment that considers potential future risks is not straightforward. Hellweg et al. applied potential long-term emissions in life cycle assessment and also gave a comprehensive introduction to life cycle assessment and compared different waste treatment technologies.[21]

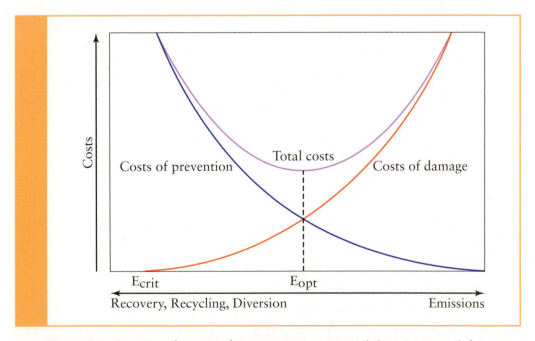

Figure 9-2 Costs as a function of increase in recovery and diversion rates (left *y*-axis) and of damage in case no measures to reduce emissions are taken (right *y*-axis). Adapted from source [20].

However, a simple example of the difficulties in life cycle analysis would be in finding the solution to the great coffee cup debate—whether to use paper coffee cups or polystyrene coffee cups. The answer most people would give is not to use either, but instead to rely on the permanent mug. But there nevertheless are times when disposable cups are necessary (e.g., in hospitals) and a decision must be made as to which type to choose.[4] So let us use life cycle analysis to make a decision.

The paper cup comes from trees, but the act of cutting trees results in environmental degradation. The foam cup comes from hydrocarbons such as oil and gas, and this also results in adverse environmental impact, including the use of nonrenewable resources. The production of the paper cup results in significant water pollution, while the production of the foam cup contributes essentially no water pollution. The production of the paper cup results in the emission of chlorine, chlorine dioxide, reduced sulfides, and particulates, while the production of the foam cup results in none of these. The paper cup does not require chlorofluorocarbons (CFCs), but neither do the newer foam cups ever since the CFCs in polystyrene were phased out. The foam cups, however, result in the emission of pentane, while the paper cup contributes none. From a materials separation perspective, the recyclability of the foam cup is much higher than the paper cup since the latter is made from several materials, including the plastic coating on the paper. However, many communities do not have recycling programs for polystyrene. Paper and foam cups both burn well, although the foam cup produces 17,200 Btu/lb (40,000 kJ/kg), while the paper cup produces only 8600 Btu/lb (20,000 kJ/kg). In the landfill, the paper cup degrades into CO_2 and CH_4 (both greenhouse gases) while the foam cup is inert. Since it is inert, it will remain in the landfill for a very long time, while the paper cup will eventually (but very slowly!) decompose. If the landfill is considered a waste storage receptacle, then the foam cup is superior, since it does not participate in the reaction, while the paper cup produces gases and probably leachate. If, on the other hand, the landfill is thought of as a treatment facility, then the foam cup is highly detrimental, since it does not biodegrade.

So which cup is better for the environment? If you wanted to do the right thing, which cup should you use? This question, like so many others in this book, is not an easy one to answer. Private individuals can, of course, practice pollution prevention by a simple expedient such as not using either plastic or paper disposable coffee cups but by using a refillable mug instead. The argument as to which kind of cup, plastic or paper, is better is then moot. It is better not to produce the waste in the first place. In addition, the coffee tastes better from a mug! We win by doing the right thing.

9-1-2 Life Cycle Management

Once the life cycle of a material or product has been analyzed, the next (engineering) step is to manage the life cycle. If the objective is to use the least energy and cause the least damage to the environment, then much of the onus is on the manufacturers of these products. The users can have the best intentions, but if the products are manufactured in such a way as to make meeting the goal impossible, then the fault is with the manufacturers. On the other hand, if the manufactured materials are easy to separate and recycle, then most likely energy is saved, and the environment is protected. This process has become known as *pollution prevention*. There are numerous examples of how industrial firms have reduced emissions or other waste production or have made it easy to recover waste products and, in the process, save money. Some automobile manufacturers, for example, are modularizing the engines so that junked parts can be easily reconditioned and reused. Printer cartridge manufacturers have found that refilling cartridges is far cheaper than remanufacturing them and now offer trade-ins. All of the efforts by industry to reduce waste (and save money in the process) will influence the solid waste stream in the future.

Bad examples abound of industrial neglect of environmental concerns for the sake of short-term economic gain. The use and manufacture of non-recyclable beverage containers, for example, is perhaps the most ubiquitous example. It might be instructive to establish some baseline of absolutely unconscionable industrial behavior in order to measure how far we have come in the pollution prevention process. The authors of this book recommend that there be an award established, called the *Trabi Award*, which could commemorate the worst, most environmentally unfriendly product ever manufactured. The Trabant, affectionately known as the Trabi, was manufactured in East Germany during the 1970s and 1980s. This homely looking car (Figure 9-3) was designed to be the East German version of the Volkswagen, "the People's Car." Its design objectives were to make it

Figure 9-3 The East German Trabant, or "Trabi." (Courtesy William A. Worrell)

as cheaply as possible, so the engineers used a two-stroke engine, which is a notoriously dirty engine, and its wide use caused massive air pollution problems. All of the components were designed at the least cost, and few survived normal use. Worst of all, the body was made of a fiberglass composite that was impossible to fix—except with duct tape! As far as the solid waste management was concerned, the Trabi had absolutely no recycling value, since it could not be melted down or reprocessed in any other way, nor could it be burned in incinerators. After the reunification of Germany, thousands of Trabis were abandoned on the streets and had to be disposed of in landfills. The Trabi is the best example of engineering design when the sole objective is production cost and environmental concerns are nonexistent. The Trabi deserves to be immortalized as an exemplar of environmentally destructive design.

9-1-3 Product Stewardship

Product stewardship, sometimes referred to as *producer responsibility*, is, according to the EPA, a product-centered approach to environmental protection. It calls on those in the product life cycle—manufacturers, retailers, users, and disposers—to share responsibility for reducing the environmental impacts of products. Instead of the traditional approach, where local government becomes responsible for a product at its end of life, product stewardship shares that responsibility among all parties. Typically, product stewardship programs focus on products that are difficult to dispose of such as batteries, fluorescent bulbs, medical waste, and paint.

On a national level, the best example of product stewardship is the Rechargeable Battery Recycling Corporation (RBRC). In 1995, members of the rechargeable battery industry established the RBRC. Joined by over 28,000 retailers, the RBRC launched a nationwide industry-funded recovery system to take back and recycle rechargeable batteries.

Product stewardship can also be implemented on a local level. In 2008, the San Luis Obispo County Integrated Waste Management Authority (IWMA) implemented an award-winning product stewardship program. The program was necessary, because on February 8, 2006, the Universal Waste Rule exemption, which allowed households and some small businesses to dispose of batteries, electronic devices, and fluorescent light bulbs in the landfill, expired. To meet these challenges, the IWMA needed a program that would be cost effective yet convenient for the public.

A retail take-back program replaced the collection of these materials at the existing permanent household hazardous waste collection facilities (PHHWCFs). While there are six PHHWCFs, these facilities are open only on Saturdays from 11 AM to 3 PM. In addition, the PHHWCFs are at locations such as landfills and wastewater treatment plants. While these facilities are adequate to accept the occasional household hazardous waste from the public, they are not convenient enough to serve as collection locations for household batteries and fluorescent tubes.

The IWMA selected the retail take-back program because it provided the public with very convenient locations to return household batteries, fluorescent tubes, latex paints, and sharps (needles). The retailers are located throughout the region and are typically open seven days per week for extended hours. In addition, the public shops at these locations when it is time to replace their household batteries and fluorescent tubes.

The IWMA's staff began setting up battery and fluorescent tube take-back retailers in December 2006. An initial list of potential take-back retailers was generated using local chamber of commerce and telephone directories. As part of meeting with each retailer and discussing the take-back program, each retailer was given battery and fluorescent tube collection containers. The size and number of the containers given to each retailer was contingent upon the size of the retailer. With full implementation 354 retail locations accept batteries and 111 retail locations accept fluorescent tubes. Over 50,000 fluorescent tubes and 1 million batteries are collected annually. Government, producers, and retailers have all contributed to the funding of these programs.

The IWMA initiated the battery and fluorescent tube retail take-back program as a free and voluntary program to the local retailers. While many retailers signed up for the program, the large national and international chains would not participate. This created an undue burden on those retailers who had agreed to participate. It was also very difficult to advertise the program to the public. The advertising program was based on the premise that you could return your old batteries and fluorescent tubes to any retailer who sold them.

Because of these problems, the IWMA adopted a mandatory ordinance in March 2008. The ordinance requires retailers to take back batteries and fluorescent tubes from the public under the following three scenarios:

1. From a consumer that buys batteries and/or fluorescent tubes (maximum amount equal to number of batteries and fluorescent tubes being purchased).
2. From a consumer who documents the previous purchase of batteries and/or fluorescent tubes (maximum amount limited to the number of batteries and fluorescent tubes previously purchased).
3. From a consumer who lives in the jurisdiction (maximum amount limited to 15 batteries and 8 fluorescent tubes per week).

Retailers are required to accept the items at no cost to the public. Within a month of adopting the ordinance, all the retailers who had previously refused to participate in the program were now participating.

9-1-4 Integrated Waste and Life Style Management

One of the major obstacles we still face in waste management is that waste problems only become visible at the end of the pipe. Only if the waste is not collected at our home or if littered trash is visible do we realize that there is obviously a problem. Our society has much of an "out of sight, out of mind" attitude. Many of the waste reduction initiatives start, therefore, at the end of the life cycle. However, upstream measures must be considered more seriously in the future, as wastes are also produced during mining and production. The contribution of refuse and recycled fractions only accounts for about 14% of the entire wastes produced from different sectors in OECD countries.[22]

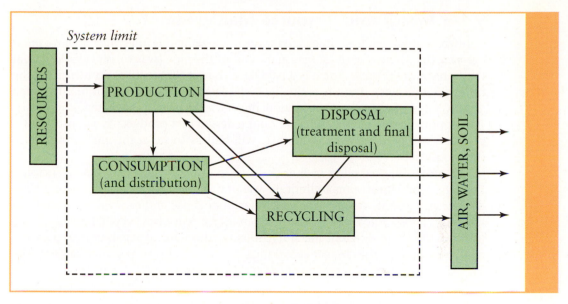

Figure 9-4 Scheme defining the integrated waste management (IWM) system and representing the relevant material flows, after Winzeler et al.[19]

Recovered and recycled waste flows need to be better integrated into all sectors of societal activities. A diagram of integrated waste management (IWM) is shown in Figure 9-4.

Another difficulty is that poorer countries have much smaller waste quantities,[22] although diversity and quantities are rapidly increasing in some developing countries. Moreover, the different waste fraction quantities differ considerably. The ideal waste management practice may therefore differ from country to country. Nevertheless, the approaches we have laid out in this textbook are generally valid and of use for the waste management engineer in the context of different waste management systems. As specific needs of the local population can be unique, bottom-up approaches can be very valuable for further developing waste management practices that fit the society. If successful, they will find further imitators worldwide.

Such a bottom-up movement is the Zero Waste Approach to Resource Management. Over the last two decades a new approach and understanding of solid waste management has been advocated by various groups. Grassroots movements introduced the zero waste approach, which was dismissed by the established waste management industry. However, those who supported zero waste continued to advocate for this approach. At one conference, lapel buttons were handed out that said, "If you are not for zero waste, how much waste are you for?" There was a blank line, and then you could write in a number. By 2015, many of those established groups were including zero waste in their discussion. A description of the zero waste approach is given in the appendix at the end of this chapter. Such an approach also calls on lifestyle changes for everyone in the society, independent of social status, education, income, or race.

9-1-5 Integrated Resource Management

A different approach from integrated waste management is *integrated resource management* (IRM). Instead of looking at the end-of-the pipe (waste), one would start at the front of the process. There is a paradigm change that focuses on increasing the resources' efficiency over the entire life cycle. This approach is not primarily aimed at a diversion of waste from the landfill, but at a reduction of the materials intensity. This approach puts more emphasize on the design of a product. As a consequence, products will be designed so that there will be less waste at the end of the life of the product.

Automobile recycling in Europe is an example of applying IRM principles. The European Union aims to limit the use of hazardous substances in new vehicle designs and also to design vehicles that facilitate reuse and recycling at the end of their life. Those automobile parts that cannot be reused or recycled are shredded in order to obtain different metal fractions for mechanical separation. Separation continues until it is no longer economical to recover and recycle the material. The EU target for materials recycling at the end of the life of a car is 85%.

9-2   FLOW CONTROL

The San Diego Materials Recovery Facility is a perfect example of how nontechnical issues often override successful engineering. This 550,000-ton-per-year mixed-waste materials recovery facility (dirty MRF) was built as a result of an agreement between a private company and the County of San Diego. When it went into operation, it met all of its performance requirements. Local communities, however, rather than send their municipal solid waste to the facility, elected to export their waste to remote landfills owned by their local haulers. Without sufficient waste to the MRF, the county closed the facility after less than two years of operation.[5] The county's inability or failure to control the solid waste stream was the principal reason why the plant closed. Restrictions or directives by local government that result in waste being transported only to designated facilities for handling, treatment, or disposal are generally referred to as *flow control*.

There are two facets to flow control: The first is the ability of government to direct MSW generated in a jurisdiction, and the second is to prevent waste generated outside your jurisdiction from entering a jurisdiction. For example, a local county may want to build a solid waste landfill for waste generated in the county but does not want to accept waste from outside the county. This example raises two legal issues that the U.S. Supreme Court has addressed in several decisions.

For many years, states and local communities struggled with the issue of legal authority to direct waste to a particular disposal or processing facility. At stake is a locality's ability to finance a solid waste disposal facility. For example, a waste-to-energy facility might cost $200 million, and this represents a large investment by the community. Before such a facility can be financed, the waste-to-energy facility owner must obtain "put-or-pay" garbage commitments from participating communities who agree to send their solid waste to that facility or pay if the waste is not delivered. Such agreements provide the security to the bond holders that the facility will be able to pay the projected debt service during the life of the bonds.

In *C&A Carbone, Inc. v. Clarkstown*, 511 U.S. 383 (1994), the Supreme Court ruled that a local ordinance directing haulers to deliver waste to a town-selected privately owned facility ran afoul of the Commerce Clause of the U.S. Constitution, which prohibits state and local legislation favoring certain private businesses over out-of-state competitors.

After *Carbone*, local government had two options to regain control of their waste. A community could enter into franchise agreement with a private hauler for the collection of solid waste. As part of the franchise agreement, the community could include the ability to direct solid waste to a specific facility. The second option is to replace the private collector system in the community with a municipal collection system. While both of these options are available, they may take as long as five years to implement. Thus, municipalities have been looking to Congress for relief but have not yet received it.

Some 13 years after *Carbone*, the high court again addressed flow control. In *United Haulers Association v. Oneida-Herkimer Solid Waste Management Authority*, 550 U.S. 330 (2007) by a 6–3 margin, the justices made a distinction between flow control to private facilities and flow control to public facilities. Important legal and policy considerations required treating public and private entities differently, the majority opinion said. "It is not the Office of the Commerce Clause to control the decision of the voters on whether government or the private sector should provide waste management services," 550 U.S., at 344. Flow control to public facilities does not discriminate against interstate commerce where all private companies—local, in-state, and out-of-state—are treated exactly the same.

The reverse side of flow control is the ability to restrict solid waste from entering a jurisdiction. In 1978, the Supreme Court ruled that the state of New Jersey unlawfully discriminated against interstate commerce by banning the disposal of out-of-state waste at all landfills. *City of Philadelphia v. New Jersey*, 437 U.S. 617.

States and localities get into trouble when they attempt disposal bans that restrict the ability of private landfills to accept waste from outside the jurisdiction or that impose higher fees on local disposal of waste generated elsewhere. Private facilities that challenge such regulations in court are likely to be successful because government officials can rarely, if ever, prove that the risks and hazards from non local wastes are worse than from wastes generated locally. See, for example, *Chemical Waste Management v. Hunt*, 504 U.S. 334 (1992); *BFI Medical Waste Systems v. Whatcom County*, 983 F.2d 911 (9th Cir. 1993). The practical implication of this restriction is that some states have become dumping grounds for other states. In 2013, 23% of the waste disposed of in Michigan came from out of state. In fact, 17% of all waste disposed of in Michigan came from Canada as allowed for under the North American Free Trade Agreement. Michigan landfills have only 28 years of remaining capacity, which is being used up by waste generated out of state. Consider the dilemma that a local recycling coordinator in Michigan must face trying to implement a recycling program to preserve landfill space when 23% of the waste is coming from out of state. However, a locality or public authority that seeks to forbid the disposal of non local wastes at its own facilities may do so. Government, acting as a private entrepreneur—that is, participating in the market—may choose its customers without creating any Commerce Clause issues. *Red River Service Corp. v. City of Minot*, 146 F.3d 583 (8th Cir. 1998).

9-3 PUBLIC OR PRIVATE OWNERSHIP AND OPERATION

The debate between public and private ownership and operation is a long one and will continue into the foreseeable future. One thing can be said with certainty: There is no one right answer that applies to all communities.

Collection of solid waste is considered an essential public service, and the question is whether this service should be provided by private haulers or public employees. The private garbage company will claim it is more efficient, less wasteful, and more motivated. The public sector will claim it is less costly because it pays no tax and does not make a profit, is more responsive to the public, and provides a living wage for its employees. There is some truth to both claims. In the final analysis, the municipality should consider cost, liability, and control when deciding on privatizing a solid waste system.[7]

A trend is to place collection services out to bid and let the municipality bid against the private companies. This approach is referred to as *competitive service delivery*. Unlike privatization, which turns the function over to the private sector, under competitive service delivery the public sector is allowed to compete for the project.[8]

While this approach appears to be good in principle, in reality it can be very difficult to implement. For example, once the public entity loses its collection system, it is almost impossible to get back into the business. The infrastructure necessary to provide the service—such as the personnel, equipment, and maintenance facilities—will no longer be available. Unlike private companies, government rules such as civil service and competitive procurement prevent local government from rapidly obtaining the resources necessary to provide collection services. In addition, public entities are subject to open-record requirements. Thus, the private companies bidding against a municipality may be able to view historical financial data prior to bidding.

Another concern is the consolidation of the private sector. As companies purchase each other, the number of viable competitors continues to decrease. This has been described as the monopolization of solid waste through mergers and acquisitions.[9] With this approach, companies can achieve *vertical integration*, whereby the company controls not only the collection of the waste but also the transfer operation, the landfill, and the recycling operation. This decrease in competition may lead to an increase in costs for collection services.

Public or private ownership of landfills is another current solid waste issue. A landfill provides an essential public health function and is needed by every municipality. The public perception based on such well-publicized incidents as "the garbage barge" is that there is a landfill shortage. At the time of the "garbage barge" incident, there was fear that landfill costs would soon exceed $100 per ton. In reality, no such shortage existed, and landfill prices dropped.

Public ownership does offer benefits to the local community. It can set its own rates, it can restrict the flow of MSW to the landfill, it can direct flow to the landfill, and it can decide on the level of service to be provided. On the other hand, a private landfill is free to compete in the open market and must offer competitive rates. One alternative that provides many benefits is to have the *operation* of a publicly owned landfill contracted out. This allows the local community to own an important asset while achieving the benefits of competition for the operation.

9-4   CONTRACTING FOR SOLID WASTE SERVICES

If a municipality does decide to contract with a private company for solid waste services, it will likely use a competitive bidding process to select the company. The municipality will prepare a *request for proposal* or **RFP**. In the RFP the municipality's level of solid waste service and terms and conditions for the contract or franchise agreement are described. While every RFP by its very nature is different, there are common elements that should be included. Those elements are described here.

Basic Service to Be Provided

The municipality must describe in detail the service included in the RFP. An example might be weekly curbside garbage collection of 12,000 residential customers with up to six 32-gallon cans of MSW per customer. This description specifies the following:

- Collection frequency—weekly
- Location of collection—curbside
- Material collected—refuse
- Number of customers—12,000
- Type of customer—residential
- Quantity of material—up to six 32-gallon cans
- Container provider—customer

Each of these factors influences the cost of providing service and must be carefully spelled out in the RFP.

Options

In addition to the basic service to be provided, the municipality can include other options that may or may not be part of the proposal. For example, a municipality may request that the company bill the customer directly, or the municipality may instead include the refuse collection bill with the water and sewer charges.

Term of Agreement

The cost of providing the requested service is directly related to the term of the agreement. Start-up costs and capital costs are amortized over the life of the agreement. Usually seven years is the minimum amount of time needed to fully amortize equipment. Thus, the longer the term of the agreement, the lower the annual cost.

Cost of Service

Every RFP must specify how the proposer is to provide the cost information. Some RFPs ask for a very detailed breakdown of costs, while others ask only for the total cost. In addition, the methodology used to adjust the costs during the term of the agreement should be specified.

Resources Provided

The RFP typically requires the proposer to provide a detailed description of the resources being provided. For example, the number and type of trucks being provided helps the municipality compare the proposals. In addition, RFPs may require the names and resumes of the key personnel who will be assigned to the project. A classic example of needing to know this information was a project in Florida in which a private company was to assume the operation of a waste-to-energy project. One company proposed a plant manager who was currently under indictment for violating an order of the fire marshal at the waste-to-energy plant that he was currently operating.

Company Experience

An RFP usually includes a requirement that the proposer include a list of similar projects and references. This allows the selecting agency to determine if the proposer is qualified to provide the required service.

Record of Violations

Proposers are asked to provide a list of violations and judgments against the company and its officers. This information is used by the municipality to determine if the company meets its standard for contracting. For example, if the company has defaulted on its last three municipal contracts or if the corporate officers have been convicted of bribing local officials, the company would probably not be selected for your contract. A record of past problems and poor performance may be sufficient reason for rejection. For example, the County of San Luis Obispo, California, rejected a large national firm because of its past practices. After being rejected, the firm appealed to the courts to overturn the decision but lost the appeal.

Financial Resources

The municipality must be assured that the proposer has the financial resources to complete the project. Thus, financial data are requested as part of the RFP. Examples might be audited annual financial statements or the most recent annual report.

Draft Agreement

A draft agreement should be included in the RFP. All proposers should be required to comment on the draft agreement and put any objections in writing prior to a selection being made. Finalizing the agreement with the selected company will be easier if the objections are known prior to the selection.

Standard Terms and Conditions

Most municipalities have a set of standard terms and conditions that are included in all RFPs. For example, the municipality has the right to reject all proposals, and the municipality will not reimburse any of the proposers for the cost of preparing their proposal.

In some cases, a municipality will decide not to use a competitive process but instead to negotiate an extension with the existing hauler or service provider. If the existing contractor has provided good service and if the price is reasonable, it can be argued that the municipality will save time and money by not going through a selection process.

While not going out to bid might save time and money in most cases, it is not always true. For example, a city in California decided not to go through a selection but instead to enter into a new 15-year agreement with the existing hauler. However, the opponents to this approach were successful in getting enough signatures to place an initiative on the ballot to require a competitive bid. This initiative became the most expensive election in the history of the city. The existing hauler spent $379,515 and the city spent $107,493 to defeat the initiative, while the proponents of competitive bidding spent $352,547.[10] In the end, the initiative was defeated, and the city entered into a 15-year agreement with the hauler as originally planned.

Agreement or Contract

If a municipality decides to contract for solid waste services, an *agreement* or *contract* will be required. Some agreements are as simple as several pages, while others require hundreds of pages. The agreement may be referred to as a *service agreement*, *franchise agreement*, or *contract* and may be either *exclusive* or *nonexclusive*. In an exclusive agreement, only the selected company can provide the specified service. Typically, residential garbage service is an exclusive service, because a municipality only wants the contracted company providing the service. In other cases (such as recycling) the agreement may be exclusive under certain circumstances but not under others. For example, the following is a section from the city of Arroyo Grande, California, recycling agreement:

> The Agreement for the Collection, processing and marketing of Recyclable Materials granted to Contractor shall be exclusive except as to the following categories of Recyclable Materials listed in this Section. The granting of this Agreement shall not preclude the categories of Recyclable Materials listed below from being delivered to and Collected and transported by others provided that nothing in this Agreement is intended to or shall be construed to excuse any person from obtaining any authorization from City which is otherwise required by law:
>
> A. Recyclable Materials separated from Solid Waste by the Waste Generator and for which Waste Generator sells or is otherwise compensated by a collector in a manner resulting in a net payment to the Waste Generator for such Recycling or related services;
>
> B. Recyclable Materials donated to a charitable, environmental or other nonprofit organization;
>
> C. Recyclable Materials which are separated at any Premises and which are transported by the owner or occupant of such Premises (or by his/her full-time employee) to a Facility;
>
> D. Other Governmental Agencies within the City which can contract for separate solid waste and recycling services; and,
>
> Contractor acknowledges and agrees that City may permit other Persons beside Contractor to Collect any or all types of the Recyclable Materials listed in this Section 4.2, without seeking or obtaining approval of Contractor under this Agreement.

This Agreement to Collect, transport, process, and market Recyclable Materials shall be interpreted to be consistent with state and federal laws, now and during the term of the Agreement, and the scope of this Agreement shall be limited by current and developing state and federal laws with regard to Recyclable Materials handling, Recyclable Materials flow control, and related doctrines. In the event that future interpretations of current law, enactment or developing legal trends limit the ability of the City to lawfully provide for the scope of services as specifically set forth herein, Contractor agrees that the scope of the Agreement will be limited to those services and materials which may be lawfully provided for under this Agreement, and that the City shall not be responsible for any lost profits and/or damages claimed by the Contractor as a result of changes in law.

Some municipalities that enter into franchise agreements with waste haulers will require that a *franchise fee* be paid by the waste hauler. This fee is in addition to the cost incurred by the selected company to provide the specified service. Franchise fees may be as little as 1 or 2% or in excess of 20%. In some cases, the franchise fee is used to support related activities (such as recycling centers) while in other cases, the fee goes to the general fund of the municipality.

9-5 FINANCING SOLID WASTE FACILITIES

Solid waste processing and disposal facilities can be either privately or publicly owned. In either case, the owner must minimize cost at a given level of service. Because solid waste facilities are most often long-term investments, the time value of money is important, and engineering economics plays a major role in deciding what kind of facility will be constructed. Funding of solid waste operations is similar to funding other utilities, such as water and sewerage services. Budgets consist of two general components: revenues and costs. Depending on the specifics of the solid waste system, such as ownership and contractual arrangements, the complexity of the financial system can vary. Some systems are complex and require financial experts, while others may require only a simple accounting system.

Solid waste operations receive revenue from the sale of services and goods. When the home or business pays its monthly garbage bill, these funds provide the revenue to cover the costs of the operation. Other sources of revenue are tipping fees at the landfill, the sale of recyclables, and the sale of electricity from a landfill-gas turbine or waste-to-energy plant. In a publicly owned system, the revenue should equal the cost. In a privately owned system, the revenue should exceed the cost, resulting in a profit.

A central idea in engineering economics is the time value of money. A dollar today does not have the same value as a dollar a year from now. Ignoring inflation for now, a dollar today can be invested in an interest-bearing account and a year from now will be worth a dollar plus the interest earned. Thus, a dollar today and a dollar a year from now cannot be added to get two dollars. They are as different as apples and oranges.

The initial cost of the facility is very important to the municipality or agency. Such a cost is known as a *capital cost* and is a one-time investment. Capital costs are paid from the proceeds of bank loans, general obligation bonds, and revenue bonds. It is common for a solid waste company to arrange for a bank loan to buy a garbage truck. The term of the loan is shorter than the life of the truck. The interest rate on the loan is based on the risk the lender perceives and includes such variables as the existence of a franchise agreement and the net worth of the company.

Municipal government can use *general obligation bonds* to finance capital projects. These bonds are based on the full faith and credit of the government. In addition, the interest paid to the bond holder is usually tax exempt, and thus, the interest rate on these bonds is low. One problem with bonds of this type is that they may require a vote of the citizens before they can be issued.

Municipal government can also use *revenue bonds* to finance capital projects. A revenue bond is guaranteed by the project. For example, a landfill may be funded by a revenue bond, and the revenue from the landfill (such as tipping fees) would be used to pay off the bond. Unlike a general obligation bond, these bonds have more risk because only the project revenue is pledged—not the full faith and credit of the municipal government. Thus, these bonds have a higher interest rate and result in a higher cost to the municipality. The interest rate is generally still tax free, and thus, the rates are lower than normal bank loans.

One variation to the revenue bond is the ability of private companies to access revenue bond financing through government. This allows a private company to own the capital project while at the same time use tax-exempt financing and receive a low interest rate. In addition, the private company can depreciate the assets and reduce its tax bill. Many waste-to-energy facilities have been built using these bonds. The companies may also receive accelerated depreciation and investment tax credits for the project. In 1984, the federal government limited the amount of private projects that could be financed using the tax-exempt bonds; now solid waste projects must compete with other beneficial projects, such as low-cost housing, for this limited amount of funding.

Another complication is that public facilities require not only the capital cost for their construction but also a yearly cost for their operation and maintenance (O&M). The latter are called the *O&M costs* and include such items as salaries, replacement parts, service, fuel, and the many other costs incurred by the facilities. Each year the elected body must approve the expenditure, so it can be difficult to guarantee multiyear financing.

Because of the time value of money, estimating the real cost of municipal facilities (such as solid waste collection and disposal operations) can be tricky. Economists and engineers use two techniques to "normalize" the dollars so that a true estimate of the cost of multiyear investment can be calculated. The first technique is to compare the costs of alternatives on the basis of *annual cost*, while the second calculates *present worth*. Both include the capital cost plus the O&M cost.

9-5-1 Calculating Annual Cost

The capital costs of competing facilities can be estimated by calculating the cost that the municipality or agency would incur if it were to pay interest on a loan of that amount. Calculating the annual cost of a capital investment is exactly like calculating the annual cost of a mortgage on a house. The owner (municipality or agency) borrows the money and then has to pay it back in a number of equal installments.

Table 9-1 6.125% Compound Interest Factors

Years	CRF	PWF	SFF
1	1.06125	0.942	1.00000
2	0.54639	1.830	0.48514
3	0.37497	2.666	0.31372
4	0.28941	3.455	0.22816
5	0.23820	4.198	0.17695
6	0.20416	4.898	0.14219
7	0.17993	5.557	0.11868
8	0.16183	6.179	0.10058
9	0.14782	6.764	0.08657
10	0.13667	7.316	0.07452
11	0.12760	7.836	0.06635
12	0.12009	8.326	0.05884
13	0.11378	8.788	0.05253
14	0.10841	9.223	0.04716
15	0.10380	9.633	0.04255

If the owner borrows X dollars and intends to pay back the loan in n installments at an interest rate of i, each installment can be calculated as

$$Y = \left[\frac{i(1 + i)^n}{(1 + i)^n - 1} \right] X$$

where

Y = the installment cost, \$
i = annual interest rate, as a fraction
n = number of installments
X = the amount borrowed, \$

The expression

$$\left[\frac{i(1 + i)^n}{(1 + i)^n - 1} \right]$$

is known as the **capital recovery factor** or CRF. The capital recovery factor does not have to be calculated, since it can be found in interest tables or is programmed into hand-held calculators. Table 9-1 shows the capital recovery factors if the interest is 6.125%.

EXAMPLE 9-1

A town wants to buy a refuse collection truck that has an expected life of 10 years. It wants to borrow the $150,000 cost of the truck and pay this back in 10 annual payments. The interest rate is 6.125%. How much are the annual installments on this capital expense?

> ### SOLUTION
>
> From Table 9-1, the capital recovery factor (CRF) for $n = 10$ is 0.13667. The annual cost to the town would then be $0.13667 \times \$150,000 = \$20,500$. That is, the town would have to pay \$20,500 each year for 10 years to pay back the loan on this truck. Note that this truck does not cost $10 \times \$20,500 = \$205,000$, because the dollars for each year are different and cannot be added.

9-5-2 Calculating Present Worth

An alternative method for estimating the actual cost of a capital investment is to figure present worth: how much one would have to invest right now, Y dollars, at some interest rate i so that one could have available X dollars every year for n years. This method is

$$Y = \left[\frac{1}{(1 + i)^n} \right] X$$

where

Y = the amount that has to be invested, \$
i = annual interest rate
n = number of years
X = amount available every year, \$

The term

$$\left[\frac{1}{(1 + i)^n} \right]$$

is called the *present worth factor* (PWF).

> ### EXAMPLE 9-2
>
> A town wants to invest money in an account drawing 6.125% interest so that it can withdraw \$20,500 every year for the next 10 years. How much must be invested?
>
> ### SOLUTION
>
> From Table 9-1, the present worth factor (PWF) for $n = 10$ is 7.316. Thus, the money required is $7.316 \times \$20,500 = \$150,000$.

9-5-3 Calculating Sinking Funds

In some situations, the municipality or agency must save money by investing it so that at some later date a specified sum of the money would be available. In solid waste engineering, this most often occurs when the landfill owner must invest money during the active life of the landfill so that, when the landfill is full, there are sufficient funds to place the final cover on the landfill. Such funds are known as *sinking funds*. It is necessary to calculate the funds Y necessary to be invested in an account that draws i percent interest so that at the end of n years the fund has X dollars in it. The calculation is

$$Y = \left[\frac{i}{(1 + i)^n - 1}\right]X$$

and the term

$$\left[\frac{i}{(1 + i)^n - 1}\right]$$

is known as the *sinking fund factor* (SFF).

EXAMPLE 9-3

A community wants to have $500,000 available at the end of 10 years by investing annually into an interest-bearing fund that yields 6.125% interest. How much would it have to invest annually?

SOLUTION

From Table 9-1, the sinking fund factor (SFF) at 10 years is 0.07452, and the required annual investment is therefore 0.07452 × $500,000 = $37,260.

Note again that 10 × $37,260 = $372,600, which is considerably less than $500,000. The reason is that the investments during the early years are drawing interest and adding to the sum available.

9-5-4 Calculating Capital Plus O&M Costs

Capital investments require not only the payment of the loan in regular installments but also the upkeep and repair of the facility. The total cost to the community is the sum of the annual payback of the capital cost plus the operating and maintenance cost.

EXAMPLE 9-4

A community wants to buy a refuse collection vehicle that has an expected life of 10 years and costs $150,000. It chooses to pay back the loan in 10 annual installments at an interest rate of 6.125%. The cost of operating the truck (gas, oil, service) is $20,000 per year. How much will this vehicle cost the community every year?

SOLUTION

From Table 9-1, the capital recovery factor for $n = 10$ is 0.13667, so the annual cost of the capital investment is $0.13667 \times \$150,000 = \$20,500$. The operating cost is \$20,000, so the total annual cost to the community is $\$20,500 + \$20,000 = \$40,500$.

9-5-5 Comparing Alternatives

One of the main uses of engineering economics is to fund the lowest-cost solution to a community need. The foregoing analysis for a single truck can be applied equally well to alternative vehicles, and the annual costs (capital plus O&M) can be compared.

EXAMPLE 9-5

As an alternative to the truck analyzed in Example 9-4, suppose the community can purchase another truck that costs \$200,000 initially but has a lower O&M cost of \$12,000. Which truck will be the most economical for the community?

SOLUTION

The annualized capital cost for this truck is $0.13667 \times \$200,000 = \$27,300$. Adding the O&M cost of \$12,000, the total annual cost to the community is \$39,300. Comparing this to the total cost calculated in Example 9-4, it appears that the \$200,000 truck is actually less expensive for the community to own and operate.

9-6 HAZARDOUS MATERIALS

Increased technological complexity and population densities have resulted in the identification of a new type of pollutant, commonly called a *hazardous substance*. The reported incidence of damage to the environment and to people by these materials has increased markedly in the last few years. The EPA maintains a list of such incidents, and some of the better documented ones have been published.[11]

The term *hazardous substance* or *hazardous waste* is difficult to define, and yet a clear definition is necessary if specialized disposal standards are to be applied to such materials. A legal definition[12] of a hazardous waste suggested by the EPA is

> … any waste or combination of wastes of a solid, liquid, contained gaseous, or semi-solid form which because of its quantity, concentration, or physical, chemical, or infectious characteristics, may (1) cause or significantly contribute to an increase in mortality or an increase in serious irreversible or

incapacitating reversible illness; or (2) pose a substantial present or potential hazard to human health or the environment when improperly treated, stored, transported or disposed of, or otherwise managed.

When deciding whether or not a specific substance is hazardous, it is useful to use a set of criteria against which the properties of the material in question can be judged. The EPA has defined hazardous materials in two ways: (1) the chemical is *listed* as a hazardous material, or (2) the material fails one of six tests that then define it as a hazardous material. The listing includes specific chemicals such as chlorinated pesticides, organic solvents, and over 50,000 others. The process used is generally that the EPA lists the material and then waits for a challenge from an interested party. If the interested party conducts tests that show the material not failing any of the six tests, it is *delisted*, making it possible to dispose of the material in ordinary landfills. All listed materials must go to specially constructed hazardous waste landfills or other treatment centers at many times the cost of nonhazardous disposal.

The six tests used to define a hazardous material are given here.

- *Radioactivity*. The stipulation is that the levels of radioactivity not exceed maximum permissible concentration levels as set by the Nuclear Regulatory Commission.

- *Bioconcentration*. This criterion captures many chemicals such as chlorinated hydrocarbon pesticides.

- *Flammability*. This standard is based on the National Fire Protection Association test for how easily a certain substance will catch fire and sustain combustion.

- *Reactivity*. Some chemicals, such as sodium, are extremely reactive if brought into contact with water.

- *Toxicity*. The criterion for toxicity is based on LD_{50} (lethal dose 50), or that dose at which 50% of the test species (e.g., rats) die when exposed to the chemical through a route other than respiration. Inhalation and dermal toxicity is next, where LC_{50} is the lethal concentration resulting in 50% mortality during an exposure time of 4 hr. Dermal irritation is measured on a Federal Drug Administration scale of 1 to 10. A grade 8 irritant causes necrosis of the skin when a 1% solution is applied. Aquatic toxicity is measured by the 96-h mean toxic limit of less than 1000 mg/liter, although the present EPA criterion for aquatic toxicity is set at 500 ppm.[13] Phytotoxicity is the ability to cause poisonous or toxic reactions in plants based on the mean inhibitory limit of 1000 ppm or less. Since the publication of these criteria, the EPA has lowered its definition of phytotoxic poisons to 100 ppm.

- *Genetic, Carcinogenic, Mutagenic, and Teratogenic Potential*. These are all measured by tests developed by the National Cancer Institute.

Note that with this system of defining what is and is not hazardous, the actual quantity of the material is not specified. This is unreasonable when disposal schemes are to be evaluated.

The most environmentally sound disposal scheme for hazardous materials (at any quantity) is destruction and conversion to nonhazardous substances. In many cases, however, this is either expensive (e.g., dilute heavy metal and pesticide wastes) or technically impossible (e.g., some radioactive waste materials).

Alternative disposal schemes are thus necessary. However, most household hazardous waste is ether treated or incinerated.

A widely used method of disposing of hazardous waste substances is the hazardous waste landfill, which is used to provide complete long-term protection for the quality of surface and subsurface waters from hazardous wastes deposited in the landfill as well as to prevent other public health and environmental problems. The hazardous waste landfill differs from the ordinary sanitary landfill, primarily in the degree of care taken to ensure minimal environmental impact. Clay liners, monitoring wells, and groundwater barriers are some of the techniques used in such landfills. The overall philosophy is strict segregation from the environment. It should be noted that just as is the case with aboveground storage, such landfills are usually not strictly disposal schemes, but rather holding operations. True disposal is still willed to future generations.

9-7 ENVIRONMENTAL JUSTICE

The EPA defines environmental justice as the fair treatment and meaningful involvement of all people regardless of race, color, national origin, or income with respect to the development, implementation, and enforcement of environmental laws, regulations, and policies. EPA has set this goal for all communities and persons across the United States. The goal will be achieved when everyone enjoys the same degree of protection from environmental and health hazards and has equal access to the decision-making process in the quest for a healthy environment in which to live, learn, and work.

In the solid waste field, the issue of environmental justice is an important factor because of the size and potential impact of solid waste facilities, such as landfills, transfer stations, and equipment yards. For example, in 2008, California adopted Senate Bill 826, which required the adoption of state minimum standards to identify and mitigate environmental justice impacts in disproportionately affected communities in which solid waste facilities are located, including providing advance notice regarding permitting or enforcement and specified mitigation measures.

Laws, such as the one passed in California, are needed because of a history of siting undesirable facilities in communities of color and low wealth. In the absence of action to promote environmental justice, the continued need for new facilities could exacerbate this environmental injustice.[17] In one classic case of taking advantage of less-advantaged people, the town council of Chapel Hill, North Carolina, made a solemn promise to a minority neighborhood that as soon as the new landfill to be sited near their homes was full, the town would find another location in another part of the county and turn the old landfill into a community park. The landfill was finally closed 20 years later (at least 10 years after it was originally supposed to have been capped). The new town administration claimed that these promises to the neighborhood were null and void because they had no legal standing. The members of the present council also argued that they personally did not make these promises and therefore they should not be bound by promises made by previous town councils![18]

9-8   THE ROLE OF THE SOLID WASTE ENGINEER

The "good old days" in solid waste management included open dumps that were routinely set on fire. These dumps polluted the groundwater, provided no landfill gas control, and had unsafe working conditions resulting in numerous injuries—but all at low cost. While the public liked the low cost for garbage disposal, they were not aware of the many problems these practices caused. The solid waste engineer has been responsible for transforming the industry into a professional field with best practices. Today, open dumps have been replaced by sanitary land-fills with gas-control leachate collection systems. Garbage incinerators have been closed, and modern waste-to-energy plants with state-of-the-art air pollution control equipment have replaced them. Collection has evolved from putting the garbage into a can to a series of recycling bins, yard waste cans, and waste oil containers. Household hazardous waste is managed separately from refuse. While all of these changes have been beneficial to society and the environment, they have resulted in an increased cost to the public. Thus, on one hand, the solid waste engineer is hailed for bringing improvements to solid waste management; on the other hand, the engineer is blamed for increasing the cost of solid waste management.

To be effective, the solid waste engineer must be competent in three areas: technology, regulations, and public communications. These three areas are like the three legs on a stool. Without all three the stool will fall over.

Technology in the solid waste field is constantly evolving. For example, new landfill liners are being developed for different applications. Materials recovery facilities are being reengineered to process different types of waste streams. Air pollution control equipment has become so extensive that it can be larger and more complex than the actual furnace in a waste-to-energy facility.

Regulations become more stringent and complex every year. Regulations pertaining to solid waste management are issued by federal, state, regional, and local agencies. In many cases, regulations from various agencies conflict with each other. For example, landfills must have caps that restrict the infiltration of water into the landfill, but at the same time, landfills must have gas extraction wells that penetrate the cap and could allow for the infiltration of surface water into the landfill.

Communication is integrally important in solid waste engineering. The engineer must be able to communicate with the public and at the same time have the technical and regulatory knowledge to develop effective solid waste systems. If the engineer cannot convey that information to the public, including such decision makers as elected officials, then projects will not be implemented. For example, an engineer must be able to explain what it means when a health risk assessment determines that the waste-to-energy plant results in a 37 in one million increased risk of cancer.[14]

Integrated solid waste management services and supporting facilities, by their very nature, are public facilities and services. The public will use them, see them, and be affected by them. Because of the public nature of these facilities and services, the solid waste engineer cannot forget that it is the elected officials who represent the public. Thus, those decisions that impact the public are best made by an informed elected official. More than one solid waste engineer has had a short career because the engineer forgot whom the public elected to make the ultimate decision.

9-9   FINAL THOUGHTS

The civility of our society depends on all of us agreeing to abide by good manners, high moral standards, and the law of the land. Given the choice, we all would want to live in a society where everyone agrees to abide by these conventions because doing so benefits everyone. This argument is applicable to professional engineering as well. To help maintain a viable engineering profession, we should demonstrate good professional manners ourselves, and if the occasion requires, admonish others for boorish behavior. We should act as role models in conducting engineering on a high moral level and promote such behavior in others. And without doubt, we should not become criminals. In short, we all have a responsibility to uphold the honor of professional engineering and to create a culture we all desire and in which we can all flourish.

But there is a larger question of why any of us should *act* in such a way. That is, if we find that having bad manners, or acting immorally, or even breaking a law is advantageous to us individually, why should we (at any given moment) not *act* in a manner that we would not necessarily want others to emulate?

The obvious answer to that question is that we don't want to get caught and suffer the consequences. Bad manners would subject us to ridicule; immoral conduct might cause us to be ostracized by others, or we might lose clients and business; and, of course, breaking a law might result in a fine or jail time. But consider now the possibility that we would not get caught and could not suffer any adverse consequences. Of course, we never know for sure that we will not get caught, but for the sake of argument, let's assume this extreme case. We have it all figured out, and it simply is impossible for us to get caught being ill-mannered, immoral, or illegal. Why should we, all things considered, still act as honorable engineers, especially if doing so might involve financial cost to us or in some other way cause us harm? The answer comes in three parts.

First, we are all members of a larger community—in this case, the engineering community—and we all benefit from this association. Acting in a manner that brings harm or discredit to this community cannot be beneficial to us in the long run. Granted, the destruction of professional engineering may be far into the future, and our small antisocial act would not be enough to destroy the profession. But we (along with all our contemporaries) have an obligation to uphold the integrity of the profession, eventually for our own good. Engineers should act honorably because the profession depends on them to do so.

Second, the antisocial act—even though we might get away with it—takes something out of us. There are circumstances where it clearly is better to lie than to tell the truth (such as saving an innocent life, for example), but all lies come at a cost to the teller.[15] There is, as it were, a reservoir of good in each human, and this can be nibbled away one justified lie at a time until the person is incapable of differentiating between lying and being truthful. This would also be true for bad manners and illegal acts. Every time we get away with something, we reduce our own standing as honorable human beings. Engaging in untruthful engineering reduces our own standing as professionals.

Finally, the reason for not being antisocial, even assuming we could get away with it, is that eventually our conscience would not stand for it. We all have a conscience within us that tells us the difference between right and wrong. Most of us, when we do antisocial things, *know* we are behaving badly and eventually regret such actions.[16]

But if by telling lies and becoming a scoundrel one gains materially, why would this be an undesirable result? If an engineer ran an engineering practice where she made it a habit to lie to clients, why is this detrimental? Why will telling lies (lies that one continues to get away with) necessarily be a bad thing?

The truth is that engineers who behave without regard to manners, morals, or laws will eventually cause harm to themselves. They *will* lose clients, their untruths *will* cause their works to fail, and they *will* have a bad conscience that will bother them. They will eventually think poorly of their own standing in the profession and regret their self-serving actions that may have been ill-mannered, immoral, or illegal.

So why be a good engineer? We might get caught if we don't behave honorably; we have a common responsibility to the professional engineering community; we lose something of our own integrity when we behave badly; and we have a conscience. But what if, in the face of these arguments, we still are not convinced? There appear to be no knock-down ethical arguments available to make us change the mind of a person set on behaving badly. We have the human option to act in any way we wish. But if we have bad manners, act immorally, or break the laws, we are not behaving honorably, and eventually, we will be harmed by our regrets for acting in this manner. That is, such behavior will always result in harm to us as professional engineers.

While the Viking society of northern Europe was in many ways cruel and crude, the Vikings had a very simple code of honor. Their goal was to live their lives so that when they died, others would say "He was a good man." The definition of what they meant by a "good man" might be quite different by contemporary standards, but the principle is important. If we live our professional engineering lives so as to uphold the exemplary values of engineering, the greatest professional honor any of us could receive would be to be remembered as a *good* engineer.

9-10 EPILOGUE

Well, dear reader, you have made it to the end of the book. The authors, with over 80 years of experience in solid waste management between them, have tried to share with you their collective knowledge. For some of you, the only future solid waste practice will be taking out the garbage. If that is the case, at least you will know that it does not just disappear when it goes into that truck. For others, this introduction may lead to a career in some aspect of solid waste management. For those readers, your journey is just beginning.

References

1. Brunner, P. H., and W. R. Ernst. 1986. "Alternative Methods for the Analysis of Municipal Solid Waste." *Waste Management and Research* 4:155.

2. "Life Cycle Analysis Measures Greenness, but Results May Not Be Black and White." 1991. *Wall Street Journal* (28 February).

3. Solano, E., R. D. Dumas, K. W. Harrison, S. Ranjithan, M. A. Barlaz, and E. D. Brill. "Integrated Solid Waste Management Using a Life-Cycle Methodology for Considering Cost, Energy, and Environmental Emissions—2. Illustrative Applications." Department of Civil Engineering, North Carolina State University.

4. Hocking, M. B. 1991. "Paper versus Polystyrene: A Complex Choice." *Science* 251, 1 (February).

5. Worrell, W. A. 1999. *The San Diego, California Mixed Waste Materials Recovery Facility: Technological Success, Political Failure.* R'99 Recovery, Recycling, Reintegration Proceeding, Volume 1. Gallon, Switzerland: EMPA.

6. *Waste News* 5, Issue 38 (February 7, 2000).

7. *County of San Diego Privatization Study.* 1991. Deloitte and Touche and R. W. Beck (October 16).

8. *The Local Government Guide to Solid Waste Competitive Service Delivery.* 1995. Public Technology Inc.

9. Biering, R. A. 1999. "The Art of Saying 'No' or Bambi Meets Godzilla." *Proceedings WasteCon 1999.* Solid Waste Association of North America.

10. *The San Diego Union-Tribune*, February 12, 2000.

11. Wentz, C. A. 1989. *Hazardous Waste Management.* New York: McGraw-Hill Publishing Co.

12. Blackman, W. C. 1993. *Basic Hazardous Waste Management.* Boca Raton, Fla: Lewis Publishers.

13. LaGrega, M. C., P. L. Buckingham, and J. C. Evans. 1994. *Hazardous Waste Management.* New York: McGraw-Hill Publishing Co.

14. *Health Risk Assessment for the Resource Recovery Facility Boiler No. 1.* 1989. Department of Solid Waste Management, Dade County, Florida. Malcome Pirnie (May).

15. Bok, S. 1978. *Lying: Moral Choice in Public and Private Life.* New York: Pantheon.

16. Pritchard, M. 1991. *On Being Responsible.* Lawrence, Kans.: University Press of Kansas.

17. Norton, J. M., S. Wing, H. J. Lipscomb, Jay S. Kaufman, Stephen W. Marshall, and A. J. Cravey. "Race, Wealth, and Solid Waste Facilities in North Carolina." Environmental Health Perspectives. September 2007.

18. Azar, S. 1998. *The Proposed Eubanks Landfill: The Ramifications of a Broken Promise.* Senior Independent Study. Durham, N.C.: Duke University Department of Civil and Environmental Engineering.

19. Winzeler, R., P. Hofer, and Morf, L., 2003. *Towards Sustainable Waste Management, in Municipal Solid Waste Treatment,* Eds: Chr. Ludwig, S. Hellweg, S. Stucki, Springer, 462–513.

20. Brunner, P. H., and H. Rechberger, 2003. *Practical Handbook of Material Flow Analysis* (Advanced Methods in Resource & Waste Management).

21. Hellweg, S., G. Doka, G. Finnveden, and K. Hungerbühler, 2003. "Ecology: Which Technologies Perform Best?," in: *Municipal Solid Waste Management,* Eds. C. Ludwig, S. Hellweg, and S. Stucki, Springer, 350–404.

22. Stucki, S., C. Ludwig, and J. Wochele, 2003. "The Diversity of Municipal Solid Waste," in: *Municipal Solid Waste Management,* Eds. C. Ludwig, S. Hellweg, and S. Stucki, Springer, 16–21.

Abbreviations Used in This Chapter

CFC = chlorofluorocarbon
CRF = capital recovery factor
EPA = Environmental Protection Agency
IRM = integrated resources management
IWM = integrated waste management

MSW = municipal solid waste
O&M = operation and maintenance
PWF = present worth factor
RFP = request for proposal
SFF = sinking fund factor

Problems

9-1. The town of Recycleville has 30,000 citizens. Each home in Recycleville generates 1.1 tons of solid waste per year and has an average of 2.4 residents per service customer. The town has no commercial businesses or schools. The local landfill has space for 100,000 tons and charges $10 per ton. Once the landfill is full, the waste will need to be sent to a remote landfill at a cost of $90 per ton.

    a. What is the 10-year average disposal cost for the town, assuming a rate of inflation of 5% per year?

    b. What will be the cost per home?

9-2. A town near Recycleville (Problem 9-1), Trashville, generates about 40,000 tons per year of solid waste. The town's landfill will be full in one year. Assume that after Trashville's landfill is full, its solid waste will come to Recycleville.

    a. If Recycleville wants to run its landfill at no cost to the citizens of Recycleville, what tipping fee should it charge Trashville?

    b. How much sooner would the landfill have to close Recycleville it accepted Trashville's refuse?

    c. If you are the mayor of Recycleville, what would you do about solid waste from Trashville?

9-3. If you are a private waste hauler trying to convince a city that you are less costly than public sector collection, how would you respond to the claim that you pay taxes, therefore your costs are higher than the public sector's?

9-4. You are preparing a request for proposal for garbage collection service. What factors would automatically disqualify you as a bidder responding to the RFP?

9-5. You have been hired to develop an integrated solid waste management plan for the city of Confusion. You present your plan to the city council, and council members decide that the town will build the new landfill but not implement your recycling component. After the meeting, the press asks you what you thought of the city council's decision. What do you tell the press?

9-6. A private company wants to construct a waste-to-energy plant that will have a capital cost of $20,000,000 and an operating cost of $300,000 per year. The company expects to receive 300 tons of refuse per day, and the facility is to have a useful life of 15 years. The community insists that it will not pay more than $40 per ton to the company for disposal of its solid waste. What interest rate must the company obtain on its $20,000,000 loan just to break even? (You will need a set of interest tables, or a calculator with interest tables, to solve this problem.)

9-7. If you didn't have the garbage collection franchise in a city, what arguments would you use to sway the city council to put the collection agreement out to bid? If you already had the agreement, what arguments would you use not to put it out to bid?

9-8. Determine the current regulatory and legal status of flow control.

9-9. A community wants to construct a hazardous waste incinerator for household hazardous wastes. The capital cost of the facility is expected to be $5,000,000. The cost of operating this facility is estimated to be about $20,000 per year plus salaries of $100,000 annually. The insurance on the facility will be $5000 per year. What is the annual cost of this facility to the community? Assume an interest rate of 6.125% and a useful life of 15 years.

9-10. A town signs a contract with a dumpster manufacturer to provide 50 dumpsters at a cost of $5000 each. The dumpsters are expected to last five years. A rival manufacturer insists that even though its dumpsters cost $8000, the expected life of 10 years makes them a better deal to the community. Is the rival manufacturer right?

9-11. A waste management agency decides to use some of the tipping fees from a landfill to buy extra land for expanding the landfill. The new land costs $2,000,000, and the existing landfill has an expected life of eight years. How much money will the agency have to put into a sinking fund annually (at 6.125% interest) to have the $2,000,000 available to buy the land?

9-12. A community wants to construct a landfill that is expected to have a capital cost of $3,000,000 and an annual operating cost of $500,000. The landfill is expected to last 15 years, and the lowest-interest loan the town can get is at 6.125%. What must the revenue stream be to finance this project?

9-13. Find the Commerce Clause in the U.S. Constitution and write it out. (*Hint:* It is only one sentence.) How would the Commerce Clause apply to waste management?

9-14. Who owns and operates your local landfill? How do the rates compare to other nearby landfills?

The Zero Waste Approach to Resource Management

Richard Anthony
Zero Waste International Alliance
www.zwia.org

The genesis of the Zero Waste movement comes from the realization that discarded materials are resources. These resources have been manufactured from a raw state with energy and labor. In the cases of metal and oil they are irreplaceable. The value of that energy and labor is still in the commodity, even after the user has discarded it.

A Zero Waste system is a resource management system. The process of wasting resources is against nature. In a Zero Waste system everything has a place before, during, and after use. There is no away. In the best-designed system, dismantling or demanufacturing would be designed into the product. The system of extraction, manufacturing, use, and disposal to incinerators or landfill will be replaced with systems that capture the materials and recycle them into a closed loop system of reuse, repair, recycle/compost, and redesign. Raw materials will be used as reserves.

This is called the "closed circle economy," and the analysis is called a "cradle to cradle" design. The recognition that disposal by burning and landfill will leave a legacy of depletion and pollution for our children will provide the basis for new analysis and new rules. These new rules will recognize the right of future generations to the planet's resources and will discourage wasteful and polluting practices.

DEFINITION OF ZERO WASTE

Toward the development of these new rules, The Zero Waste International Alliance (www.zwia.org) has peer reviewed and approved of the following definition of Zero Waste.

"Zero Waste is a goal that is ethical, economical, efficient, and visionary to guide people in changing their lifestyles and practices to emulate sustainable natural cycles, where all discarded materials are designed to become resources for others to use."

"Zero Waste means designing and managing products and processes to systematically avoid and eliminate the volume and toxicity of waste and materials, conserve and recover all resources, and not burn or bury them."

"Implementing Zero Waste will eliminate all discharges to land, water, or air that are a threat to planetary, human, animal or plant health."

ZERO WASTE AND GLOBAL WARMING

Landfills are one of the largest sources of greenhouse gases (GHGs) in any community. According to the USEPA, on a pound per pound basis the comparative impact of methane on climate change is 28 to 36 times greater than carbon dioxide over a 100-year period (see Chapter 7). In most California city's climate plans, the landfill is among the top three GHG-emitting entities (traffic and farms round out the top three). In 2014, California Air Resource Board (CARB) officials recommended that compostable organic materials be managed with aerobic and anaerobic digestion technologies. A NASA photo (Figure 1) shows methane hovering over the north and south poles of Earth.

Figure 2, which shows a "waste berg," illustrates that there are 71 tons "upstream" of wasted materials and energy for every 1 ton of municipal solid

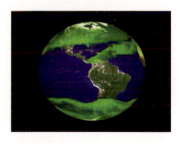

Figure 1 NASA photo showing the location on methane in the atmosphere. Source: GISS, NASA.

Figure 2 Waste berg. Source: adike/Shutterstock.com.

waste discarded. Thus source reduction programs, not only eliminate municipal solid waste, they also eliminates significant upstream wasted material and energy.

If every discard in California was recycled or composted, the savings would be the equivalent of eliminating all auto exhaust in California (EPA Waste Assessment Model). In 2011 California passed legislation establishing a 75% reduction goal and recommending to cities that source separation be required.

ZERO WASTE COMMUNITIES AND BUSINESSES

The Zero Waste movement is becoming increasingly prevalent in the United States, but, New Zealand and Japan were the first nations to come up with Zero Waste campaigns. Today cities all over the world have adopted Zero Waste goals. These include San Francisco, Los Angeles, San Diego, Fresno, Oceanside, and San Luis Obispo and dozens more in California. Other cities around the United States include Fort Collins, Colorado; Austin, Texas; and Chicago, Illinois. International examples include over 66% of New Zealand cities; Buenos Aires, Argentina; 400+ cities in Italy; Nagoya, Japan; and Vancouver, Canada.

In the business sector, Zero Waste is a cost-cutting and efficiency measure tied into international management policies. Hundreds of businesses are achieving 90% diversion from disposal at landfills and/or incinerators. New international measurements are being developed by Quality Control Rating Services (Underwriter Laboratories, UL). Today companies like Walmart, Toyota, Sierra Nevada Brewery, Frito–Lay, Vons/Safeway, and hundreds of other have adopted sustainability plans that call for Zero Waste practices.

BASIC PRINCIPLES

Different from the Integrated Waste Management approach, the Zero Waste approach considers all discarded resources as commodities. Unwanted discards can be separated at the source, stored separately, separately collected, processed, and sent to markets for reuse and recycling/composting. Ninety percent of our daily discards could be managed this way in a community collection program. The handling of the residual (less than 10%) can be discussed in the public forum on whether to require a product redesign or local ban.

There are five basic principles that are the pillars of the Zero Waste approach.

- The first prinicple is that resources are finite.
  The process of wasting resources is against nature. Therefore the ultimate option is to control population and recycle resources to survive. Because the human species is driven to survive, the reasons and the answers can be seen in nature. Where there are limits in materials and space, contradictions to the flow of nature are obvious.

- The second principle is that there is no away.
  The notion of Zero Waste is as much as a principle of survival for the human species as it is a matter of fact in nature. A close examination of natural systems reveals

that there is very little waste in nature. Everything is connected to everything else. Every discard is another's feedstock. When the planet is seen as a finite sphere in space, there can be no away on planet Earth. Everything that is sent away must go someplace.

- The third principle is that today's wasting and pollution rob the future of resources that future generations will need.
  Even though disposal by landfilling or burning and landfilling of the ash may be cost effective under today's rules, the legacy of depletion and pollution for our children will provide the basis for new rules. These new rules will recognize the future's right to the planet's resources and will discourage waste.

- The fourth principle is highest and best use.
  There is a hierarchy of use of materials that involves the highest and best use of materials in the areas of energy and resources. This hierarchy includes reduce, reuse, and recycle, where recycling includes repair and composting. The "three R's" are used to teach pollution prevention. The first R in the "three R's" (reduce, reuse, and recycle) refers to source reduction, or the area of discard management that addresses overpackaging and single-use products. The "three R's" are taught as a means to demonstrate how product design can lead to decreasing waste. Consumers are encouraged to consider buying products that can be reused, repaired, recycled, and composted.

- The fifth principle is mandatory source separation.
  Thus, Zero Waste theory calls for disposal systems that place disposal cost responsibility on the manufacturers and encourages them to redesign products for recyclable ability. The discard management service provider, whether government or private contractor, is mandated to collect source-separated material from clearly labeled and conveniently located storage containers and deliver them to processing centers that will sort, process, and reintroduce these materials back into the use system.

RESOURCE USE

As the world population and living standards increase, world resources are used at increased rates as well. The impact of this increasing demand on the remaining of the planet's finite resources—like petroleum and metals—and its biodiversity of animals, birds, flowers, fish, and trees—is leading to the depletion of these resources leading to their depletion and, in some cases, extinction.

Figure 3, from the 1972 book, *The Limits to Growth*, shows poululation growth from 1900 to 2100. Near the year 2000, the lines cross, showing the population of the planet increasing while resources are depleted. This projection is a reality today; new recycling mills and factories in India and China are using Western discards as material feedstock. As the resources become depleted and the demand increases, the value of these discards increases as well. The monetary value of today's recycled materials market is 100 times higher than what it was in 1970.

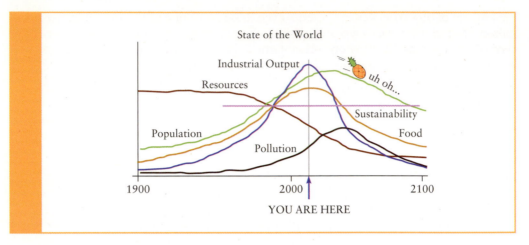

Figure 3 Relationship between population and resouces.

GREENHOUSE GAS AND OTHER POLLUTION REDUCTION

Table 1 shows the reductions in energy use, air pollution, water pollution, mining waste, and water use when recycled resources are substituted for virgin materials. In the case of aluminum, no new bauxite has to be mined and added to the aluminum melt to recycle it into another product. Today, steel is the most recycled metal.

Table 1, which was created by the United States Environmental Protection Agency (EPA), is the basis of the Waste Assessment Model. For each commodity, the reduction by recycling in energy use, air pollution, water pollution, mining waste, and water use is given.

The Zero Waste approach to managing resources considers the Earth's need and demand for these materials and the energy and greenhouse gas savings that occur when discards are managed as commodities.

ZERO WASTE MANAGEMENT

Today, whether it's your home, your business, or your community, there are basic approaches to handling discards. The Zero Waste approach looks at discards, both upstream and downstream of what some would call the waste stream. Efforts to reduce

Table 1 Pollution, Energy, and Waste Reductions with Recycling

	Aluminum	Steel	Paper	Glass
Energy Use	90–97%	47–74%	23–74%	4–32%
Air Pollution	95%	85%	74%	20%
Water Pollution	97%	76%	35%	
Mining Wastes	99%	97%		80%
Water Use		40%	58%	50%

this stream of discarded materials can happen before purchase and use (upstream, or prevention) or after use (downstream, end of pipe or recovery). In other words, the Zero Waste approach is about the prevention of waste and the recovery of discards.

Up stream prevention programs include the following:

- Clean Production
  - The goal of the final product is that the manufacturing of product will not hurt the product, the profit, and the planet (triple bottom line). Factory managers are responsible for not incurring additional cost by creating and disposing of waste and/or toxic materials.

- Product Redesign
  - If any of the triple bottom-line parameters (actions that hurt product, planet, or profit) are exceeded, the product needs to be redesigned. In our energy-conscious economy, the product must be repairable and/or recyclable/compostable, and there will be no toxins used or created by the process that cannot be reused and recycled by the manufacturer.

- Product Stewardship
  - A company takes pride in its products and creates sustainable materials because is the right thing to do for the planet. In the past, product stewardship was voluntary, but today companies are being required to take the lead in making their products and packages conform to international requirements.

Downstream recovery programs focus on capturing commodities at the point of discard.

- Reuse materials should be handled through box truck collection and drop-off centers, where materials can be evaluated for reuse and repair. If not reusable and repairable, materials are dismantled into the basic recycling categories. All reusable and durable goods should be collected separately and processed as commodities.

- Composting compostable organics is the basic way of handling organic discards. These materials can be collected as yard trimmings and food scraps and composted in the back yard and/or collected separately and sent to farms and facilities that have composting and anaerobic digestion capabilities. The result will be healthy farms with less water and petrochemical demands and no local, human-created sources of leachate and methane emissions from the landfill.

- All containers and paper products must be recycled. These containers include metal, glass, paper, and plastic.

- Resource recovery parks are the new transfer stations where commodity clusters can be collected separately and then transferred to the processing facilities. Special discards like rocks and wood have recovery areas at these parks.

MARKET CATEGORIES

All discards can be sorted into the following 12 market categories.

The percentage of each of the above market categories in the discards varies by community. Figure 4 is a typical composition of these market categories in a community's discards.

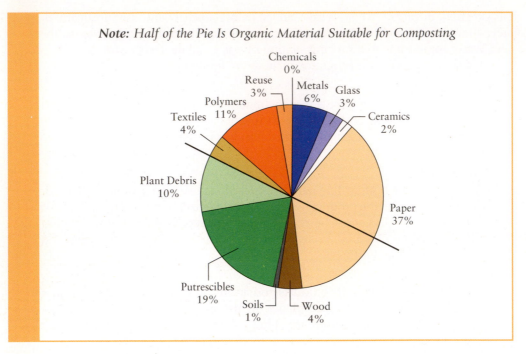

Figure 4 Market categories.

The following are the twelve market categories:
1. **Reusable** are materials that can be reused, repaired, and /or dismantled for recycling.
2. **Paper** includes cardboard, newspaper, writing paper, tissues, and towels.
3. **Plant debris** includes all yard trimmings.
4. **Putrescible** are food scraps and organics that putrefy.
5. **Wood** includes painted and unpainted, although some painted wood is problematic because of lead based paint.
6. **Ceramics** are rocks, concrete, and asphalt.
7. **Glass** includes containers but not leaded glass.
8. **Polymers** are plastic and can be sorted into resin codes for high resale value.
9. **Soil** includes dirt.
10. **Metal** includes ferrous magnetic metals like steel and iron and nonferrous metals like copper, aluminum, brass, gold, etc.
11. **Textiles** are reusable but in a class of their own and include cloth and woolen natural fibers as well as synthetic fibers.
12. **Chemicals** are hazardous when disposed of but can by recycled for further use.

Each of the materials in these categories has a positive monetary value. Credit for this system of market categories must be given to Dr. Dan Knapp of Urban Ore in Berkeley, California.

MASTER CATEGORY CLUSTERS

Often, when people in rural areas have discards, they try to use all of them. Both the cost and distance involved in removal and the saving and practical use of resources are key drivers. Typically reusable products are taken to the church or a local charity or given to the family and neighbors for reuse and repair (reuse). Excess food goes to family, workers, neighbors, animals, and ultimately the land (compost). Bottles and cans can are a problem but are used for storing material and ultimately recycled at rural transfer stations (recycling). Paper and wrappers are burned as fuel. Special discards like pesticide containers must be hauled away.

Based on practical experience in marketing commodities, in a Zero Waste system, discards should be sorted into four clusters, reusable and repairable products, recyclable materials, compostable organics, and landfill (for legacy and badly designed materials). Containers can be varied and different colors. As of 2015, state-of-the-art programs use wheel carts for storage and collection. These vary in size.

The blue cart (recycle) is for paper and containers (paper, metals, glass, and polymers). The comingled four commodities can be sorted with magnets and hand sorters at a materials recovery facility (MFR), baled, and sold.

The organic cart should be green and can collect food, vegetative debris, dirty paper, paper, plant debris, putrescibles, wood, and soil. The comingled organic commodities are taken to a composting facility to be processed into soil.

Discarded items, sometimes called bulky waste or items set out for charity pickup, need to be collected in a box truck and taken to a warehouse for further sorting. These materials include furniture, appliances, clothing, toys, tools, reusable goods, and textiles. This collection can be made on call.

Transfer stations and landfills need to be converted to resource recovery parks to handle self-hauled loads and special discards. These materials include chemicals, construction and demolition materials, wood, ceramics, soils, and, in case of self-hauled reuse, recycle and compost categories.

CLUSTERS AND FACILITIES

Table 2 shows the facilities needed for each cluster.

Reusable materials need a warehouse for sorting and dismantling. An unloading area, baler, and loading dock are necessary.

Recycling materials need a materials recovery facility (MFR). This facility takes in comingled containers and papers and with magnets, lasers, blowers, and hand sorting creates bales for global markets. Many MRFs are completely hand-sorted.

Compostable materials need land for grinding the materials, trommel sorting, windrowing the material, wetting the materials, and turning the soil for aerobic composting. CO_2 and soil are the products. In an anaerobic system the material is reduced without air and creates methane, which is extracted for reuse. The final digestate needs to be composted and stabilized and can be used as a soil amendment.

Table 2

Clusters	Facilities	

Recyclables:
    Paper and containers made of paper, metals, glass, polymers
Organics:
    Food, vegetative debris, food dirty paper, paper, plant debris, putrescibles, wood
Reused Products:
    Furniture, appliances, clothing, toys, tools, reusable goods, textiles
Special Discards:
    Chemicals, construction and demolition materials, wood, ceramics, soils
	Recyclables:
	    Papers, plastic, glass and metal containers (Materials Recovery Facility, MRF)
	Organics:
	    Food, vegetable debris, and food paper, putrescibles, untreated wood and sheetrock [Composting or Anaerobic Digestion (AD)]
	Reuse & Repair:
	    Reuse, repair, dismantling, reconditioning, remanufacturing, manufacturing and resale of furniture, large and small appliances, electronics, textiles, toys, tools, metal and ceramic plumbing, fixtures, lighting, lumber and other used building materials (Repair and Dismantling)
		Construction and Demolition (C and D):
		    Rock, soils, concrete, asphalt, brick, land clearing debris, and mixed construction and demolition materials (Sorting and Grinding)
		Regulated Materials:
		    Used motor oil, paint, pesticides, cleaners, and other chemicals (Reuse, Take Back, and HHW Disposal)

Construction and Demolition (C and D) material can be as much as one-third of all discards disposed of at a landfill. Comingled C and D materials can be sorted on a belt to recover metal, wood, and rock.

Toxic materials need to be taken back to manufacturers or collected and safely packed for proper disposal.

REVENUE AND JOBS FROM DISCARDS

To estimate the value of discards, take the annual tons disposed of at the landfill and/or incinerator and apply the percentages in the pie chart (Figure 4). You can look at local studies to see if there is a waste characterization analysis. These studies typically sort samples of discards in 75 or more categories. These can be combined into the 12 market categories.

Once the annual amounts of materials discarded by category are calculated, their value can be estimated. The value of baled material is based on official published market prices. These prices are posted in market journals, newsletters, and Internet sites. The price for finished compost is used for organic materials.

There is a collection and processing cost for discards, but when compared to the cost of polluting (landfill and incineration), these materials have net positive end value. Table 3 shows the market values for each commodity class. One-third of the value is from reusable materials. This is a good illustration of why we should manage these materials away from landfills. Other commodities like paper, metal, and polymers have a global value as well.

Table 3 **Commodity Value and Jobs Estimate for the State of Deleware**

Market Categories	Jobs	Tons per Year	Market Price $/T (est.)	Total Value of Discards in Delaware ($)
1. Reuse	350	28,000	550	15,400,000
2. Paper	65	370,000	20	7,400,000
3. Plant Trimmings	30	100,000	7	700,000
4. Putrescibles	85	190,000	7	1,330,000
5. Wood	24	40,000	4	320,000
6. Ceramics	7	20,000	4	80,000
7. Soils	20	10,000	7	70,000
8. Metals	35	60,000	40	2,400,000
9. Glass	75	30,000	10	300,000
10. Polymers	1,020	110,000	100	11,000,000
11. Textiles	340	40,000	200	8,000,000
12. Chemicals	4	2,000	15	30,000
Total	2,055	1,000,000		47,030,000

The Washington, D.C. based Institute for Local Self Reliance has calculated the number of jobs created in the processing of discards. The Institute claims that, on average, one job could be created for every 10,000 tons disposed of at the landfill or incinerator. For the one million tons listed in Table 3, there was nearly $50,000,000 was lost annually to the landfill and more than 2000 potential jobs.

Table 3 shows the value of commodities in terms of revenue and job creations. The value of a ton of baled paper, metal, and polymers (plastic) is a global price and is basically the same worldwide. Reusable and organic materials are locally used products. The value and amount of labor used for reuse and repair varies because labor rates are different around the world.

RESIDUALS

After all the commodities are recovery, less than 10% of discards remain and are considered residuals. Diapers often comprise as much as 6% of residuals. Diapers are a product that should be redesigned so that they would no longer be unrecoverable. Counties, cities, and joint powers agencies have the power to make rules in the name of health and safety. Products like diapers should be compostable.

Residuals also include composite packaging material, wrappers, and legacy waste like lead painted wood. Some of this waste can be redesigned, and the remainder can be buried in a secure landfill with no unprocessed organics mixed in. The goal would be in the future to find a use for all discards so that residuals could be eliminated.

SUMMARY

The elements of a Zero Waste system include:

1. Producers taking responsibility for the impact of their product on the environment.
2. Producers designing products for the environment.
3. Clean production systems at factories that create neither wasted materials nor toxic discharges.
4. Retail stores taking back products that are not recyclable or compostable.
5. Consumers purchasing products that are environmentally friendly.
6. Resource recovery parks replacing transfer stations and landfills.
7. Rules changing to require separation, ban organics from landfill, banning C and D materials unless there is a plan to take back where no recycling system or composting system is in place.
8. Tax rules changing so that resources, not labor, are taxed.
9. Many new jobs being created in reuse, repair, recycling and composting.

APPENDIX A

The Phantom Solid Waste Problem

A-1 BACKGROUND

The City Manager of the City of Phantom has asked your firm to evaluate its options for integrated solid waste management. He expects a preliminary study from your firm detailing the possible options for further study.

The impetus for this study is the expiration of a contract with Trash County Landfill, which is presently receiving all of Phantom's refuse under a grandfather contract. Trash County has told Phantom that it will begin to charge \$80/ton in the year $X + 3$ when the present contract expires. This potential threat has promoted the study you are conducting. (*Note:* $X =$ the present year.)

In addition, the state has passed new regulations to address greenhouse gas reduction requirements and recycling mandates. Every city must reduce the amount of waste being landfilled by 50% in $X + 3$ years or pay a fine of \$10,000 per day.

The populations to be served and the sources and quantities of municipal solid waste (MSW) are estimated first. Unfortunately, there is no information on solid waste generation in Phantom, and the best available information from other sources must be used. The first task is to estimate the solid waste production, both quantity and composition, generated in Phantom.

The second task involves designing an integrated waste management system for the residential and commercial solid waste. Potential components of the intergratred waste management system include:

1. Recycling and composting facilities
2. Waste-to-energy facility, with the production of steam and conversion to electricity that will be bought by the power company. A landfill for the unacceptable items and the incinerator ash is also necessary.
3. Sanitary landfill

You have decided that your report should be organized as a series of chapters, which will be incorporated into the final report:

Chapter 1 Sources, Composition, and Quantities
Chapter 2 Collection
Chapter 3 Recycling
Chapter 4 Composting
Chapter 5 Waste-to-Energy (Mass Burning)
Chapter 6 Siting the Sanitary Landfill
Chapter 7 Design of the Sanitary Landfill
Chapter 8 Technical, Economic, and Environmental Evaluation

That is, once you have developed some figures on the amount and type of solid waste generated, you will be comparing the various solid waste management options. Remember that one of your alternatives is accepting the high tipping fee by the Trash County Landfill and delivering all of the nonrecycled waste to that landfill.

You will complete each chapter, developing the numerical and policy conclusions as necessary. You will work as groups, and all recommendations, calculations, and policies will be a group effort. However, the final product must be your own writing. Each chapter will have two sections: (1) the text: the written report and all graphs, tables, or other illustrations necessary for the reader to understand your conclusions, and (2) the appendix: calculations and other supporting material, including working graphs. When the final report is prepared, the calculations will all go into a common appendix. Begin each chapter with the proper headings. You are encouraged to consult *Report Writing for Environmental Engineers and Scientists* for questions of style and format (available from Lakeshore Press, P.O. Box 92, Woodsville, NH 03785). http://lakeshorepressbooks.com/store/engineering-books

Individual chapters are due as indicated on the syllabus. A late penalty of 5% per day will be charged for any fraction of a day late. The day begins when the reports are collected at the beginning of the class period.

A-1-1   Useful Data

Population of the City of Phantom

Year	Population
1960	95,722
1970	105,634
1980	115,689
1990	125,728
2000	135,880
2010	145,220

About 15% of the people in Phantom live in apartments served by dumpsters collected by the City of Phantom.

A-1-2 Existing Solid Waste Collection Program and Method of Disposal

The City engineer estimates that at present there are 1000 commercial accounts (including apartments) serviced by the City, and that the residential collection, based on billing records, serves 52,000 homes.

The City does not have any idea how many tons of refuse are actually being disposed of because the City is working under a grandfather clause with Trash County that allows the City to deliver all of its refuse to the landfill at a fixed annual cost. This contract, however, is expected to expire by the year $X + 3$. The expected expiration of the (very favorable) contract is what prompted this study.

A-1-3 Existing Recycling Program

The Boy Scouts collect newsprint once a month. They take it to Bud's Scrap Yard and receive $10/ton. They collect about 50 tons/year.

The Kiwanis Club has a drop-off location for aluminum cans and clear glass. The Club transports the materials to Bud's. Latest data show that the Club collects 15 tons of aluminum and 35 tons of clear glass per year. The Club pays for the drop-off containers and the transportation and keeps the income. Bud's Scrap Yard pays $600 per ton for the aluminum and $10 per ton for the clear glass.

The City has thus far not spent any money on developing its own recycling program, although it is State-mandated to do so.

A-1-4 City Organization

The City has an elected city council, which elects its own chair, who then serves as the mayor. The City is divided into five districts, and one councilperson is elected from each district. The council hires the city manager. He is the one who commissioned the study you are doing.

CHAPTER 1

A-2   SOURCES, COMPOSITION, AND QUANTITIES

We assume here (as the worst-case assumption) that there are no data on solid waste production or solid waste composition.

Three sources of solid waste generation data that may be useful to you are (1) national average (a crude and usually unacceptable estimate when applied to specific localities), (2) data from other nearby communities, and (3) data obtained from statewide surveys.

The municipal solid waste (MSW) can be identified as coming from distinct sources:

1. Residential (single-family homes)
2. Public institutional (schools, government offices, for example)
3. Commercial (including apartments)

Constuction and demolition debris is not considered in this study. Other than the C&D waste, all other components of MSW will need to be collected and disposed of, and therefore the quantities must be estimated.

Use the population estimates for the year $X + 3$ in your calculations since this would be the earliest any one of the alternatives you evaluate would be placed into operation. Based on the census data, estimate the population to be served by the City's solid waste collection and disposal program.

You can estimate that the public institutional waste is 10% of the total solid waste and the commercial waste is 20% of the total solid waste. The commercial establishments in Phantom are served by the municipal collection system since there is no reasonably priced alternative. The City charges a fee for every dumpster load collected. Residents currently provide their own garbage cans and pay $15 per month to set out up to six garbage cans.

National solid waste production averages are based on a sum of residential, public institutional, and commercial waste since these often cannot be distinguished during collection. Such figures are available in various publications both on a national level and state level.

For this chapter,

1. Estimate the population for the year $X + 3$. This is your design population. (*Note:* X = present year.)
2. Estimate the generation and composition of the three different types of municipal solid waste for the year $X + 3$. The three types of solid waste are
   a. residential,
   b. public institutional,
   c. commercial

Express the composition as percent by weight of various identifiable materials, such as newsprint, aluminum, organics, etc.

Consider and select alternative tabular and/or graphical means to present your results. The graphics should be informative, accurate, and easy to read. You will use the information from these figures for subsequent chapters. More than one graph may be useful and appropriate. This graphical presentation must include composition as well as quantities. The quality of the graphic will be a consideration in grading this report.

CHAPTER 2

A-3 COLLECTION

The City currently collects residental garbage with a fleet of old rear loader trucks. Citizens place their garbage at the curb in their own cans. The City wants to continue collecting MSW using city-owned trucks and city employees, but City officials recognize the need to cut labor costs and injuries. Commercial establishments will also be collected using new city trucks and city employees. At present, the City services 1000 dumpsters located in the commercial zone.

Garbage, recyclables, and/or organics will be picked up from individual residences. Apartment buildings, schools, and other municipal facilities and commercial businesses will be collected using dumpsters. Two types of trucks—one type used for residential pickup and another type used for collecting apartments, institutions, and commercial establishments—must be purchased.

You may use the following cost data, or develop your own.

Collection Vehicles for Small Containers (such as residential)

Size (yd³)	Type	Capital Cost ($)	Compaction Ratio	Truck Operating Cost (not labor) ($/hour)
25	Rear loader	250,000	1:4	15
20	Rear loader	200,000	1:4	13
20	Side loader	250,000	1:4	15

Collection Vehicles for Dumpsters (such as commercial)

| 25 | Front loader | 250,000 | 1:3 | 15 |
| 30 | Front loader | 300,000 | 1:3 | 18 |

Assume a crew of three persons per residential rear loader truck and one person for side loader, each paid at $35/hour (including fringe benefits, insurance, etc.). Dumpster collection vehicles are managed by only one person. Expected vehicle life is 10 years. Assume an overhead factor for salaries and benefits of support staff at 30% of labor cost.

Estimate that for individual households, one collection vehicle can service 80 residences per hour if the system uses waste wheelers placed on the curb and 50 per hour if the residents set their cans on the curb. If waste wheelers are to be purchased for individual households, estimate their cost as $50 per waste wheeler. Waste wheelers have an expected life of 10 years.

Estimate that a collection vehicle requires one hour each way to travel to and from the disposal site and 30 minutes each way to travel to and from the MRF and compost sites.

Estimate that a dumpster has a capacity of 3 yd³ and that it takes 10 minutes to empty a dumpster into the truck, including the travel to the next dumpster site. Assume that each dumpster costs $800 and has a useful life of 5 years.

For this chapter,

1. Design a plan for the collection of the garbage, recyclables, and/or organics.

   a. Dumpsters and vehicles:

      What fraction of solid waste will go into the dumpsters? Where will the dumpsters be placed to service the public institutions, the commercial establishments, and the apartments? How many commercial dumpsters will be provided? How many collection vehicles will be needed to service the dumpsters?

   b. Residential collection and vehicles:

      Establish a routing system for residential solid waste collection, including the type and number of trucks, frequency of collection, and number of personnel required (with job classifications). If waste wheelers are to be used, who will pay for them? Who will pay for lost/damaged waste wheelers? Amortize any capital costs over the expected life.

2. Calculate the expected gross cost for both residential and commercial/public institutional solid waste collection for the year $X + 3$. Include in your calculations the cost of labor, the vehicles and collection containers (dumpsters and waste wheelers if you choose to use them) by annualizing the cost using 5.0% interest. (*Note:* If you borrow $1, you pay back $1.05 the next year. If you borrow $1 and pay it back in 10 equal installments over the next 10 years, you pay $0.1295 every year.) The annual cost of collection for Phantom, as $/year, is represented by the symbol G_r for residential and G_c for commercial.

3. Decide what you will charge residential and commercial establishments for collecting per month. How did you decide on this figure? Using the estimated number of accounts, calculate the income to the town from charging residential and commercial establishments, as $/month. This income is represented by the symbols F_r and F_c.

4. Calculate the cost of collection on a tonnage basis by dividing the net cost of collection previously calculated by the total quantity of refuse collected by the city for residential and commercial tonnage.

This cost, as $/ton, is used in the calculations in subsequent chapters and is represented by the symbol C_r and C_c.

CHAPTER 3

A-4 RECYCLING

The State has mandated that all cities must divert 50% of their waste from the landfill by year $X + 3$. Regardless of what other options are adopted by the City, a recycling program is a must. It is unlikely that the City will be able to depend on the Boy Scouts and the Kiwanians for very long into the future, although the Scouts and Kiwanis might be allowed to continue their programs. This is a problem that will complicate your recommendations.

A survey of potential markets for clean recycled materials indicates the following:

Material	Price paid at the Market Site ($/tons)
Aluminum	1200
Steel cans	50
Glass	
Clear	30
Mixed	20
Paper	
Corrugated	60
Newsprint	80
Office	100
Mixed	30
Plastic (No. 1 and 2)	200

The estimation of collection cost can be difficult since the costs will depend on the level of participation, the materials collected, and the collection systems used. Use the information from Chapter 2 to estimate collection costs. For this chapter,

1. Assume initially that all of the materials listed above are candidates for recycling. For the year $X + 3$, estimate the fraction of each material produced by the town (from Chapter 1) that might be recycled (residential, commercial, and public institutional). Estimate the quantities of each material to be collected, and sum these. Will the Boy Scouts and Kiwanians be allowed to continue their recycling programs?
2. Calculate the cost of collection for the materials to be recycled by using the information in Chapter 2.

3. Design a materials recovery facility (MRF). Specify the unit operations to be used and decide on the capacity of those unit operations. Draw a schematic of the process train and indicate what materials are to be recovered. (A schematic is a flow diagram, with the various processes identified by boxes and the flow of material shown by arrows.)

4. Draw a floor plan of the facility, including the receiving area and the building. Be sure to indicate all storage and conveying equipment. Show how the material moves through your facility. The floor plan should be one step better than a schematic. It should include the various processing operations in rough size and space according to their operation. All material transfer operations must be identified. Assume a capital cost of $25,000 per ton of designed capacity.

5. Estimate the quantity of materials to be recovered. Use the prices listed as market values to calculate the expected income from the sale of the recovered materials. Assume an operating and maintenance cost of $60 per ton processed.

6. Estimate the amount of waste destined for the landfill as all materials are not recovered.

7. Calculate the annual cost of the MRF and the cost per ton.

$$\text{Annual cost of the MRF option (\$/year)} = L - M + C(Q_{prod}) + D(Q_{rej}).$$

where

L = annualized cost of equipment and building (capital plus O&M), $/year

M = annual income (or loss) from sale of recovered materials,

C = annual cost of collection, $/ton (from Chapter 2)

Q_{prod} = quantity of recyables brought to the MRF, tons/year

D = cost of landfilling rejected material, $/ton (assume $100/ton for transport and disposal)

Q_{rej} = quantity of materials not recovered (rejected) in MRF, tons/year

CHAPTER 4

A-5   COMPOSTING

As discussed above, the State has mandated that all cities must divert 50% of their waste from the landfill by year $X + 3$. Organics is a large fraction of the waste stream that must be considered. The current market value for compost is $40 per cubic yard. Use the information from Chapter 2 to estimate collection costs. For this chapter,

1. Assume initially that all the organic material can be composted. For the year $X + 3$, estimate the fraction of organic material produced (residential, commercial, and public institutional). Estimate the quantities of each material to be collected (some people might have home compositng systems).

2. Calculate the cost of collection for the materials by using the information in Chapter 2.
3. Design a composting facility. Specify the unit operations to be used and decide on the capacity of those unit operations. Draw a schematic of the process train and indicate what materials are to be recovered. (A schematic is a flow diagram, with the various processes identified by boxes and the flow of material shown by arrows.)
4. Draw a site plan of the facility, including the receiving area and composting area. Be sure to indicate all storage and conveying equipment. Show how the organics move through your facility. The floor plan should be one step better than a schematic. It should include the various processing operations in rough size and space according to their operation. Assume a capital cost of $5000 per ton of designed capacity and an operating cost of $30 per ton.
5. Estimate the quantity of materials to be recovered. Use the $40 per cubic yard to calculate the expected income from the sale of the compost. Assume an operating and maintenance cost of $30 per ton processed.
6. Estimate the amount of waste destined for the landfill (not all materials are compostible).
7. Calculate the net composting cost and the cost per ton processed.

$$\text{Annual cost of the compost option (\$/year)} = L - M + C(Q_{prod}) + D(Q_{rej})$$

where

L = annualized cost of equipment and building plus operating costs (capital plus O&M), $/year
M = annual income from sale of compost,
C = annual cost of collection, $/ton (from Chapter 2)
Q_{prod} = quantity of organics brought to the compost site, tons/year
D = cost of landfilling rejected material, $/ton (assume $100/ton for transport and disposal)
Q_{rej} = quantity of materials not recovered (rejected), tons/year

CHAPTER 5

A-6 WASTE TO ENERGY (MASS BURNING)

The State has considered waste-to-energy production to be recycling. This option meets the State mandate for recycling, and thus it might not be necessary to have the City conduct a separate recycling and composting program.

Mass burning is the combustion of unprocessed refuse. Burn units are sized on the basis of 24-hour operation and are designed to operate continuously. Typically, such units can operate as much as 350 days out of the year, with 15 days per year for repair and maintenance. Debt service should be calculated at an annual rate of 5.0% over 25 years.

The annual cost can be estimated as follows:

Capital cost—$150,000/ton of daily design capacity

Operation (salaries, fuel, O&M)—$130/ton burned

For example, if the capacity is 300 tons per day, the capital cost would be 300 tons X $150,000 = $45 million.

Estimate that a waste-to-energy plant will be able to reduce the volume of the refuse fed to it by 90%, but the remaining 10% of the waste (as ash) must go to the landfill. The density of the ash when placed in a landfill is about 2000 lb/cubic yard. During the time that the unit is down for repairs, all of the refuse must go to the landfill.

The facility is to produce steam and run a turbine, which will produce electricity to be sold to the power company at an income of $0.05/kWh. Assume that the plant will produce 600 kWh of electricity per ton of refuse delivered to the waste-to-energy facility.

For this chapter,

1. Select the type of mass-burn facility to be used. Draw a picture showing its major components. Note how the steam is produced and where the turbine is.
2. Specify what waste (if any) will be excluded from the facility.
3. Estimate the quantity of refuse to be processed under three scenarios: (1) all waste, (2) waste minus organics, and (3) waste minus organics and recycalbles. Calculate the income derived from the sale of the electricity.
4. Estimate the amount of waste destined for the landfill.
5. Estimate the total annual cost and cost per ton received of the mass-burn option as

Annual cost of the waste-to-energy option ($/year) $= N - P + D(Q_{ash}) + C(Q_{prod})$

where

N = annualized cost of the mass-burn facility (capital plus O&M), $/year
P = annual income from the sale of electricity, $/year
D = cost of landfilling, $/ton (assume $120/ton for transport and disposal)
Q_{ash} = quantity of ash and unburned refuse landfilled, tons/year
C = annual cost of collection, $/ton (from Chapter 2)
Q_{prod} = quantity of waste brought to the WTE plant, tons/year

 CHAPTER 6

A-7 SITING THE SANITARY LANDFILL

The City has decided that it might want to operate its own landfill. Many considerations must go into selecting a new site, including transportation, proximity to the town, social and political constraints, environmental constraints, and soil types.

For this chapter,

1. Estimate the solid waste production (minus recycled and organic waste) for the City of Phantom for the years $X + 13$ and $X + 23$. (Make sure you explain to the reader why you have selected these two years.)

2. Set the expected life of the landfill. Too many years, and the land will be too expensive. Too few years, and the City will be looking for a new landfill in a few years. Using the solid waste production figures from Chapter 1, as well as the estimates for the years $X + 13$ and $X + 23$, estimate the land area needed. Assume a compacted refuse density of 1000 lb/yd³. Estimate earth cover as 20% of the refuse volume. Make sure enough earth exists to provide the cover. If not enough is available on site, indicate where it will be obtained.

3. Select a site. Anticipate the opposition to your site. Who will be against it, and why?

How can this criticism be countered?

CHAPTER 7

A-8 DESIGN OF THE SANITARY LANDFILL

At present, State law requires that all new landfills meet RCRA standards (Subtitle D landfills) and have, at a minimum, the following:

   a. Double liner (clay plus synthetic)

   b. Leachate collection and treatment (or disposal)

   c. Gas venting or collection and use

   d. A plan and fund for closure

Assume land costs at between $50,000 (one hour drive from the city) and $100,000 per acre (30 minute drive from the city), depending on proximity to town, availability of transportation, and productivity. Assume $50,000 per mile for the construction of an unpaved road, and add $50,000 per mile for the paving. Assume site clearing, grading, fencing, etc. as $50,000 per acre. A scale costs $200,000 and is expected to last 20 years. Assume landfill cell construction, including liners, pipelines for gas control, leachate-collection system at $200,000 per acre, and operating costs at $20 per ton of refuse landfilled. Assume closure costs as $100,000 per acre. Closure cost has to be handled as a sinking fund and the cost attached to the cost of running the landfill. That is, a fund must be established so that at the time the landfill is full, enough money is available in this fund to close the landfill at a cost of $100,000 per acre. You must use the interest table to make this calculation. (see chapter 9, Table 9-1)

For this chapter,

1. Using the land area established in Chapter 6, calculate the cost of the land and cell construction and annualize these capital costs using the expected life of the facility at 5.0% interest.

2. Calculate the annualized cost of the scale.

3. Calculate the annual cost of the closure. (Note: You must establish a sinking fund over the life of the landfill, and each year some funds will be placed

into this fund. For example, if the interest rate is 5.0% and you want to have $1 at the end of ten years, you must invest $0.0795 every year.) You estimate the cost of closure, then calculate the annual contribution to this fund so that the money will be available when needed to close the landfill.

4. Estimate the annual operating cost of the landfill, using the figures for the first year the landfill would be in operation ($X + 3$).
5. Add these four annual costs. This is your gross annual cost of the landfill for the year $X + 3$.
6. Establish the tipping fee charged, expressed as $/ton of refuse received.

This is the net cost of the landfill, under this option, to the people of Phantom.

CHAPTER 8

A-9 TECHNICAL, ECONOMIC, AND ENVIRONMENTAL EVALUATION

In this chapter, collect all of the cost data generated in the earlier chapters and present comparisons in tabular or graphical form.

Discuss the technical considerations (dependability, reserve capacity, what future solid waste management systems might be available, etc.). Present a brief discussion of the environmental concerns and ramifications of the various alternatives, listing each and responding to what are perceived to be the major public concerns. Summarize the costs associated with the alternatives and the overall fiscal impact on the businesses and residents. Remember that the City is under a State mandate to have some type of recycling effort, so this must be included in your plan. Remember also that one of your alternatives is to recommend that the City sign a long-term contract with the Trash County Landfill to send all of the refuse to that landfill at a cost of $80/ton. Should the City simply do that and save itself all the trouble of running a solid waste disposal program? What are the pros and cons of this?

Finally (!) choose one recommended alternative and justify your choice.

APPENDIX B

Bulk Densities of Refuse Components

This appendix is based on Appendix J of *Conducting a Diversion Study—A Guide for California Jurisdictions* published in March 2001 by the California Integrated Waste Management Board. The table is a compilation of data from various sources listed at the end of this appendix.

Construction and Demolition

Material	Size	Study	Pounds
Ashes, dry	1 cubic foot	FEECO	35–40
Ashes, wet	1 cubic foot	FEECO	45–50
Asphalt, crushed	1 cubic foot	FEECO	45
Asphalt/paving, crushed	1 cubic yard	Tellus	1,380
Asphalt/shingles, comp, loose	1 cubic yard	Tellus	418.5
Asphalt/tar roofing	1 cubic yard	Tellus	2,919
Bone meal, raw	1 cubic foot	FEECO	54.9
Brick, common hard	1 cubic foot	FEECO	112–125
Brick, whole	1 cubic yard	Tellus	3,024
Cement, bulk	1 cubic foot	FEECO	100
Cement, mortar	1 cubic foot	FEECO	145
Ceramic tile, loose 6" × 6"	1 cubic yard	Tellus	1,214
Chalk, lumpy	1 cubic foot	FEECO	75–85
Charcoal	1 cubic foot	FEECO	15–30
Clay, kaolin	1 cubic foot	FEECO	22–33
Clay, potter's dry	1 cubic foot	FEECO	119
Concrete, cinder	1 cubic foot	FEECO	90–110
Concrete, scrap, loose	1 cubic yard	Tellus	1,855
Cork, dry	1 cubic foot	FEECO	15
Earth, common, dry	1 cubic foot	FEECO	70–80
Earth, loose	1 cubic foot	FEECO	76
Earth, moist, loose	1 cubic foot	FEECO	78
Earth, mud	1 cubic foot	FEECO	104–112
Earth, wet, containing clay	1 cubic foot	FEECO	100–110

Construction and Demolition (*Continued*)

Material	Size	Study	Pounds
Fiberglass insulation, loose	1 cubic yard	Tellus	17
Fines, loose	1 cubic yard	Tellus	2,700
Glass, broken	1 cubic foot	FEECO	80–100
Glass, plate	1 cubic foot	FEECO	172
Glass, window	1 cubic foot	FEECO	157
Granite, broken or crushed	1 cubic foot	FEECO	95–100
Granite, solid	1 cubic foot	FEECO	130–166
Gravel, dry	1 cubic foot	FEECO	100
Gravel, loose	1 cubic yard	Tellus	2,565
Gravel, wet	1 cubic foot	FEECO	100–120
Gypsum, pulverized	1 cubic foot	FEECO	60–80
Gypsum, solid	1 cubic foot	FEECO	142
Lime, hydrated	1 cubic foot	FEECO	30
Limestone, crushed	1 cubic foot	FEECO	85–90
Limestone, finely ground	1 cubic foot	FEECO	99.8
Limestone, solid	1 cubic foot	FEECO	165
Mortar, hardened	1 cubic foot	FEECO	100
Mortar, wet	1 cubic foot	FEECO	150
Mud, dry close	1 cubic foot	FEECO	110
Mud, wet fluid	1 cubic foot	FEECO	120
Pebbles	1 cubic foot	FEECO	90–100
Pumice, ground	1 cubic foot	FEECO	40–45
Pumice, stone	1 cubic foot	FEECO	39
Quartz, sand	1 cubic foot	FEECO	70–80
Quartz, solid	1 cubic foot	FEECO	165
Rock, loose	1 cubic yard	Tellus	2,570
Rock, soft	1 cubic foot	FEECO	100–110
Sand, dry	1 cubic foot	FEECO	90–110
Sand, loose	1 cubic yard	Tellus	2,441
Sand, moist	1 cubic foot	FEECO	100–110
Sand, wet	1 cubic foot	FEECO	110–130
Sewage, sludge	1 cubic foot	FEECO	40–50
Sewage, sludge dried	1 cubic foot	FEECO	35
Sheetrock scrap, loose	1 cubic yard	Tellus	393.5
Slag, crushed	1 cubic yard	Tellus	1,998
Slag, loose	1 cubic yard	Tellus	2,970
Slag, solid	1 cubic foot	FEECO	160–180
Slate, fine ground	1 cubic foot	FEECO	80–90
Slate, granulated	1 cubic foot	FEECO	95

(Continued)

Construction and Demolition (*Continued*)

Material	Size	Study	Pounds
Slate, solid	1 cubic foot	FEECO	165–175
Sludge, raw sewage	1 cubic foot	FEECO	64
Soap, chips	1 cubic foot	FEECO	15–25
Soap, powder	1 cubic foot	FEECO	20–25
Soap, solid	1 cubic foot	FEECO	50
Soil/sandy loam, loose	1 cubic yard	Tellus	2,392
Stone or gravel	1 cubic foot	FEECO	95–100
Stone, crushed	1 cubic foot	FEECO	100
Stone, crushed, size reduced	1 cubic yard	Tellus	2,700
Stone, large	1 cubic foot	FEECO	100
Wax	1 cubic foot	FEECO	60.5
Wood ashes	1 cubic foot	FEECO	48

Glass

Item	Size	Study	Pounds
Glass, broken	1 cubic foot	FEECO	80–100.00
Glass, broken	1 cubic yard	FEECO	2,160.00
Glass, crushed	1 cubic foot	FEECO	40–50.0
Glass, plate	1 cubic foot	FEECO	172
Glass, window	1 cubic foot	FEECO	157
Glass, 1 gallon jug	each	USEPA	2.10–2.80
Glass, beer bottle	each	USEPA	0.53
Glass, beverage—8 oz	1 bottle	USEPA	0.5
Glass, beverage—8 oz	1 case = 24 bottles	USEPA	12
Glass, beverage—12 oz	1 case = 24 bottles	USEPA	22
Glass, wine bottle	each	USEPA	1.08

Plastic

Item	Size	Study	Pounds
Film plastic/mixed, loose	1 cubic yard	Tellus	22.55
HDPE Film plastics, semi-compacted	1 cubic yard	Tellus	75.96
LDPE Film plastics, semi-compacted	1 cubic yard	Tellus	72.32

Plastic (*Continued*)

Item	Size	Study	Pounds
HDPE, common beverage containers			
Plastic, HDPE juice, 8 oz.	8 oz.	USEPA	0.1
Plastic, 1 gallon HDPE jug	1 gallon	USEPA	0.33
Plastic, 1 gallon HDPE beverage cont.	milk/juice	USEPA	0.19–0.25
PETE, common beverage containers			
Plastic, 1 liter PETE beverage bottle w/o cap	1 liter	USEPA corr.	0.09
Plastic, PETE water bottle (over 1.5 liters)	50 oz.	USEPA	0.12
Plastic, PETE, 2 liter	1 bottle	USEPA	0.13
Plastic containers			
Plastic, 1 gallon container—mayo	1 gallon	USEPA	0.42
Plastic, 1/2 gallon plastic beverage cont.	1/2 gallon	USEPA	0.09
Plastic, beverage container	12 oz.	USEPA	0.05

Miscellaneous Plastic Items

Item	Size	Study	Pounds
Plastic, bubble wrap	33 gallons	USEPA	3
Plastic, bucket	25 gallons	USEPA	1.1
Plastic, bucket w/metal handle	5 gallons	USEPA	1.9
Plastic, cake decorator's boxes	each	USEPA	0.63
Plastic, grocery bag	100 bags	USEPA corr.	0.77
Plastic, HDPE 10–12 fluid oz.	.	USEPA	0.05
Plastic, HDPE beverage case	.	USEPA	1.2
Plastic, HDPE bread case	.	USEPA	1.5
Plastic, HDPE gallon containers (not beverage)	1 gallon	USEPA	0.06
Plastic, HDPE (auto) oil container	1 quart size	USEPA	0.2
Plastic, pot	1 qt size	USEPA	0.25
Plastic, mixed HDPE & PETE	1 cubic yard	USEPA	32
Plastic, pallet, 48" × 48"		USEPA	40

(Continued)

Miscellaneous Plastic Items (*Continued*)

Item	Size	Study	Pounds
Plastic, sheeting	square yard	USEPA	1
Plastic, whole, uncompacted PETE	1 cubic yard	USEPA	30–40
Polyethylene, resin pellets	1 cubic foot	FEECO	30–35
Polystyrene beads	1 cubic foot	FEECO	40
Polystyrene, packaging	33 gallon	USEPA	1.5
Styrofoam kernels	1 cubic yard	Tellus	6.27
Polystyrene, blown formed foam	1 cubic yard	Tellus	9.62
Polystyrene, rigid, whole	1 cubic yard	Tellus	21.76
PVC, loose	1 cubic yard	Tellus	341.12

Paper

Item	Size	Study	Pounds
Books, hardback, loose	1 cubic yard	Tellus	529.29
Books, paperback, loose	1 cubic yard	Tellus	427.5
Egg flats	one dozen	USEPA	0.12
Egg flats	12" × 12"	USEPA	0.5
Paper sacks	25# size	USEPA	0.5
Paper sacks	50# dry goods	USEPA	1
Calendars/books	1 cubic foot	FEECO	50
Catalogs	100 pages ledger	USEPA	1
Computer printout, loose	1 cubic yard	USEPA	655
Mixed paper, loose (construction, fax, manila, some chipboard)	1 cubic yard	USEPA	363.5
Mixed paper, compacted (construction, fax, manila, some chipboard)	1 cubic yard	USEPA	755
Office paper (white, color, CPO, junk mail)	13 gallons	USEPA	10.01
Office paper (white, color, CPO, junk mail)	33 gallons	USEPA	25.41
Office paper (white, color, CPO, junk mail)	55 gallons	USEPA	42.35
Shredded paper	33 gallons	USEPA	8
White ledger paper	12" stack	USEPA	12

Paper (*Continued*)

Item	Size	Study	Pounds
White ledger #20, 8.5" × 11"	1 ream (500 sheets)	USEPA	5
White ledger #30, 8.5" × 14"	1 ream (500 sheets)	USEPA	6.4
White ledger w/o CPO, loose	1 cubic yard	Tellus	363.51
White ledger, uncompacted stacked	1 cubic yard	USEPA	400
White ledger, compacted stacked	1 cubic yard	USEPA	800
Colored message pads	1 carton (144 pads)	USEPA	22
Padded envelope	9" × 12"	USEPA	0.88
Magazines, 8.5" × 11"	10 units	USEPA	3
Manila envelope	1 cubic foot	FEECO	37
Newspapers	12" stack	USEPA	35
Newspapers	1 cubic foot	FEECO	38
Newspapers, loose	1 cubic yard	USEPA	400
Newspapers, stacked	1 cubic yard	USEPA	875
Phone book	Ventura	USEPA	4
Paper pulp, stock	1 cubic foot	FEECO	60–62.00
Tab cards, uncompacted	1 cubic yard	USEPA	605
Tab cards, compacted	1 cubic yard	USEPA	1,275.00
Yellow legal pads	1 case (72 pads)	USEPA	38

Chipboard

Item	Size	Study	Pounds
Chipboard, beverage case	4 pack	USEPA	0.1
Chipboard, beverage case	6 pack	USEPA	0.2
Chipboard, cereal box	average	USEPA	0.15
Chipboard, fabric bolt		USEPA	0.69
Paperboard/boxboard/ chipboard, whole	1 cubic yard	Tellus	21.5

OCC

Item	Size	Study	Pounds
OCC, beverage case	4 six-packs full case	USEPA corr.	0.99
OCC, box, large	48" × 48" × 60"	USEPA	4
OCC, box, medium	24" × 24" × 30"	USEPA	2.2
OCC, box, small	12" × 12" × 15"	USEPA	1.1
OCC, flattened boxes, loose	1 cubic yard	Tellus	50.08
OCC, stacked	1 cubic yard	USEPA	50

(Continued)

OCC (*Continued*)

Item	Size	Study	Pounds
OCC, whole boxes	1 cubic yard	Tellus	16.64
OCC, uncompacted	1 cubic yard	USEPA	100
OCC, compacted	1 cubic yard	USEPA	400

Organics

Item	Size	Study	Pounds
Yard trims, mixed	1 cubic yard	USEPA	108
Yard trims, mixed	40 cubic yards	USEPA	4,320
Grass	33 gallons	USEPA	25
Grass	3 cubic yards	USEPA	840
Grass & leaves	3 cubic yards	USEPA	325
Large limbs & stumps	1 cubic yard	Tellus	1,080
Leaves, dry	1 cubic yard	Tellus	343.7
Leaves	33 gallons	USEPA	12
Leaves	3 cubic yards	USEPA	200–250
Pine needles, loose	1 cubic yard	Tellus	74.42
Prunings, dry	1 cubic yard	Tellus	36.9
Prunings, green	1 cubic yard	Tellus	46.69
Prunings, shredded	1 cubic yard	Tellus	527

Other Organic Material

Item	Size	Study	Pounds
Hay, baled	1 cubic foot	FEECO	24
Hay, loose	1 cubic foot	FEECO	5
Straw, baled	1 cubic foot	FEECO	24
Straw, loose	1 cubic foot	FEECO	3
Compost	1 cubic foot	FEECO	30–50
Compost, loose	1 cubic yard	Tellus	463.39

Food

Item	Size	Study	Pounds
Bread, bulk	1 cubic foot	FEECO	18
Fat	1 cubic foot	FEECO	57
Fats, solid/liquid (cooking oil)	1 gallon	USEPA	7.45
Fats, solid/liquid (cooking oil)	55 gallon drum	USEPA	410
Fish, scrap	1 cubic foot	FEECO	40–50

Food (*Continued*)

Item	Size	Study	Pounds
Meat, ground	1 cubic foot	FEECO	50–55
Oil, olive	1 cubic foot	FEECO	57.1
Oyster shells, whole	1 cubic foot	FEECO	75–80
Produce waste, mixed, loose	1 cubic yard	Tellus	1,443

Manure

Item	Size	Study	Pounds
Manure	1 cubic foot	FEECO	25
Manure, cattle	1 cubic yard	Tellus	1,628
Manure, dried poultry	1 cubic foot	FEECO	41.2
Manure, dried sheep & cattle	1 cubic foot	FEECO	24.3
Manure, horse	1 cubic yard	Tellus	1,252

Wood

Item	Size	Study	Pounds
Cork, dry	1 cubic foot	FEECO	15
Pallet, wood or plastic	average 48" × 48"	USEPA	40
Particle board, loose	1 cubic yard	Tellus	425.14
Plywood, sheet 2' × 4'	1 cubic yard	Tellus	776.3
Roofing/shake shingle, bundle	1 cubic yard	Tellus	435.3
Sawdust, loose	1 cubic yard	Tellus	375
Shavings, loose	1 cubic yard	Tellus	440
Wood chips, shredded	1 cubic yard	USEPA	500
Wood scrap, loose	1 cubic yard	Tellus	329.5
Wood, bark, refuse	1 cubic foot	FEECO	30-50
Wood, pulp, moist	1 cubic foot	FEECO	45–65
Wood, shavings	1 cubic foot	FEECO	15

Miscellaneous

Item	Size	Study	Pounds
Toner cartridge	.	USEPA	2.5

Rubber

Item	Size	Study	Pounds
Tire, bus	.	USEPA	75
Tire, car	.	USEPA	20

(Continued)

Rubber (*Continued*)

Item	Size	Study	Pounds
Tire, truck	.	USEPA	60–100
Rubber, car bumper	.	USEPA	15
Rubber, manufactured	1 cubic foot	FEECO	95
Rubber, pelletized	1 cubic foot	FEECO	50–55

Textiles

Item	Size	Study	Pounds
Clothing, used, mixed	cubic yard	Tellus	225
Fabric, canvas	square yard	USEPA	1
Leather, dry	1 cubic foot	FEECO	54
Leather, scrap, semi-compacted	1 cubic yard	Tellus	303
Rope	1 cubic foot	FEECO	42
String	yard	USEPA	1 gram
Used clothing, mixed, loose	1 cubic yard	Tellus	225
Used clothing, compacted	1 cubic yard	Tellus	540
Wool	1 cubic foot	FEECO	15–30
Carpet & padding, loose	1 cubic yard	Tellus	84.4

Metals

Aluminum

Item	Size	Study	Pounds
Aluminum foil, loose	1 cubic yard	Tellus	48.1
Aluminum scrap, cubed	1 cubic yard	Tellus	424
Aluminum scrap, whole	1 cubic yard	Tellus	175
Aluminum cans, uncrushed	1 case = 24 cans	USEPA	0.89
Aluminum cans, crushed	13 gallons	USEPA corr.	7.02
Aluminum cans, crushed	33 gallons	USEPA	17.82
Aluminum cans, crushed	39 gallons	USEPA	31.06
Aluminum cans, crushed & uncrushed mix	1 cubic yard	Tellus	91.4
Aluminum cans, uncrushed	1 full grocery bag	USEPA	1.5
Aluminum cans, uncrushed	13 gallons	USEPA	2.21
Aluminum cans, uncrushed	33 gallons	USEPA	5.61
Aluminum cans, uncrushed	39 gallons	USEPA	6.63
Aluminum cans (whole)	1 cubic yard	USEPA	65
Aluminum, chips	1 cubic foot	FEECO	15–20
Aluminum/tin cans commingled–uncrushed	33 gallons	USEPA	11.55

Ferrous

Item	Size	Study	Pounds
Metal scrap	55 gallons	USEPA	226.5
Metal scrap	cubic yard	USEPA	906
Metal, car bumper	each	USEPA	40
Paint can	5 gallons	USEPA	2.21
Radiator, ferrous	each	USEPA	20
Hanger (adult)	each	CIWMB	0.14
Hanger (child)	each	CIWMB	0.09
Tin can, ferrous	#2.5	USEPA	0.13
Tin can, ferrous	#5	USEPA	0.28
Tin can, ferrous	#10	USEPA	0.77
Tin coated steel cans	1 cubic yard	USEPA	850
Tin coated steel cans	1 case (6 #10 cans)	USEPA	22
Tin, tuna can (3/4 of #10)	each	USEPA	0.58
Tin, cat food can, ferrous	8 oz.	USEPA	0.14
Tin, dog food can, large	22 oz.	USEPA	0.22
Tin, dog food can, Vet	15.5 oz.	USEPA	0.11
Tin, cast	1 cubic foot	FEECO	455
Cast iron chips or borings	1 cubic foot	FEECO	130–200
Iron, cast ductile	1 cubic foot	FEECO	444
Iron, ore	1 cubic foot	FEECO	100–200
Iron, wrought	1 cubic foot	FEECO	480
Steel, shavings	1 cubic foot	FEECO	58–65
Steel, solid	1 cubic foot	FEECO	487
Steel, trimmings	1 cubic foot	FEECO	75–150
Brass, cast	1 cubic foot	FEECO	519
Brass, scrap	1 cubic yard	Tellus	906.43
Bronze	1 cubic foot	FEECO	552
Copper fittings, loose	1 cubic yard	Tellus	1,047.62
Copper pipe, whole	1 cubic yard	Tellus	210.94
Copper, cast	1 cubic foot	FEECO	542
Copper, ore	1 cubic foot	FEECO	120–150
Copper, scrap	1 cubic yard	Tellus	1,093.52
Copper, wire, whole	1 cubic yard	Tellus	337.5
Chrome ore (chromite)	1 cubic foot	FEECO	125–140
Lead, commercial	1 cubic foot	FEECO	710
Lead, ores	1 cubic foot	FEECO	200–270
Lead, scrap	1 cubic yard	Tellus	1,603.84
Nickel, ore	1 cubic foot	FEECO	150
Nickel, rolled	1 cubic foot	FEECO	541

Furniture

Item	Type	Material	Size	Pounds
Desk	Executive, single pedestal	Wood	72" × 36"	246.33
Desk	Executive, double pedestal	Wood	.	345
Desk	Double pedestal	Laminate	72" × 36"	299.5
Desk	Double pedestal	Laminate	60" × 30"	231
Desk	Single pedestal	Laminate	72" × 36"	250
Desk	Single pedestal	Laminate	42" × 24"	146
Desk	Double pedestal	Metal	72" × 36"	224.67
Desk	Double pedestal	Metal	60" × 30"	184.75
Desk	Double pedestal	Metal	54" × 24"	124
Desk	Single pedestal	Metal	72" × 36"	189
Desk	Single pedestal	Metal	48" × 30"	133.67
Desk	Single pedestal	Metal	42" × 24"	146
Desk	Single pedestal	Metal	40" × 20"	82
Desk	Small modular panel system	.	.	422
Desk	Large modular panel system	.	.	650
Workstation	with return	Laminate	60" × 30"	329.33
Workstation	with return	Metal	60" × 30"	230.67
Bridge	Executive	Wood	.	76.67
Bridge	.	Laminate	.	140
Credenza	.	Wood	.	250.78
Credenza	.	Laminate	.	230.14
Credenza	with knee space	Metal	60" × 24"	156.67
Round conference table	.	Wood	42" diameter	91.5
Bookcase	3 shelves	Wood	36" wide	90
Bookcase	4 shelves	Wood	36" wide	110.9
Bookcase	5 shelves	Wood	36" wide	138.8
Bookcase	6 shelves	Wood	36" wide	134.6
Bookcase	7 shelves	Wood	34" wide	138.5
Bookcase	4 shelves	Laminate	.	85
Bookcase	5 shelves	Laminate	.	110
Bookcase	2 shelves	Metal	34"–36"	44.5
Bookcase	3 shelves	Metal	34"–36"	57.5
Bookcase	4 shelves	Metal	34"–36"	70.5
Bookcase	5 shelves	Metal	34"–36"	89
Bookcase	6 shelves	Metal	34"–36"	101
File cabinet	2 drawer, lateral	Wood	.	155.14
File cabinet	2 drawer, lateral	Laminate	.	171.5
File cabinet	2 drawer, lateral	Metal	30"–42"	230.67

Furniture (*Continued*)

Item	Type	Material	Size	Pounds
File cabinet	4 drawer, lateral	Metal	36"	207.33
File cabinet	2 drawer, vertical	Metal	Letter size	60.6
File cabinet	4 drawer, vertical	Metal	Letter size	107.6
File cabinet	2 drawer, vertical	Metal	Legal size	71.5
File cabinet	4 drawer, vertical	Metal	Legal size	123.5
Chair	Executive desk	.	.	51.167
Chair	Guest arm	.	.	38.2
Chair	Swivel arm	.	.	45.25
Chair	Secretary with no arms	.	.	31.76
Chair	Stacking	.	.	15.83
Personal computer	CPU (Central Processing Unit)	.	.	26
Computer monitor	.	.	.	30
Computer printer	.	.	.	25.33

* This data is from the U.S. EPA Business Users Guide Study.

Source Acronyms Used:

CIWMB	California Integrated Waste Management Board
FEECO	FEECO Incorporated
Tellus	Tellus Institute, Boston, Massachusetts
USEPA	United States Environmental Protection Agency *Business Users Guide*

Source: Based on Appendix J of Conducting a Diversion Study - A guide for California Jurisdictios, March 2001, California Integrated Waste Management Board. The table is a compilation of data from various sources listed at the end of the appendix.

References

EPA Municipal and Industrial Solid Waste, Washington, D.C. and University of California Los Angeles Extension Recycling and Municipal Solid Waste Management Program. 1996. *Business Waste Prevention Quantification Methodologies—Business Users Guide* (November). Los Angeles, California.

EPA. 1997. *Measuring Recycling: A Guide For State and Local Governments* (September). Publication number EPA530-R-97-011.

Cal Recovery Inc., Tellus Institute, and ACT... now. 1991. *Conversion Factors for Individual Material Types*. Submitted to California Integrated Waste Management Board (December).

FEECO International, Inc. *FEECO International Handbook*, 8th Printing (Section 22-45 to 22-510). Green Bay, Wisconsin.

Harivandi, M. Ali, Victor A. Gibeault, and Trevor O'Shaughnessy. 1996. "Grass-cycling in California." *California Turfgrass Culture* 46: 1–2.

Alameda County Waste Management Authority Home Composting Survey. 1992. *Source Reduction Through Home Composting*. Summary published in *Biocycle* (April).

Conversions

Multiply	by	to obtain
acres	0.404	ha
acres	43,560	ft^2
acres	4,047	m^2
acres	4,840	yd^2
acre-ft	1233	m^3
atmospheres	14.7	lb/in^2
atmospheres	29.95	in. mercury
atmospheres	33.9	ft of water
atmospheres	10,330	kg/m^2
Btu	252	cal
Btu	1.053	kJ
Btu	1,053	J
Btu	2.93×10^{-4}	kWh
Btu/ft^3	8,905	cal/m^3
Btu/lb	2.32	kJ/kg
Btu/lb	0.555	cal/g
Btu/s	1.05	kW
Btu/ton	278	cal/tonne
Btu/ton	0.00116	kJ/kg
calories	4.18	J
calories	0.0039	Btu
calories	1.16×10^{-6}	kWh
calories/g	1.80	Btu/lb
$calories/m^3$	0.000112	Btu/ft^3
calories/tonne	0.00360	Btu/ton
centimeters	0.393	in.
cubic ft	1728	in^3
cubic ft	7.48	gal
cubic ft	0.0283	m^3
cubic ft	28.3	L
cubic ft/lb	0.0623	m^3/kg
cubic ft/s	0.0283	m^3/s
cubic ft/s	449	gal/min
cubic ft of water	61.7	lb of water
cubic in. of water	0.0361	lb water
cubic m	35.3	ft^3
cubic m	264	gal

Multiply	by	to obtain
cubic m	1.31	yd^3
cubic m/day	264	gal/day
cubic m/hr	4.4	gal/min
cubic m/hr	0.00638	million gal/day
cubic m/s	1	cumec
cubic m/s	35.31	ft^3/s
cubic m/s	15,850	gal/min
cubic m/s	22.8	mil gal/day
cumec	1	m^3/s
cubic yards	0.765	m^3
cubic yards	202	gal
C-ration	100	rations
decacards	52	cards
feet	0.305	m
feet/min	0.00508	m/s
feet/s	0.305	m/s
fish	10^{-6}	microfiche
foot lb (force)	1.357	J
foot lb (force)	1.357	Nm
gallons	0.00378	m^3
gallons	3.78	L
gallons/day	43.8×10^{-6}	L/s
gallons/day/ft^2	0.0407	m^3/day/m^2
gallons/min	0.00223	ft^3/s
gallons/min	0.0631	L/s
gallons/min	0.227	m^3/hr
gallons/min	6.31×10^{-5}	m^3/s
gallons/min/ft^2	2.42	m^3/hr/m^2
gallons of water	8.34	lb water
grams	0.0022	lb
grams/cm^3	1,000	kg/m^3
hectares	2.47	acre
hectares	1.076×10^5	ft^2
horsepower	0.745	kW
horsepower	33,000	ft-lb/min
inches	2.54	cm
inches of mercury	0.49	lb/in^2
inches of mercury	0.00338	N/m^2
inches of water	249	N/m^2
joules	0.239	cal
joules	9.48×10^{-4}	Btu
joules	0.738	ft-lb
joules	2.78×10^{-7}	kWh
joules	1	Nm
joules/g	0.430	Btu/lb
joules/s	1	W
kilocalories	3.968	Btu
kilocalories/kg	1.80	Btu/lb

(Continued)

Multiply	by	to obtain
kilograms	2.20	lb
kilograms	0.0011	tons
kilograms/ha	0.893	lb/acre
kilograms/hr	2.2	lb/hr
kilograms/m^3	0.0624	lb/ft^3
kilograms/m^3	1.68	lb/yd^3
kilograms/m^3	1.69	lb/yd^3
kilograms/tonne	2.0	lb/ton
kilojoules	9.49	Btu
kilojoules/kg	0.431	Btu/lb
kilometers	0.622	mile
kilometer/hr	0.622	miles/hr
kilowatts	1.341	horsepower
kilowatt-hr	3600	kJ
lite year	365	days drinking low-calorie beer
liters	0.0353	ft^3
liters	0.264	gal
liters/s	15.8	gal/min
liters/s	0.0228	million gal/day
megaphone	10^{12}	microphone
meters	3.28	ft
meters	1.094	yd
meters/s	3.28	ft/s
meters/s	196.8	ft/min
microscope	10^{-6}	mouthwash
miles	1.61	km
miles/hr	0.447	m/s
miles/hr	88	ft/min
miles/hr	1.609	km/hr
milligrams/L	0.001	kg/m^3
million gal	3,785	m^3
million gal/day	43.8	L/s
million gal/day	3785	m^3/day
million gal/day	157	m^3/hr
million gal/day	0.0438	m^3/s
newtons	0.225	lb(force)
newtons/m^2	2.94×10^{-4}	in. mercury
newtons/m^2	1.4×10^{-4}	lb/in^2
newtons/m^2	10	poise
newton-m	1	J
pounds (mass)	0.454	kg
pounds (mass)	454	g
pounds (mass)/acre	1.12	kg/ha
pounds(mass)/ft^3	16.04	kg/m^3
pounds (mass)/ton	0.50	kg/tonne
pounds (mass)/yd^3	0.593	kg/m^3
pounds (force)	4.45	N

Multiply	by	to obtain
pounds (force)/in^2	0.068	atmospheres
pounds (force)/in^2	2.04	in. of mercury
pounds (force)/in^2	6895	N/m^2
pounds (force)/in^2	6.89	kPa
pound cake	454	Graham crackers
pounds of water	0.01602	ft^3
pounds of water	27.68	in^3
pounds of water	0.1198	gal
square ft	0.0929	m^2
square m	10.74	ft^2
square m	1.196	yd^2
square miles	2.59	km^2
tons (2000 lb)	0.907	tonnes (1000 kg)
tons	907	kg
tons/acre	2.24	tonnes/ha
tonnes (1000 kg)	1.10	ton (2000 lb)
tonnes/ha	0.446	tons/acre
two kilomockingbirds	2000	mockingbird
watts	1	J/s
yard	0.914	m

Composition and Analysis of Waste, Raw Material, and Fuels

by Jörg Wochele
PAUL SCHERRER INSTITUT

Abstract

CAWAF is a database of elemental Composition and Analysys of Waste, raw material, And Fuels for investigations of thermal treatment. The data collection consists of more than 150 datasets of main, minor, and trace elements, ash, water content, and heating values. Diagrams allow material characterization and comparisons of the different materials.

Contents

D-1   INTRODUCTION

Heating values, proximate, and ultimate analysis are the common data of fuels for combustion calculation and material characterization. This kind of information is also valuable for waste, biomasses, raw, and other materials to calculate and investigate their behavior by thermal treatment (pyrolysis, gasification, combustion, tempering, calcinations, annealing, etc.). In this appendix, data of different materials and of different sources were brought together on the same common base (element amounts on mass basis) including elemental analysis of main elements (C H O N S Cl), mineral elements (Si, Fe, Al, Ca, Mg, Na, K, Ti), heavy metals (Zn, Cu, Pb, Cd, Hg, Cr, Ni), and minor and trace elements (As, F, I, Br, etc.) for quick access and to allow investigations of very different questions.

Depending on the focus, every science (combustion, mineralogy, glass, cement, waste, environmental, system analysis, or system design) has its own preferences in presenting the data; sometimes small pieces of information are excluded in one presentation but are included in presentations in other neighboring fields. Water content or humidity is such a detail; it is often missing in publications of costly prepared analysis. Lack of clear declaration of the base (wet, dry, ash free, only ash (wet, dry)) is another problem.

The main purpose of this appendix is to present heavy metal and trace elements content in thermal treating materials. Whereas data of the main elements from ultimate analysis are available in numerous publications, minor and trace elements are quite scarce and not often given together with the main elements. Therefore, also incomplete datasets and secondary sources are included, as a guideline when no other data are available and in the hope that, together with other datasets, we can guess or reconstruct the missing values.

For an overview, to compare the very different materials, and for evaluating missing values, H/C-O/C diagrams, ternary C H O diagrams, and accumulation diagrams referring to Earth's crust are included.

Every material dataset (waste, fuels) has a name and a @-number for identification. The material datasets are divided in different tables (Tables D-5-1 to D-5-5). The entries are grouped in

- solid fuels (coal, peat, wood)

- solid waste

- solid products (raw materials, plastics, etc.)

- liquid waste, sludge, etc.

- liquids (fuels, alcohols, etc.)

- others (gaseous fuels, etc.)

This data compilation grew from our data needs during years of investigating the heavy metals in thermal treatment and has become a considerable data collection. An effort was made to depict the loose collected material in a readable form in the hope that the material presented here will be useful for others.

D-2 DEFINITIONS

The data derived from the different analyses were interpreted and linked in the following way (mass basis; e.g., g/kg fuel):

Waste or Fuel

organic substance		minerals	water	
combustibles		ash	water	
fixed C	volatiles	ash	water	
C H O N S Cl		ash	water	on wet basis
Si Fe Ca Al Mg	Na K Ti			on dry basis
... Zn Cu Pb Cd Hg	Cr Ni ...			on dry basis

raw, as received	(ar) or (wet) or (raw)			100%	**wet basis**
dry	(dry)		100%		**dry basis**
dry, ash free	(daf)	100%			

ar, wet, raw: $100\% = w + a + C + H + O + N + S + Cl$ [wt/%]
dry: $100\% = a + C + H + O + N + S + Cl$ [wt/%]
daf: $100\% = C + H + O + N + S + Cl$ [wt/%]
C O H: $100\% = C + H + O$ [wt/%] or [mol/mol]

    a = ash
   w = water, humidity
 wet = with water content
   ar = as received
  dry = dried matter
  daf = dried and ash free
CHO = carbon C, oxygen O, hydrogen H; $C + H + O = 100\%$

D-2-1 Fuel

Fuel is taken for short for waste, fuel, raw materials, products, materials, goods (usually on a wet mass basis), for organics (C H O N S Cl), and on a dry mass basis for mineral and trace elements. In some sources chlorine is not counted to organics.

D-2-2 Ash, Minerals

In general, in this discussion no distinctions are made between ash content and mineral matter, because of lack of information (for exceptions, see Table 4, "remarks"). Elemental composition of fuel may be calculated from given ash analysis and marked with #2.

D-2-3 Water

The water content (humidity) can be based on "raw," "as received," "or water content of substance in practical application" (e.g., sludge). Natural drying during storage of goods takes place, and therefore the value can be different from the analyzed material. In general, water content is given here as received or water of the analyzed good. (In German: Analyse-feucht = ar.) In goods, water content may vary considerably.

D-3   CONVERSIONS

Conversions from one system to another are simple, but it is convenient to have them at hand:

Elements are: C H O N S Cl .. Si Fe .. P Br .. Zn Cu .. As

a = ash

w = water, humidity

Conversion from … to …

ar → dry:	$z_{dry} = z_{ar}/(1-w)$	[-] (wt);	z = a, element.
ar → daf:	$z_{daf} = z_{ar}/(1-w-a_{ar})$	[-] (wt);	z = element.
ar → CHO:	$zCHO = z_{ar}/(C_{ar}+H_{ar}+O_{ar})$	[-] (wt);	z = C, H, O.

dry → ar:	$z_{ar} = z_{dry} * (1-w)$	[-] (wt);	z = a, element.
dry → daf:	$z_{daf} = z_{dry}/(1-a_{dry})$	[-] (wt);	z = element.
dry → CHO:	$z_{CHO} = z_{dry}/(C_{dry} + H_{dry} + O_{dry})$	[-] (wt);	z = C, H, O.

daf → ar:	$z_{ar} = z_{daf} * (1-w-a_{ar})$	[-] (wt);	z = a, element.
daf → dry:	$z_{daf} = z_{daf} * (1-a_{dry})$	[-] (wt);	z = element.
daf → CHO:	$z_{CHO} = z_{daf}/(C_{daf}+H_{daf}+O_{daf})$	[-] (wt);	z = C, H, O.

CHO → ar:	$z_{ar} = z_{CHO} * (1-a_{ar}-w-N_{ar}-S_{ar}-Cl_{ar})$	[-] (wt);	z = a, element.
CHO → dry:	$z_{dry} = z_{CHO} * (1-a_{dry}-N_{dry}-S_{dry}-Cl_{dry})$	[-] (wt);	z = element.
CHO → daf:	$z_{daf} = z_{CHO} * (1 - N_{daf}-S_{daf}-Cl_{daf})$	[-] (wt);	z = C, H, O.

H/C [-] (wt) →   H/C [-] (mol) = H/C [g/kg fuel/g/kg fuel] * 11.9169 (MM-C/MM-H)

O/C [-] (wt) →   O/C [-] (mol) = O/C [g/kg fuel/g/kg fuel] * 0.75072 (MM-C/MM-O)

Chemical formula normalized to C = 1: $C_1H_xO_yN_zS_vCl_w$

C [mol] =1

H [mol] = H [g] / C [g] * 11.9169

O [mol] = O [g] / C [g] * 0.75072

N [mol] = N [g] / C [g] * 0.85752

S [mol]   = S [g] / C [g] * 0.37457

Cl [mol] = Cl [g] / C [g] * 0.33879

Element of dry basis z_{dry} (as in Tables D-5-2, D-5-3, D-5-4) → element of dry ash $z_{dry\ ash}$ (#2)

$$z_{dry\ ash} = z_{dry} / a \qquad\qquad z = element$$

Element of dry basis z_{dry} (as in Tables D-5-2, D-5-3, D-5-4) $\rightarrow$ oxides of dry ash $ox_{dry\ ash}$ (#2)

$$ox_{dry\ ash} = z_{dry} / a * F$$

z = element

ox = oxid of z

F = factor, see Table D-1

Table D-1 **Molar Mass of Mineral Elements and Their Oxides as well as Conversion Factors F**

Element	Molar Mass Elements [g/mol]	Oxid	Molar Mass Oxid [g/mol]	Conversion Factor F (element → oxid)
Al	26.98	Al_2O_3	101.96	1.890
Ba	137.327	BaO	153.326	1.117
Ca	40.08	CaO	56.08	1.399
Cd	112.411	CdO	128.410	1.142
Cr	51.996	Cr_2O_3	151.989	2.923
Cu	63.546	CuO	79.545	1.252
Fe	55.85	Fe_2O_3	159.70	1.430
K	39.10	K_2O	94.20	1.205
Mg	24.31	MgO	40.31	1.658
Mn	54.04	MnO_2	86.93	1.609
Na	23.0	Na_2O	62.0	1.348
P	30.974	P_2O_5	141.94	4.583
Pb	207.2	PbO	223.199	1.077
S	32.066	SO_3	80.064	2.497
Si	28.09	SiO_2	60.09	2.139
Sr	87.620	SrO	103.619	1.183
Ti	47.88	TiO_2	79.88	1.668
Zn	65.39	ZnO	81.39	1.245

D-4 DIAGRAMS

Different diagrams for characterization/classification of the materials are in use; quantities are measured either in mass or in moles:

D-4-1 H/C-O/C-Diagrams

- are independent of the basis (wet, dry, daf)
- give orientation about carbon content
- show the carbonating degree
- are based on wt or mol ratios

D-4-2 Ternary-Diagrams: C H O Diagram

C H O diagram

- wt-% and mol-% diagrams

- C + H + O = 100%

- describes the organic part of the material

- shows the carbonating degree

D-4-3 Accumulation Factor Diagram

For mineral elements (Si, Al, Fe, Ca, Mg, Na, K, Ti) the ratio to Earth's crust is as given in Figure 4. Earth's crust's elemental composition is @-number 155.

$$AF = C_z(fuel) / C_z(Earth's\ crust) \qquad\qquad C\ [g/kg\ dry] \qquad z = element$$

The diagrams presented here (Figures D-1 through D-4) are built from the datasets of Tables D-5-1 through D-5-4 at end of this report.

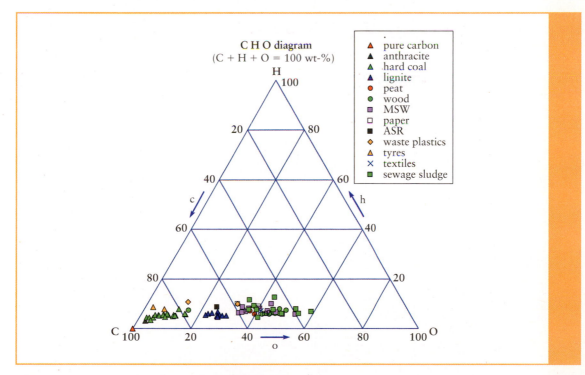

Figure D-1 C H O diagram in wt-% for waste and fuels.

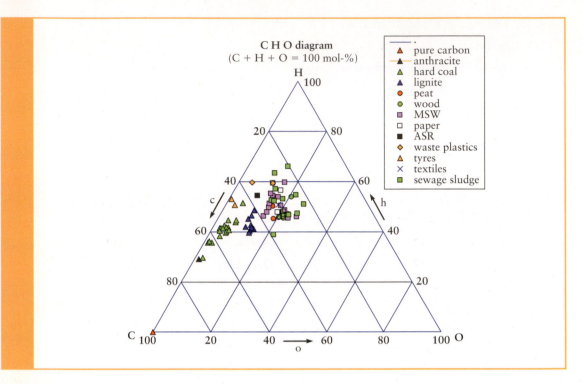

Figure D-2 C H O diagram in mol-% for waste and fuels.

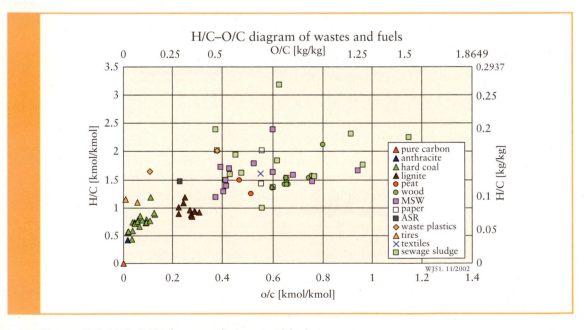

Figure D-3 H/C–O/C diagram of waste and fuels.

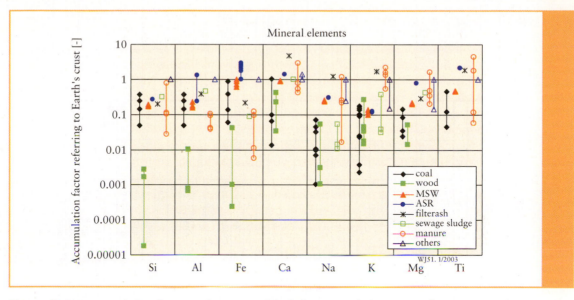

Figure D-4 Accumulation factors of waste and fuels for mineral elements referring to Earth's crust (Earth's crust: @-number = 155).

D-5 TABLES

– Table D-5-1: Ultimate and proximate analysis, heating value [C H O N S Cl, water, ash, Hu, Ho] (wet mass basis)

– Table D-5-2: Elemental analysis of mineral elements [Si Fe Ca Al Mg Na K] (dry mass basis)

– Table D-5-3: Heavy metals, minor elements [P F Br I Zn Cu Pb Cd Hg Cr Ni As] (dry mass basis)

– Table D-5-4: Remarks, further elements (dry mass basis)

– Table D-5-5: References

Note that Table 1 is on a wet mass basis, whereas the other tables are on a dry mass basis. In the tables, fuel is used as abbreviation for waste, fuel, raw materials, products, materials, goods.

Many of the table entries have their roots in the German language. The German terms used in the tables are defined therein. In Table D-5-1 the datasets are in order of the groups (solid fuels, solid waste, etc.), and in Table D-5-4 the order is based on the @-number.

-PSI-

Composition and Analysis of Wastes, Raw Materials and Fuels

Table D-5-1: proximate and ultimate analysis, heating values

CAWAF Tab 1, page 1
WJ51, 05.05.03

CAWAF1.xls

last= 157

solid fuels (coal, peat, wood, energy crop)

@-No.	waste or fuel		ref.	remarks	proximate analysis (massfraction, sum=1)		ultimate analysis (massfraction, sum=1) (wet mass basis)								lower heating value Hu	upper heating value Ho
					fixed carbon	volatile matter	ash,inert a	water w	carbon c	hydrogen h	oxigen o	nitrogen n	sulfur s	chlorine cl		
					kg/kg fuel	kg/kg fuel	kg/kg fuel	kg/kg fuel	kg/kg fuel	kg/kg fuel	kg/kg fuel	kg/kg fuel	kg/kg fuel	kg/kg fuel	MJ/kg fuel	MJ/kg fuel
26	carbon (pure)	Kohlenstoff (rein)	/1/				0	0	1	0	0	0	0	0	33.82	33.82
12	pure carbon	reiner Kohlenstoff	/2/				0	0	1	0	0	0	0	0	33.82	33.82
32	anthracite	Anthrazit (Ruhr)	/1/				0.05	0.03	0.85	0.03	0.02	0.01	0.01		31.4	
14	anthracite	Anthrazit (Ruhr)	/2/				0.03	0.05	0.85	0.03	0.02	0.01	0.01		31.4	32.13
33	coke (blast furn.)	Koks (Hochofen,Ruhr)	/1/				0.09	0.05	0.83	0.005	0.005	0.01	0.01		28.9	
13	coke	Koks (Gaskoks)	/2/				0.12	0.015	0.84	0.003	0.01	0.005	0.007		29.3	29.4
123	coke from..to	Koks (VDZ)	/26/	*123												
124	coke	Koks (ZKG)	/26/	*124												
31	coal	Magerkohle (Ruhr)	/1/				0.05	0.03	0.84	0.04	0.02	0.01	0.01		31.4	
30	coal	Esskohle (Ruhr)	/1/				0.05	0.03	0.82	0.04	0.04	0.01	0.01		31.8	
29	bituminous coal	Fettkohle (Ruhr)	/1/				0.05	0.03	0.81	0.05	0.04	0.01	0.01		31.8	
55	bituminous coal	(Goettelborn, D)	/11/	*55	0.522	0.3	0.089	0.089	0.6723	0.0446	0.0832	0.0128	0.0073	0.001813	27.03	
57	bituminous coal		/12/		0.491	0.265	0.154	0.09	0.6237	0.0378	0.0748	0.0129	0.0068	0.000053		25
69	bituminous coal (Pittsburgg)		/17/	*69	0.5277	0.3774	0.0877	0.0072	0.7492	0.0496	0.0609	0.0148	0.0306			
142	bituminous		/33/	*142												
154	coal		/35/	*154 #2	0.58	0.23	0.14	0.05					0.000376			
28	coal	Gaskohle (Ruhr)	/1/				0.05	0.03	0.8	0.05	0.05	0.01	0.01		31.4	
27	coal	Gasflammkohle (Ruhr)	/1/				0.05	0.03	0.77	0.05	0.08	0.01	0.01		30.1	
34	coal	Pechkohle (Oberbayern)	/1/				0.11	0.1	0.58	0.043	0.1	0.012	0.055		22.93	
47	black coal (typical)		/7/				0.138	0.108	0.59	0.045	0.0998	0.01	0.009	0.0002	23.6	25.2
48	black coal from		/7/	*48			0.08	0.08	0.55	0.034	0.07	0.003	0.005	0	23.5	25
	ff. to						0.15	0.12	0.8	0.08	0.12	0.02	0.015	0.001	27.5	29
140	hard coal	Steinkohle	/32/	*140										0.0009		
15	hard coal from	Steinkohle	/2/				0.03	0.03	0.7	0.034	0.018	0.011	0.006		28.05	28.87
	ff. to						0.08	0.05	0.83	0.053	0.12	0.011	0.009		33.07	34.36
61	polish coal		/13/				0.0686	0.086	0.7084	0.0512	0.0636	0.0145	0.007	0.0008		
53	silantek coal		/9/		0.615	0.242	0.131	0.012	0.7414	0.0442	0.0465	0.0183	0.0066		31.474	28.93

-PSI-

Composition and Analysis of Wastes, Raw Materials and Fuels

Table D-5-1: proximate and ultimate analysis, heating values

CAWAF Tab 1, page 2
WJ51, 05.05.03

@-No.	waste or fuel	ref.	remarks	proximate analysis (massfraction, sum=1)			ultimate analysis (massfraction, sum = 1) (wet mass basis)							lower heating value Hu	upper heating value Ho
last= 157				fixed carbon	volatile matter	ash.inert a	water w	carbon c	hydrogen h	oxigen o	nitrogen n	sulfur s	chlorine cl	MJ/kg fuel	MJ/kg fuel
				kg/kg fuel	kg/kg fuel	kg/kg fuel	kg/kg fuel	kg/kg fuel	kg/kg fuel	kg/kg fuel	kg/kg fuel	kg/kg fuel	kg/kg fuel	MJ/kg fuel	MJ/kg fuel
79	hard coal Steinkohle (D) Trockenfeuerung	/3/	*79			0.058	0.109	0.71	0.045	0.054	0.014	0.01			
80	hard coal (slag-type furnace) (Steinkohle (D) Schmelzfeuerung)	/3/	*80			0.15	0.094	0.642	0.036	0.055	0.013	0.01			
157	coal German-Anna	/37/	#1 #3	0.796	0.12	0.084	0	0.8262	0.0302	0.0366	0.0092	0.0073			33
141	lignite/sub-bituminous	/33/	*141												
35	lignite Braunkohle, roh (Rheinland)	/1/				0.05	0.5	0.3	0.03	0.1	0.01	0.01		9.63	
16	lignite from Rohbraunkohle	/2/				0.03	0.5	0.25	0.02	0.09	0.003	0.002		7.53	9.19
	ff. to					0.057	0.6	0.32	0.025	0.124	0.004	0.01		10.46	12.47
81	lignite Braunkohle (D)	/3/	*81			0.093	0.538	0.245	0.019	0.101	0.002	0.002			
70	pulverized lignite Braunkohlestaub (D)	/18/		0.2991	0.3807	0.1002	0.22	0.4623	0.0339	0.1714	0.0055	0.0066		16.989	
71	pulverized lignite Trocken-Braunkohlestaub (D)	/18/		0.3375	0.4295	0.113	0.12	0.5216	0.0383	0.1934	0.0062	0.0075		19.48	
17	lignite briquets Braunkohlebriketts (Mitteldeutschland)	/2/				0.089	0.15	0.53	0.045	0.159	0.006	0.021		20.93	22.28
36	lignite briquets Braunkohlebriketts (Rheinand)	/1/				0.05	0.15	0.55	0.05	0.18	0.01	0.01		19.25	
68	lignite (South Banko)	/17/	*68	0.3962	0.504	0.0245	0.0753	0.6423	0.0483	0.194	0.0106	0.005			
156	German Lignite	/37/	#1 #3	0.4603	0.4947	0.045	0	0.6389	0.0497	0.2454	0.0057	0.0048			25.1
73	lignite Braunkohle	/20/		0.304	0.379	0.039	0.278	0.466	0.033	0.173	0.005	0.006		16.918	
83	coal from Kohle	/21/	*83 *83a									0.002	0.00002		
	ff. to											0.065	0.006		
121	coal from..to Kohle (NFP)	/26/	*121												
122	coal from..to Kohle (KKL)	/26/	*122												
37	peat (air-dry) Torf, lufttrocken	/1/				0.05	0.25	0.38	0.04	0.26	0.01	0.01		13.8	
18	peat (air-dry) Torf, luftgetrocknet	/2/				0.07	0.2	0.4	0.05	0.25	0.02	0.01		15.49	17.07
38	wood (air-dry) Holz, lufttrocken	/1/				0.01	0.15	0.42	0.05	0.37				14.65	
19	wood (air-dry) Holz, luftgetrocknet	/2/				0.005	0.15	0.44	0.05	0.35	0.005	0		15.49	16.95
105	wood unbelasteted Holz	/25/				0.0032	0.447	0.2754	0.0355	0.2386	0.00006	0.00006	0.00022		
41	natural wood, dry	/4/	*41 #1			0.004	0	0.499	0.061	0.4345	0.0015	0.00006	0.00005		
67	Robinia pseudoacacia (wood)	/16/	*67 #1	0.193	0.792	0.015	0	0.439	0.078	0.4678	0.0002	0	0	16.6	
74	wood chips Holzschnitzel	/20/	*74	0.219	0.552	0.009	0.22	0.357	0.047	0.359	0.007	0.001		11.197	

-PSI-

Composition and Analysis of Wastes, Raw Materials and Fuels

Table D-5-1: proximate and ultimate analysis, heating values

CAWAF Tab 1, page 3
WJ51, 05.05.03

last= 157

@-No.	waste or fuel	ref.	remarks	\multicolumn{4}{c}{proximate analysis (massfraction, sum=1)}				\multicolumn{6}{c}{ultimate analysis (massfraction, sum=1) (wet mass basis)}						lower heating value Hu	upper heating value Ho
				fixed carbon	volatile matter	ash,inert a	water w	carbon c	hydrogen h	oxigen o	nitrogen n	sulfur s	chlorine cl	MJ/kg fuel	MJ/kg fuel
				kg/kg fuel	kg/kg fuel	kg/kg fuel	kg/kg fuel	kg/kg fuel	kg/kg fuel	kg/kg fuel	kg/kg fuel	kg/kg fuel	kg/kg fuel		
58	wood pellets	/12/		0.1715	0.7312	0.0073	0.09	0.45507	0.0543	0.3918	0.00126	0.00018	0.00009	17	
42	urban waste wood, dry	/4/	*42 #1			0.053	0	0.443	0.057	0.438	0.008	0.001385	0.000849		
98	wood in MSW Holz	/24/	*98 *94			0.047	0.0302		0.0475					12.84	20.63
106	waste wood Altholz	/25/	*106			0.1606	0.05	0.3933	0.0466	0.3427	0.0029	0.0001	0.0038		
128	waste wood Altholz (Infras)	/26/	*128											17	
129	waste wood mean, max Altholz (AHK)	/26/	*129											17	
144	Biomass (wood,straw,grass) from..to	/33/	*144												
145	Biomass (straw/grass) from..to	/33/													
146	Biomass (wood) from..to	/33/													
148	pine shavings	/35/	*148 #2	0.0836	0.4658	0.0006	0.45					1.20E-07			
149	reed canary grass	/35/	*149 #2	0.069	0.265	0.014	0.652					1.10E-06			
138	corn	/31/	*138	0.1554	0.692	0.0136	0.139	0.3624	0.0623	0.4101	0.0108	0.0018		13.653	15.361
139	soybean	/31/	*139	0.0824	0.7413	0.0513	0.125	0.4165	0.0648	0.2851	0.0548	0.0026		17.287	19.012
	solid wastes														
1	municipal solid waste Hausmüll	/2/				0.194	0.241	0.284	0.039	0.227	0.009	0.001	0.005	10.6	12.03
2	municipal solid waste Hausmüll (Kap.2)	/2/				0.273	0.347	0.222	0.026	0.123	0.009			8	9.4
11	municipal solid waste from Hausmüll ff.	/2/	*11			0.25	0.15	0.28	0.04	0.16	0.002	0.003	0.004	7	
	to					0.35	0.35	0.4	0.05	0.22	0.013	0.005	0.01	15	
39	municipal solid waste from Hausmüll ff.	/5/	*39			0.2	0.25	0.2	0.02	0.1	0.005	0.005		6	
	to					0.35	0.3	0.25	0.05	0.2	0.005	0.005		10	
51	high volatile solid waste	/8/	*51	0.0707	0.4153	0.029	0.485	0.2266	0.028	0.2288	0.0013	0.0013			
64	houshold waste Hausmüll	/14/	*64			0.1413	0.29	0.3	0.045	0.21	0.005	0.0014	0.0073	10.5	
65	municipal solid waste Siedlungsabfall	/14/	*65			0.1118	0.24	0.38	0.055	0.2	0.005	0.0013	0.0069	12.5	
110	municipal solid waste Siedlungsabfall	/27/	*110			0.1266	0.25	0.37	0.04	0.2	0.005	0.0013	0.0071		
94	municipal solid waste Wiener Hausmüll	/24/	*94			0.2747	0.301	0.2512	0.0377		0.0013	0.0013	0.0071	8.49	14.07
72	municipal solid waste Europ.Hausmüll	/19/	*72			0.1736	0.224		0.0349	0.3155	0.0013				
75	municipal solid waste Hausmüll aufbereitet	/20/	*75			0.447	0.056	0.237	0.0316	0.2144	0.001	0.004		8.697	
6	news paper Zeitungen	/2/			0.417	0.016	0.25	0.365	0.047	0.318	0.001	0.002	0.001	14.46	16.09

-PSI-   Composition and Analysis of Wastes, Raw Materials and Fuels

CAWAF Tab 1, page 4
WJ51, 05.05.03

Table D-5-1: proximate and ultimate analysis, heating values

last= 157

@-No.	waste or fuel		remarks	ref.	proximate analysis (massfraction, sum=1)		ultimate analysis (massfraction, sum = 1)					(wet mass basis)			lower heating value Hu	upper heating value Ho
					fixed carbon	volatile matter	ash.inert a	water w	carbon c	hydrogen h	oxigen o	nitrogen n	sulfur s	chlorine cl		
					kg/kg fuel	kg/kg fuel	kg/kg fuel	kg/kg fuel	kg/kg fuel	kg/kg fuel	kg/kg fuel	kg/kg fuel	kg/kg fuel	kg/kg fuel	MJ/kg fuel	MJ/kg fuel
62	paper (dry)		*62 #1	/13/			0.103	0	0.466	0.079	0.3455	0.0033	0.0016	0.0016	20.3	20.3
95	paper in MSW	Papier	*95 *94	/24/			0.1209	0.249		0.0443					10.1	15.35
127	waste paper	Altpapier	*127	/26/												15
132	waste paper from	Altpapier	*132	/28/			0.02	0.04	0.35	0.04	0.28	0.001	0.001		13.5	
	ff. to						0.25	0.1	0.5	0.06	0.37	0.003	0.003		17.5	
96	cardboard in MSW	Pappe	*96 *94	/24/			0.1297	0.259		0.0474		0.003	0.0012	0.0013	11.65	17.6
63	80%paper/20%plastic waste			/13/			0.078	0.046	0.519	0.088	0.2635	0.003	0.003	0.03		22.9
7	plastic wastes	Kunststoffabfälle		/2/			0.086	0.15	0.563	0.078	0.081	0.009			26.92	28.9
97	plastics in MSW	Kunststoff	*97 *94	/24/			0.0847	0.136		0.1253					30.3	38.56
103	plastic films in MSW	Kunststoff-Folien	*103 *94	/24/			0.0909	0.203		0.1188					29.56	40.86
104	plastic solid, MSW	Kunststoff-Formkörper	*104 *94	/24/			0.086	0.087		0.142					30.63	36.88
8	rubber, leather	Gummi, Leder		/2/			0.225	0.1	0.43	0.054	0.116	0.013	0.012	0.05	19.59	21.01
99	rubber, leather in MSW	Gummi, Leder	*99 *94	/24/			0.0721	0.073		0.2292					21.35	24.57
5	car tyres	Autoreifen		/2/			0.026	0.022	0.794	0.073	0.059	0.012	0.014		35.96	37.64
54	scrap tyre	Altreifen		/10/	0.1744	0.6249	0.1908	0.0099	0.7079	0.0681	0.007	0.006	0.0103		29.866	
125	waste tyres from..to	Altreifen	*125	/26/											29	
9	textiles	Textilien		/2/			0.02	0.25	0.371	0.05	0.272	0.031	0.003	0.003	15.32	17.03
100	textiles in MSW	Textilien	*100 *94	/24/			0.0653	0.266		0.047					13.1	19.9
101	vegetable waste in MSW	Vegetabilien	*101 *94	/24/			0.1129	0.645		0.0153					4.33	14.74
10	garden wastes	Gartenabfälle		/2/			0.101	0.45	0.233	0.029	0.175	0.009	0.002	0.001	9.3	11.04
49	chicken litter from		*49	/7/	0.13	0.48	0.09	0.15				0.01	0.008	0.002	15	
	ff. to				0.16	0.62	0.14	0.5				0.06	0.015	0.009	16	
50	straw from		*50	/7/	0.14	0.57	0.03	0.07				0.003	0.0004	0.002	17	
	ff. to				0.18	0.75	0.12	0.3				0.009	0.004	0.0075	20	
102	minerals,scrap in MSW	Mineralien, Schrott	*102 *94	/24/			1	0							-0.9	
66	ASR	RESH (ALSTOM)	*66	/15/			0.453	0.05	0.324	0.04	0.1	0.01	0.003	0.02	15.5	
133	ASR	RESH (CT ENVIRONMENT)	*133	/29/				0.102	0.271	0.0332		0.0018	0.0036	0.0125		
135	ASR	RESH	*135	/30/				0.008	0.377				0.004	0.0159		
136	ASR	RESH	*136	/30/				0.039						0.00644		
137	ASR from	RESH 1985	*137 #1	/30/				0	0.4			0.004	0.004	0.01		

Composition and Analysis of Wastes, Raw Materials and Fuels

Table D-5-1: proximate and ultimate analysis, heating values

CAWAF Tab 1, page 5
WJ51, 05.05.03

-PSI-

@-No.	waste or fuel	ref.	remarks	proximate analysis (massfraction, sum=1)				ultimate analysis (massfraction, sum = 1) (wet mass basis)							lower heating value Hu	upper heating value Ho
				fixed carbon	volatile matter	ash,inert a	water w	water w	carbon c	hydrogen h	oxigen o	nitrogen n	sulfur s	chlorine cl	MJ/kg fuel	MJ/kg fuel
				kg/kg fuel	kg/kg fuel	kg/kg fuel	kg/kg fuel	kg/kg fuel	kg/kg fuel	kg/kg fuel	kg/kg fuel	kg/kg fuel	kg/kg fuel	kg/kg fuel		
last= 157																
ff. to						0.4	0	0					0.01	0.05		
134	MSW filterash KVA-Filterasche	/29/	*134				0.005		0.009				0.0338	0.1134		0
solid products (raw materials, plastics,....)																
76	PE polyethylene (C2H4)n	/22/				0	0	0	0.8563	0.1437	0	0	0	0	46.37	
77	PVC polyviylchloride (C2H3Cl)n	/22/				0	0	0	0.3844	0.0483	0	0	0	0.5673	18.33	
87	PP polypropylene (C3H6)n	/22/				0	0	0	0.85629	0.14371	0	0	0	0	46.41	
88	acetate (C4H6O2)n	/22/				0	0	0	0.55807	0.07024	0.37169	0	0	0	28.08	
78	latex (rubber) (C4H8O2S)n	/22/				0	0	0	0.3998	0.0671	0.2663	0	0.2668	0	19.456	
89	cellulose (C6H10O5)n	/22/				0	0	0	0.44446	0.06216	0.49338	0	0	0	16.75	
90	nylon (C6H11ON)n	/22/				0	0	0	0.63686	0.09798	0.14139	0.12378	0	0	31.73	
91	polystyrene (C8H8)n	/22/				0	0	0	0.92258	0.07742	0	0	0	0	41.5	
92	polyetyelene tetraphathalic (C10H8O4)n	/22/				0	0	0	0.62502	0.04196	0.33302	0	0	0	29.55	
93	polyurethane (C12H22O4N4)n	/22/				0	0	0	0.50338	0.07744	0.22351	0.19567	0	0	26.06	
107	limestone Kalk	/26/	*107													
108	clay Ton	/26/	*108													
109	marl Mergel	/26/	*109													
111	clinker (mean) (Zement-) Klinker (HMB)	/26/	*111													
112	clinker form..to (Zement-) Klinker (HMB)	/26/	*112													
113	clinker (mean) (Zement-) Klinker (USA)	/26/	*113													
114	clinker form..to (Zement-) Klinker (USA)	/26/	*114													
115	cement raw powder (mean) Zement-Rohr	/26/	*115													
116	cement raw powder from..to Zement-Rohr	/26/	*116													
117	cement raw powder from..to Zement-Rohr	/26/	*117													
118	cement raw powder (mean) Zement-Rohm	/26/	*118													
119	cement raw powder Zement-Rohmehl (BCl	/26/	*119													
liquid wastes, sludges,....																
3	sewage sludge Klärschlamm luftgetretrockt	/2/				0.344	0.04		0.237	0.046	0.289	0.031	0.013		12.11	12.95

-PSI-

Composition and Analysis of Wastes, Raw Materials and Fuels

Table D-5-1: proximate and ultimate analysis, heating values

CAWAF Tab 1, page 6
WJ51, 05.05.03

@-No.	waste or fuel		ref.	remarks	proximate analysis (massfraction, sum=1)		ultimate analysis (massfraction, sum = 1)					(wet mass basis)			lower heating value Hu	upper heating value Ho
last= 157					fixed carbon	volatile matter	ash,inert a	water w	carbon c	hydrogen h	oxigen o	nitrogen n	sulfur s	chlorine cl		
					kg/kg fuel	kg/kg fuel	kg/kg fuel	kg/kg fuel	kg/kg fuel	kg/kg fuel	kg/kg fuel	kg/kg fuel	kg/kg fuel	kg/kg fuel	MJ/kg fuel	MJ/kg fuel
4	sewage sludge Klärschlamm entwässert		/2/				0.092	0.742	0.078	0.012	0.064	0.008	0.004		1.21	12.95
40	sewage sludge from Klärschlamm		/5/	*40			0.1	0.1	0.1	0.02	0.05	0.01	0.05		2	
ff.	to						0.4	0.8	0.3	0.08	0.25	0.01	0.05		10	
43	sewage sludge A, Thailand, district plant		/6/		0.0209	0.445	0.3803	0.1538	0.1666	0.0314	0.2548	0.0056	0.0076		10.77	
44	sewage sludge B, Thailand, textil plant		/6/		0.031	0.5953	0.3529	0.0208	0.2836	0.0372	0.2897	0.0098	0.006		16.24	
45	sewage sludge C, Thailand, beer plant		/6/		0.0692	0.4063	0.4883	0.0362	0.1911	0.0283	0.2454	0.0072	0.0036		14.85	
46	sewage sludge D, Thailand, public plant		/6/		0.18	0.6015	0.1232	0.0953	0.4173	0.0348	0.309	0.0086	0.0118		23.57	
56	sewage sludge (Backnang, D)		/11/	*56	0.03	0.445	0.474	0.051	0.2448	0.0399	0.1479	0.0323	0.0095	0.000567	9.48	
59	sewage sludge, dry		/12/		0.0452	0.457	0.3078	0.19	0.26717	0.03566	0.15392	0.03566	0.00954	0.000251	10	
60	sewage sludge, wet		/12/		0.0086	0.1354	0.156	0.7	0.07502	0.01022	0.04755	0.00878	0.0023	0.00013	1.1	
126	sewage sludge (CH, 1989) Klärschlamm		/26/	*126											7.6	
143	sewage sludge		/33/	*143												
150	sheep manure		/35/	*150 #2	0.073	0.34	0.109	0.478					0.0024			
151	dairy free-stall manure		/35/	*151 #2	0.021	0.091	0.185	0.703					0.00003			
152	dairy tie-stall manure		/35/	*152 #2	0.022	0.091	0.189	0.698					0.000106			
153	poultry litter		/35/	*153 #2	0.077	0.553	0.17	0.2					0.0022			
52	crude oil terminal sludge (Malaysia) from		/9/	*52	0.0007	0.048	0.443	0.0745	0.0383	0.0059	0.0457	0.0057	0.0067		1.323	
ff.	to				0.0017	0.4816	0.7676	0.1827	0.34	0.0476	0.0731	0.0093	0.0127		17.162	
	liquids (fuels, alcohols,....)															
20	methanole	Methanol	/2/				0	0	0.375	0.125	0.5	0	0	0	19.51	22.31
21	ethanole	Äthanol	/2/				0	0	0.52	0.13	0.35	0	0	0	26.96	29.89
22	benzene pure	Benzol, rein	/2/				0	0	0.922	0.078	0	0	0	0	40.23	41.94
23	gasoline	Benzin (Mittelwert)	/2/				0	0	0.85	0.15	0	0	0	0	42.5	46.7
24	light heating oil	Heizöl EL	/2/	*24			0	0.004	0.86	0.13	0.004	0	0.002		42.7	45.4
25	heavy fuel oil	Heizöl S	/2/	*25			0	0.004	0.84	0.11	0.015	0.003	0.028		40.2	42.3
120	heavy fuel oil	Heizöl S	/26/	*120												
130	waste oil	Altöl Messg.1	/26/	*130											39.5	
131	waste oil	Altöl Messg.2	/26/	*131											39.5	

-PSI-

Composition and Analysis of Wastes, Raw Materials and Fuels

Table D-5-1: proximate and ultimate analysis, heating values

CAWAF Tab 1, page 7
WJ51, 05.05.03

@-No.	waste or fuel		ref.	remarks	proximate analysis (massfraction, sum=1)		ultimate analysis (massfraction, sum = 1) (wet mass basis)								lower heating value	upper heating value
					fixed carbon	volatile matter	ash,inert	water	carbon	hydrogen	oxigen	nitrogen	sulfur	chlorine	Hu	Ho
last= 157							a	w	c	h	o	n	s	cl		
					kg/kg fuel	kg/kg fuel	kg/kg fuel	kg/kg fuel	kg/kg fuel	kg/kg fuel	kg/kg fuel	kg/kg fuel	kg/kg fuel	kg/kg fuel	MJ/kg fuel	MJ/kg fuel
		others														
84	methane (CH4)	Methan	/23/				0	0	0.74869	0.25131	0	0	0	0	50.011	55.496
85	ethane (C2H6)	Äthan	/23/				0	0	0.79889	0.20111	0	0	0	0	47.499	51.874
82	propane (C3H8)	Propan	/23/				0	0	0.81715	0.18285	0	0	0	0	46.369	50.369
86	n-butane (C4H10)	n-Butan	/23/				0	0	0.82659	0.17341	0	0	0	0	45.741	49.53
87	i-butane (C4H10)	i-Butan	/23/				0	0	0.82659	0.17341	0	0	0	0	45.636	49.404
147	human body		/34/	*147			0.0318	0.7	0.1851	0.0171	0.0279	0.033	0.003	0.0021		
155	earth's crust	Erdkruste	/36/	*155 #1				0	0.0002	0.0014	0.466	0.00002	0.00026	0.00013		

-PSI-   **Composition and Analysis of Wastes, Raw Materials and Fuels**

Table D-5-2: mineral elements

CAWAF Tab2, page 1
WJ51, 05.05.03

CAWAF1.xls

@-No.	waste or fuel	ref.	remarks	Si	Fe	Ca	Al	Mg	Na	K	Ti
				g/kg fuel dry	g/kg fuel dry	g/kg fuel dry	g/kg fuel dry	g/kg fuel dry	g/kg fuel dry	g/kg fuel dry	g/kg fuel dry
							dry mass basis				
solid fuels (coal, peat, wood, energy crop)											
55	bituminous coal (Goettelborn, D)	/11/	*55	19.76	6.92	3.51	11.75	1.76	0.285	2.31	0.527
142	bituminous from	/33/	*142						0.3	0.6	
ff.	to								1.25	4.6	
154	coal	/35/	*154	33.13	18.85	2.4	19.71	0.727	0.273	2.708	
140	hard coal Steinkohle	/32/	*140						0.9	3.5	
141	lignite/sub-bituminous from	/33/	*141						0.03	0.06	
ff.	to								0.2	0.65	
83	coal from Kohle	/21/	*83 *83a	6	3	0.5	4	0.5	0.3	0.1	0.2
ff.	to			70	43	37	30	3	2	4	2
41	natural wood, dry	/4/	*41 #1	0.47	0.05	1.21	0.064	0.3	0.03	0.38	
67	Robinia pseudoacacia (wood)	/16/	*67 #1	0.005	0.012	8.17	0.055	1.07	0.087	1.15	
42	urban waste wood, dry	/4/	*42 #1	0.75	2.1	15	0.84	1.06		0.89	
146	Biomass (wood) from	/33/							0.03	0.45	
ff.	to								1.5	7	
145	Biomass (straw/grass) from	/33/							0.07	5	
ff.	to								13.7	20	
148	pine shavings	/35/	*148 #2	0.267	0.042	0.063	0.071	0.02	0.01	0.041	0.007
149	reed canary grass	/35/	*149 #2	8.24	0.421	2.81	0.36	1.31	0.71	6.16	1.23
solid wastes											
11	municipal solid waste from Hausmüll	/2/	*11		30						
ff.	to				50						
64	houshold waste Hausmüll	/14/	*64	53.5	43.7	33.8	18.3	4.5	6.8	3.5	2.1
65	municipal solid waste	/14/	*65	48.7	35.5	31.6	14.5	4.2	6.7	3	2
110	municipal solid waste Siedlungsabfall	/27/	*110	46.3	35	33.8	12.5		7.1	2.6	
66	ASR RESH (ALSTOM)	/15/	*66		118		18.9		8.4	3.2	
133	ASR RESH (CT ENVIRONMENT)	/29/	*133	75	146	51	106	16		3	9
135	ASR RESH	/30/	*135	140	140						

Composition and Analysis of Wastes, Raw Materials and Fuels
Table D-5-2: mineral elements

CAWAF Tab2, page 2
WJ51, 05.05.03

-PSI-

@-No.	waste or fuel		ref.	remarks	dry mass basis							
					Si	Fe	Ca	Al	Mg	Na	K	Ti
					g/kg fuel dry	g/kg fuel dry	g/kg fuel dry	g/kg fuel dry	g/kg fuel dry	g/kg fuel dry	g/kg fuel dry	g/kg fuel dry
136	ASR	RESH	/30/	*136		105						
137	ASR from RESH 1985 ff. to		/30/	*137 #1		50 90						
134	MSW Filterash	KVA-Filterasche	/29/	*134	55	11	176	31	6	34	44	8
	liquid wastes, sludges,....											
56	sewage sludge (Backnang, D)		/11/	*56	87.2	4.42	35.8	36.2	8.74	1.48	9.59	
143	sewage sludge from ff. to		/33/	*143						0.3 0.4	0.8 1	
150	sheep manure		/35/	*150 #2	28.6	0.285	19.1	3.41	7.24	7.19	40.6	0.251
151	dairy free-stall manure		/35/	*151 #2	218.4	0.562	28.4	3.16	9.96	6.1	34.9	7.69
152	dairy tie-stall manure		/35/	*152 #2	7.6	6	104	7.47	33.6	32.6	55.5	19.04
153	poultry litter		/35/	*153 #2	31.3	4.8	15.4	8.22	4.11	0.454	14.02	0.52
	others											
147	human body		/34/	*147			50		3	7	3.9	
155	earth crust	Erdkruste	/36/	*155 #1	277.2	50	36.3	81.3	20.9	28.3	25.9	4.4

-PSI-

Composition and Analysis of Wastes, Raw Materials and Fuels
Table D-5-3: heavy metals and trace elements

CAWAF Tab3, page 1
WJ51, 05.05.03

CAWAF1.xls

@-No.	waste or fuel	ref.	remarks	P	F	Br	I	Zn	Pb	Cu	Cd	Hg	Cr	Ni	As
				mg/kg dry	mg/kg dry	mg/kg dry	mg/kg dry	mg/kg dry	mg/kg dry	mg/kg dry	mg/kg dry	mg/kg dry	mg/kg dry	mg/kg dry	mg/kg dry
												dry mass basis			
solid fuels (coal, peat, wood, energy crop)															
123	coke from — Koks (VDZ)	/26/	*123					16	6		0.04		5	300	
ff.	to							220	102	3.1			104	355	
124	coke — Koks (ZKG)	/26/	*124									0.16	11	208	0.6
55	bituminous coal (Goettelborn, D)	/11/	*55	21											
142	bituminous from	/33/	*142					8	1	5	0.1	0.03	4	6	1
ff.	to							180	250	154	2	2	75	55	50
154	coal	/35/	*154 #2	128											
61	polish coal	/13/							24.1		0.15	31.7		72.2	0.57
79	hard coal Steinkohle (D) Trockenfeuerung	/3/	*79					55	28	22	0.3	0.17	17	33	9
80	hard coal (slag-type furnace) Steinkohle (D) Schmelzfeuerung	/3/	*80					55	44	33	0.4	0.33	22	33	9.9
140	hard coal Steinkohle	/32/	*140		100										
81	lignite Braunkohle (D)	/3/	*81					8.4	1.4	1.8	0.06	0.2	5.8	2.36	2
141	lignite/sub-bituminous from	/33/	*141					6	4	5	0.05	0.05	10	4	1
ff.	to							90	35	70	3	2	100	40	30
83	coal from Kohle	/21/	*83 *83a	5	100	4	0.0002	5	1	5	0.05	0.02	4	3	0.5
ff.	to			1400		50	0.014	5400	200	60	65	2	60	80	90
121	coal from Kohle (NFP)	/26/	*121					16	11	30	0.1	0.1	5	20	9
ff.	to							220	270	44	10	3.3	80	80	50
122	coal from Kohle (KKL)	/26/	*122					18	18	10	0.1	0.5	3.5	15	2
ff.	to							209	273	60	10	1.4	81	95	50
41	natural wood, dry	/4/	*41 #1					11.1	0.4	1.2	0.1	0.01	2.4	0.3	0.1
105	wood unbelasteted Holz	/25/						20	25	14	0.4		2	0.6	
42	urban waste wood, dry	/4/	*42 #1					535	314	27	3.4	0.3	32	6	4.1
106	waste wood Altholz	/25/	*106		0.2			167	93.5	10.6	0.38	0.32	14	2.1	
128	waste wood Altholz (Infras)	/26/	*128					729	512	12	1.46	0.075	16.4	3.36	0.806
129	waste wood mean Altholz (AHK)	/26/	*129					1544	1031	1432	3.45	0.31	53	40	1.6
129	ff. Max							3143	1776	6844	8	1	192	264	3.6
144	Biomass (wood,straw,grass) from	/33/	*144					20	1	1	0.1	0.01	0.15	0.2	0.1
ff.	to							150	5	90	0.3	0.05	25	10	6

-PSI-

Composition and Analysis of Wastes, Raw Materials and Fuels

Table D-5-3: heavy metals and trace elements

CAWAF Tab3, page 2
WJ51, 05.05.03

@-No.	waste or fuel		ref.	remarks	P	F	Br	I	Zn	Pb	Cu	Cd	Hg	Cr	Ni	As
					mg/kg dry	mg/kg dry	mg/kg dry	mg/kg dry	dry mass basis mg/kg dry	mg/kg dry	mg/kg dry	mg/kg dry	mg/kg dry	mg/kg dry	mg/kg dry	mg/kg dry
148	pine shavings		/35/	*148 #2	3.14											
149	reed canary grass		/35/	*149 #2	1240											
	solid wastes															
11	municipal solid waste	Hausmüll from	/2/	*11					470	390	60	1	0.5	30		
	ff.	to							6500	1800	2100	33	11	2700		
64	houshold waste	Hausmüll	/14/	*64	1350	300	155		1830	746	1150	15.5	5.92	437	141	3.8
65	municipal solid waste		/14/	*65	908	237	184		2110	921	1030	14.5	5.53	474	145	3.55
110	municipal solid waste	Siedlungsabfall	/27/	*110	800	253			1867	933	573	14.7	4			
127	waste paper	Altpapier	/26/	*127					90	66	26	0.9	0.03	34	13	
125	waste tyres from	Altreifen	/26/	*125					3000	3	65	0.1		11	17	
	ff.	to							3000	760	65	10		97	77	
49	chicken litter	from	/7/	*49	3											
	ff.	to			40											
50	straw	from	/7/	*50	0.1											
	ff.	to			3.5											
66	ASR	RESH (ALSTOM)	/15/	*66		421					18090					
133	ASR	RESH (CT ENVIRONMENT)	/29/	*133	1000				22000	2000	21000	50	3	2000	1000	
135	ASR	RESH	/30/	*135		300			20000	8100	25000	150	4.9	1800	1200	
136	ASR	RESH	/30/	*136					16000	10000	10000	60	5	900	600	
137	ASR from	RESH 1985	/30/	*137 #1		300			10000	2500	10000	20	1			
	ff.	to				800			15000	5000	15000	80	3			
134	MSW Filterash	KVA-Filterasche	/29/	*134	3000				29000	10000	2000	630	1	1000		
	solid products (raw materials, plastics,....)															
107	limestone from	Kalk	/26/	*107					22	0.4		0.035	0.03	1.2	1.5	0.2
	ff.	to							24	13		0.1		16	7.5	12
108	clay from	Ton	/26/	*108					59	13		0.016	0.45	90	67	13
	ff.	to							115	22		0.3		109	71	23
109	marl	Mergel	/26/	*109					84	12		0.5	0.02	31	21	9
111	clinker (mean)	(Zement-) Klinker (HMB)	/26/	*111					60	25	40	0.2	0.2	30	30	15
112	clinker from	(Zement-) Klinker (HMB)	/26/	*112					29	1	5	0.01	0.02	11	12	2
	ff.	to							531	105	136	1.5	1.2	319	397	87

-PSI-

Composition and Analysis of Wastes, Raw Materials and Fuels

Table D-5-3: heavy metals and trace elements

CAWAF Tab3, page 3
WJ51, 05.05.03

@-No.	waste or fuel	ref.	remarks	dry mass basis											
				P mg/kg dry	F mg/kg dry	Br mg/kg dry	I mg/kg dry	Zn mg/kg dry	Pb mg/kg dry	Cu mg/kg dry	Cd mg/kg dry	Hg mg/kg dry	Cr mg/kg dry	Ni mg/kg dry	As mg/kg dry
113	clinker (mean) (Zement-) Klinker (USA)	/26/	*113						12		0.34	0.014	76	31	19
114	clinker from (Zement-) Klinker (USA)	/26/	*114						1		0.03	0.001	25	10	5
114 ff.	to								12		1.12	0.039	422	129	71
115	cement raw powder (mean) Zement-Rohm	/26/	*115					30	12	17	0.2	0.07	27	22	11
116	cement raw powder from Zement-Rohmeh	/26/	*116					15	4		0.04	0.02	23	18	3
116 ff.	to							90	23		1	0.6	39	30	28
117	cement raw powder from Zement-Rohmeh	/26/	*117					31	4		0.04			18	3
117 ff.	to							47	15		0.15			23	15
118	cement raw powder (mean) Zement-Rohm	/26/	*118										28	30	
119	cement raw powder Zement-Rohmehl (BCl)	/26/	*119					30	17	24	0.03	0.01	59	18	31
	liquid wastes, sludges,....														
56	sewage sludge (Backnang, D)	/11/	*56	20600								1.59			
126	sewage sludge (CH, 1989) Klärschlamm from	/26/	*126					186	22	40	1	1	15	6	
126 ff.	to							2600	1200	1292	19	5	888	171	
143	sewage sludge from	/33/	*143					1600	120	230	1	1	50	25	2
143 ff.	to							800	150	310	5	3	200	45	15
52	crude oil terminal sludge (Malaysia) from	/9/	*52						211	850	120				
52 ff.	to								251	890	290				
150	sheep manure	/35/	*150 #2	4200											
151	dairy free-stall manure	/35/	*151 #2	3940											
152	dairy tie-stall manure	/35/	*152 #2	20000											
153	poultry litter	/35/	*153 #2	5190											
	liquids (fuels, alcohols,...)														
120	heavy fuel oil Heizöl S	/26/	*120						3.5	3	2	0.006	1	19	8
130	waste oil Altöl Messg.1	/26/	*130					730	420	54	1	0.02	12	5.7	5
131	waste oil Altöl Messg.2	/26/	*131					670	180	100	0.68	0.001	11	3.2	1.3

-PSI-

Composition and Analysis of Wastes, Raw Materials and Fuels

Table D-5-3: heavy metals and trace elements

CAWAF Tab3, page 4
WJ51, 05.05.03

@-No.	waste or fuel	ref.	remarks	P mg/kg dry	F mg/kg dry	Br mg/kg dry	I mg/kg dry	Zn mg/kg dry	Pb mg/kg dry	Cu mg/kg dry	Cd mg/kg dry	Hg mg/kg dry	Cr mg/kg dry	Ni mg/kg dry	As mg/kg dry
	others														
147	human body	/34/	*147	3300											
155	earth crust (Erdkruste)	/36/	*155 #1	1050	625	2.5	0.5	70	13	55	0.2	0.08	100	75	1.8

-PSI-

Composition and Analysis of Wastes, Raw Materials and Fuels

CAWAF Tab4: page1
WJ51, 05.05.03

Table D-5-4: remarks

CAWAF1.xls

#	Name	Ref	Marker	Remark
#1			#1	#1 fuel dry mass basis (no water content given)
#2			#2	#2 mineral-elements calculatet from ash analysis
#3			#3	#3 NREL (see References)
11	Hausmüll from … to	/2/	*11	*11 dioxin [ng/kg fuel wet] = 10 … 256
24	Heizöl EL	/2/	*24	*24 s: <=0.002 [kg/kg fuel wet]
25	Heizöl S	/2/	*25	*25 s: <=0.028 [kg/kg fuel wet]
39	Hausmüll	/5/	*39	*39 [kg/kg fuel wet]; combustibles: 0.4…0.5; n <0.005; s <0.005.
40	Klärschlamm	/5/	*40	*40 [kg/kg fuel wet]; combustibles: 0.2…0.5; n <0.05; s <0.01.
41	natural wood, dry	/4/	*41	*41 BIOMETH analysis; [g/kg fuel dry] : Sn = n.v.g.; Co = 0.0001.
42	urban waste wood, dry	/4/	*42	*42 BIOMETH analysis; [g/kg fuel dry] : Sn = 0.006; Co = 0.001.
48	black coal	/7/	*48	*48 (UK), as received
49	chicken litter	/7/	*49	*49 (UK), as received
50	straw	/7/	*50	*50 (UK), as received
51	high volatile solid waste	/8/	*51	*51 simulated high-moisture and high-volatile solid waste, avr. size = 12.0 mm
52	crude oil terminal sludge (Malaysia) fr.to	/9/	*52	*52 considered as low radioactiv waste, conc: U-238= 0.0012-0.0026 g/kg fuel, Th-232= 0.0035-0.0111 g/kg fuel; Ra-226=25-96 Bq/kg; Ra-228= 26-123 Bq/kg.
56	sewage sludge (Backnang, D)	/11/	*56	*56 Ti = 1.99 [g/kg fuel]
62	paper (dry)	/13/	*62	*62 Fit WJ51 from analysis of given paper/plastic waste mixtures
64	houshold waste	/14/	*64	*64 [g/kg fuel] : Ba=0.37, Mn=0.27, Sn=0.074, MO=0.009, Sb=0.031,Co=0.0037, Li=0.01.
65	municipal solid waste	/14/	*65	*65 [g/kg fuel] : Ba=0.43, Mn=0.22, Sn=0.071, MO=0.011, Sb=0.041,Co=0.0034, Li=0.009.
66	ASR RESH (ALSTOM)	/15/	*66	*66 ASR = automotiv shredder rest
67	Robinia pseudoacacia (wood)	/16/	*67	*67 no water content given, branch inkl. bark, fuel density=380 kg/m³, char density=100 kg/m³
68	lignite (South Banko)	/17/	*68	*68 ASTM volatile matter
69	bit.coal (Pittsburgg)	/17/	*69	*69 ASTM volatile matter
72	municipal solid waste Europ.Hausmüll	/19/	*72	*72 TAMARA-experiments with 25% BRAM and 75% breackup, homogeized houshold waste. Representative for Central Europe.
74	wood chips Holzschnitzel	/20/	*74	*74 wood chips from pine (Holzschnitzel aus Kiefernholz)
75	municipal solid waste Hausmüll aufbereitet	/20/	*75	*75 MSW: Hausmüll aufbereitet: zerkleinert, getrocknet, von Metallteilen befreit, pelletiert.
79	Steinkohle (D) Trockenfeuerung	/3/	*79	*79 hard coal: Steinkohle (D), Trockenfeuerung; ca. average value of in German used coal; mg/kg fuel(ar): Be=1.5; Co=9.8; Sb=0.9; Se=2.0; Te=0.2; Tl=0.51; U=1.0; V=29.4.
80	Steinkohle (D) Schmelzfeuerung	/3/	*80	*80 hard coal: Steinkohe (D), Schmelzfeuerung; ca. average value of in German used coal; mg/kg fuel(ar): Be=2.0; Co=11.8; Sb=0.9; Se=2.0; Te=0.2; Tl=0.51; U=1.27; V=51.6.
81	Braunkohle (D)	/3/	*81	*81 lignite: Braunkohle (D); ca. average value of in German used coal; mg/kg fuel(ar): Be=0.17; Co=0.00092; Sb=0.065; Se=0.23; Te=0.009; Tl=0.009; U=0.46; V=3.74.

Composition and Analysis of Wastes, Raw Materials and Fuels

Table D-5-4: remarks

CAWAF Tab4: page2
WJ51, 05.05.03

-PSI-

#			/source/	*83	*83a	*94	col
83	Kohle		/21/	*83			
83	Kohle		/21/		*83a		
94	municipal solid waste	Wiener Hausmüll	/24/			*94	
95	paper in MSW	Papier	/24/				*95 *94
96	cardboard in MSW	Pappe	/24/				*96 *94
97	plastics in MSW	Kunststoff	/24/				*97 *94
98	wood in MSW	Holz	/24/				*98 *94
99	rubber, leather in MSW	Gummi, Leder	/24/				*99 *94
100	textiles in MSW	Textilien	/24/				*100 *94
101	vegetable waste in MSW	Vegetabilien	/24/				*101 *94
102	minerals,scrap in MSW	Mineralien, Schrott	/24/				*102 *94
103	plastic films in MSW	Kunststoff-Folien	/24/				*103 *94
104	plastic solid, MSW	Kunststoff-Formkörper	/24/				*104 *94
106	waste wood	Altholz	/25/				*106
107	limestone	Kalk	/26/				*107
108	clay	Ton	/26/				*108
109	marl	Mergel	/26/				*109
110	municipal solid waste	Siedlungsabfall	/27/				*110
111	clinker (mean)	(Zement-) Klinker (HMB)	/26/				*111
112	clinker form...to	(Zement-) Klinker (HMB)	/26/				*112
113	clinker (mean)	(Zement-) Klinker (USA)	/26/				*113
114	clinker form...to	(Zement-) Klinker (USA)	/26/				*114
115	cement raw powder (mean)	Zement-Rohm	/26/				*115
116	cement raw powder from...to	Zement-Rohm	/26/				*116
117	cement raw powder from...to	Zement-Rohm	/26/				*117
118	cement raw powder (mean)	Zement-Rohm	/26/				*118
119	cement raw powder	Zement-Rohmehl (BCU	/26/				*119
120	heavy fuel oil	Heizöl S	/26/				*120
121	coal from...to	Kohle (NFP)	/26/				*121
122	coal from...to	Kohle (KKL)	/26/				*122
123	coke from...to	Koks (VDZ)	/26/				*123

*83 /21/, Bandbreiten der Spurenelementgehalte typischer Kohlen [mg/kg coal] :   Be=0.2-4.0; B=11-360; Sc=0.1-100; V=5-80; Mn=6-180; Co=1-40; Ga=1-8; Ge=1-43; Se=0.5-8; Rb=0.3-60; Sr=10-500; Y=0.1-60; Zr=8-130; Mo=1-30; Ag=0.1-2; Sn=1-50; Sb=0.1-10; Cs=0.02-6; Ba=2-500; La=2-30; Hf=0.1-2; Ta=0.04-1; W=0.1-4; Tl=0.2-2; Ce=2-50; Sm=0.2-5; Eu=0.1-1; Th=0.6-9; U=0.3-5.

*83a: mean value of hard coal (Steinkohle) /21/p.17: F =100 [mg/kg coal].

*94 Wiener Abfall 1983, components= paper 24.3 %wt, cardboard 13.9 %wt, plastics 8.4 %wt, rubber,lether 1.2 %wt, wood 2.3 %wt, textiles 5.3 %wt, vegetable waste 19.6 %wt, minerals, scrap 25 %wt. See also @ 95-104.

*95 component of Wiener MSW, see *94.

*96 component of Wiener MSW, see *94.

*97 component of Wiener MSW, see *94, consits of 43 %wt of *103 and 57 %wt of *104.

*98 component of Wiener MSW, see *94.

*99 component of Wiener MSW, see *94.

*100 component of Wiener MSW, see *94.

*101 component of Wiener MSW, see *94.

*102 component of Wiener MSW, see *94.

*103 component of Wiener MSW, see *94, *97.

*104 component of Wiener MSW, see *94, *97.

*106 Mn= 0.0798 g/kg fuel(wet).

*107 Tl= 0.05-0.5 mg/kg.

*108 Tl=0.7-1.6 mg/kg.

*109 Tl = 1.0, Mn= 0.557 mg/kg.

*110 MSWI St-Gallen (CH).

*111 Tl=0.2, V=30 mg/kg; HMB = Mitteilung Zementindustrie.

*112 Tl=0.01-1.2, V=10-100, Co=10-21, Sb=0.1-0.95 mg/kg; HMB = Mitteilung Zementindustrie.

*113 Tl=1.08, Ba=280 mg/kg.

*114 Tl=0.01-2.68, Ba 91-1402, Sb=0.7-4.0 mg/kg.

*115 all mean values; Tl=0.5, Sb=0.08, V=50, Co=11 mg/kg; HMB= Mitteilung Zementindustrie.

*116 Tl=0.1-6, V=32-102 mg/kg; HMB= Zementindustrie (CH).

*117 Tl=0.21-0.78 mg/kg; VDZ= Mitteilung Zementindustrie.

*118 Tl= 0.4, Sb=0.08, V=48, Co=11, Ba=118 mg/kg; ZKG= D-Journal: 'Zement, Kalk, Gips'.

*119 Tl=0.1, Co=6.6 mg/kg; BCU = Mitteilung Zementindustrie.

*120 V=21, Co=0.8 mg/kg.

*121 Tl=0.2-4, Mn=109-125, Co=6-10 mg/kg; NFP= Nationales Forschungsprojekt (CH), Klärschlammverwertung.

*122 Tl<=3-4, Sb=0.4-2, Co=27-182, Co= <=7-30, Ba=45-355 mg/kg; KKL= Kautz,Kirsch,Laufhütte: Spurengehalt in Kohle, 1975.

*123 Tl=0.04-3.1 mg/kg; VDZ = Mitteilung Zementindustrie.

-PSI-

Composition and Analysis of Wastes, Raw Materials and Fuels

CAWAF Tab4: page3
WJ51, 05.05.03

Table D-5-4: remarks

	Name		Ref	No.	Remark
124	coke	Koks (ZKG)	/26/	*124	*124 Tl=0.1, Sb=0.04, V=776, Co=2.6, Ba=8.4 mg/kg; ZKG= D-Journal: 'Zement, Kalk, Gips'.
125	waste tyres	Altreifen	/26/	*125	*125 Tl=0.2-0.3 mg/kg; Mitteilung Zementwerk Eclépens (CH)
126	sewage sludge (CH, 1989)	Klärschlamm	/26/	*126	*126 Hu=7.6 MJ/kg for 90%dry matter and 40% org. substance
127	waste paper	Altpapier	/26/	*127	*127 Tl < 2 mg/kg; Mitteilung Holderbank Zement AG,1990.
128	waste wood	Altholz (Infras)	/26/	*128	*128 Tl=0 mg/kg; Lit.: INFRAS (CH)
129	waste wood from..to	Altholz (AHK)	/26/	*129	*129 AHK= Altholzkonzept des Kt. Zürich, Bau- und Möbelholzproben, 1990
130	waste oil	Altöl Messg.1	/26/	*130	*130 Tl < 5 mg/kg; Mitteilung Amt. für Umweltschutz, Kt.St Gallen (CH)
131	waste oil	Altöl Messg.2	/26/	*131	*131 Tl < 1 mg/kg; Mitteilung Amt. für Umweltschutz, Kt.St Gallen (CH)
132	waste paper from..to	Altpapier	/28/	*132	*132 ash = annealing residue
133	ASR RESH (CT ENVIRONMENT)		/29/	*133	*133 Ba=6.3, Mn=0.9, Sn=0.08, Sb=0.27g/kg, Hu,trocken = 14.831 MJ/kg, Schüttdichte =400 kg/m3; ASR = automotiv shredder rest.
134	MSW Filterash KVA-Filterasche		/29/	*134	*134 Ba=2, Mn=1, Sn=3, Sb=2 g/kg; Hu,trocken = 0, Schüttdichte =640 kg/m3.
135	ASR RESH		/30/	*135	*135 RESH Thommen (CH), 1994; Glühverlust 50.9%, ASR = automotiv shredder rest.
136	ASR RESH		/30/	*136	*136 RESH (CH), Studie EWI 1993. Glühverlust 46.4%, ASR = automotiv shredder rest.
137	ASR RESH 1985		/30/	*137	*137 RESH (CH), 1985, Glühverlust = 50-55 %, ASR = automotiv shredder rest.
138	corn		/31/	*138	*138 apparent density = 1.35 g/cm3
139	soybean		/31/	*139	*139 apparent density =1.21 g/cm3.
140	hard coal	Steinkohle	/32/	*140	*140 mean values.
141	lignite/sub-bituminous		/33/	*141	*141 Mn=10-110, Co=5-20, V=10-60 mg/kg.
142	bituminous		/33/	*142	*142 Mn=20-200, Co=1-15, V=10-60 mg/kg.
143	sewage sludge		/33/	*143	*143 Mn=280, Co=5-10, V=45 mg/kg.
144	Biomass (wood,straw,grass) from..to		/33/	*144	*144 Mn=20-420, Co=0.1-0.5, V=0.1-3 mg/kg. (wood and straw/grass)
145	Biomass (straw/grass) from..to		/33/		
146	Biomass (wood) from..to		/33/		
147	human body		/34/	*147	*147 Trace elements: B F Si V Cr Mn Fe Co Cu Zn Se Mo Sn I.
148	pine shavings		/35/	*148 #2	*148 Hu(dry basis)=19.476 MJ/kg, Ba=0.806, Mn=1.83,Sr=4.06 mg/kg fuel.
149	reed canary grass		/35/	*149 #2	*149 Hu(dry basis)=16.838 MJ/kg, Ba=6.27, Mn=9.57, Sr=13.0 mg/kg fuel.
150	sheep manure		/35/	*150 #2	*150 Hu(dry basis)=16.038 MJ/kg, Ba=48.8, Mn=115.2, Sr= 27.6 mg/kg fuel.
151	dairy free-stall manure		/35/	*151 #2	*151 Hu(dry basis)=8.836 MJ/kg, Ba=33.1, Mn=195.4, Sr= 156.4 mg/kg fuel.
152	dairy tie-stall manure		/35/	*152 #2	*152 Hu(dry basis)=19.080 MJ/kg, Ba=33.8, Mn=164.4, Sr=175.7 mg/kg fuel.
153	poultry litter		/35/	*153 #2	*153 Hu(dry basis)=14.88 MJ/kg, Ba=95.34, Mn=476.6, Sr=54.02 mg/kg fuel.
154	coal		/35/	*154 #2	*154 Hu(dry basis)=30.238 MJ/kg.

page 482 Appendix D

-PSI-

Composition and Analysis of Wastes, Raw Materials and Fuels

Table D-5-4: remarks

CAWAF Tab4: page4
WJ51, 05.05.03

				*155 #1	remarks
155	earth crust	Erdkruste	/36/	*155 #1	*155 mean earth's crust concentration of continental crust and crust rocks, [mg/kg] Ag=0.07, Au=0.004, B=10, Ba=425, Be=2.8, Bi=0.2, Ce=60, Co=25, Cs=3, Dy=3.0, Eu=1.2, Er=2.8, Ga=15, Gd=5.4, Ge=1.5, Hf=3.0, Ho=1.2, In=0.1, Ir=0.001, La=30, Li=20,Lu=0.5, Mn=950, Mo=1.5, Nb=20, Nd=28, Os=0.005, Pd=0.01, Pr=8.2, Pt=0.01, Rb=90, Re=0.001, Rh=0.005, Ru=0.01, Sb=0.2, Sc=22, Se=0.05, Sm=6.0, Sn=2, Sr=375, Ta=2, Tb=0.9, Te=0.01, Th=7.2, Tm=0.5, U=1.8, V=135, W=1.5, Y=33, Yb=3.4, Zr=165 [mg/kg].
156	German Lignite		/37/	#1 #3	
157	coal German-Anna		/37/	#1 #3	

-PSI-

Composition and Analysis of Wastes, Raw Materials and Fuels

Table D-5-5: References

CAWAF Tab5, page 1
WJ51, 05.05.03

CAWAF1.xls

#3	
	NREL (Nat. Renewable Energy Lab, Golden, CO, USA), *An Atlas of Thermal Data For Biomass and Other Fuels*, NREL/TP-433-7965, June 1995, (only ultimate analysis, dry).
/1/	Recknagel, Sprenger, *Taschenbuch für Heizung und Klimatechnik*, Oldenburg, 62. Jhrg., 83/84.
/2/	Scholz, Beckmann, *Abfallbehandlung in thermischen Verfahren*.
/3/	P.Markewitz, *Theoretische Untersuchungen zum Verhalten von Schwermetallen in Kohlekraftwerken*, Forschungszentrum Jülich (D), 7/1991.
/4/	C. von Scala, *The influence of contaminants on the gasification of waste wood*, Diss 1998, ETH-Zürich.
/5/	K.Görner, *Charakterisierung, Modellbildung und Simulation technischer erbrennungssysteme*, Habil. 1992, Uni.Stuttgart.
/6/	C.Treeyawetchakun, V.Meeyoo, P.Rangsunvigit, P.C.Wright, *Kinetik study of the pyrolysis of Thailand Municipal and Industrieal Sludges by TGA*, Incineration 2001, Brussels, 2-4 July 2001.
/7/	S.Mould, P.Thornley, *Environmental and Economic Evaluation of two UK Bio-Mass Fuelled Power Plants*, Incineration 2001, Brussels, 2-4 July 2001.
/8/	Y.B.Yang, Y.R.Goh, V.Nasserzadeh, J.Swithenbank, *Mathematical Modelling of Solid Combustion on a Travelling Bed and the Effect of Channelling*, Incineration 2001, Brussels, 2-4 July 2001.
/9/	A.H.Shamsuddin, K.Sopian, N.Y.Muhd Noor, A.Ilmi, H.A.Mohamad Puad, (Malaysia), *A Study on Combustion of Terminal Oil Sludge in a Fluidesed Bed Combustor*, Incineration 2001, Brussels, 2-4 July 2001.
/10/	Y.Chi, S.Li, J.Huang, R.Li, X.Li, D.Yan, Z.Ma, J.Yan, K.Cen, (Uni Hangzhou, China), *Clean Fuels from Pyrolysis of Scrap Tyre in Rotary Kiln*, Incineration 2001, Brussels, 2-4 July 2001.
/11/	M.Hocquel, S.Unterberger, K.R.G.Hein, (Uni Stuttgart), *Understanding mercury behaviour - a contribution to higher removal efficiencies*, Incineration 2001, Brussels, 2-4 July 2001.
/12/	L.E.Amand, H.Miettinen-Westberg, M.Karlsson, (Uni Göteborg, S), B.Leckner, J.Werther, et al. (Uni Hamburg, D), *Co-Combustion of Sewage Sludge with Wood/Coal in a Circulating Fluidised Bed Boiler*, Incineration 2001, Brussels, 2-4 July 2001.
/13/	S.Chaiklangmuang, J.M.Jones, M.Pourkashanian, A.B.Ross, A.Williams, (Uni Leeds, UK), *Combustion of Paper and Plastic Wastes*, Incineration 2001, Brussels, 2-4 July 2001.
/14/	H.Belevi, *Environmental Engineering of MSW Incineration*, v/d/f 1999.
/15/	Dokumentation ALSTOM Power zu IGEA-Anlage, 7.2001.
/16/	R.Chirone, P.Salatino, F.Scala, *The relevance of attrition to the fate of ashes during fluidzed-bed combustion of biomass*, in Proceedings of the Combustion Institute, Vol 28, 2000/pp. 2279-2286.
/17/	S.T.Perry, E.M.Hambly, T.H.Fletcher, M.S.Solum, R.J.Pugmire, *Solid-State 13C NMR Characterisation of matched tars and chars from rapid coal devolatilization*, in Proceedings of the Combustion Institute, Vol 28, 2000/pp. 2313-1319.
/18/	S.Vockrodt, H.Müller, M.Pollack, C.Götte, (D), *Auswirkungen der Mitverbrennung von Trockenbraunkohle auf das Emissions- und Verschmutzungsverhalten eines Grossdampferzeugers*, in VDI Bericht 1534, Braunschweig 3/2000.
/19/	F.Rückert, U.Schnell, K.R.G.Hein, B.Peters, H.Hunsinger (Karlsruhe), *Simulation verschiedener Feuerungsprnzipien einer Rostfeurungsanlage (TAMARA)*, in VDI-Bericht 1534, Braunschweig 3/2000.

-PSI-

Composition and Analysis of Wastes, Raw Materials and Fuels

CAWAF Tab5, page 2
WJ51, 05.05.03

Table D-5-5: References

/20/	W.Bernstein, T.Brunner, A.Hiller, J.Albrecht, N.Quang (D), *Die Mitverbrennung von Abfallstoffen in der zurkulierenden Wirbelschicht - Experiment und Modellierung.*
/21/	in: N.Krzikalla, *Das Verhalten ausgewählter Spurenelemente in GuD-Kraftwerken mit integrierter Kohlevergasung* , Forschungszentrum Jülich , Jül-2756, 1993, from: D.W.Koppenaal, *Trace element studies on coal gasification process streams* , Diss. University Of Missouri, Columbia (USA),1978.
/22/	*? Chemical Compositions and Heating Value of the Examined Materials (Table 2).*
/23/	Dubbel, *Taschenbuch für den Maschinenbau* , Springer Verlag, 1974.
/24/	B.Bilitewski, G.Hardtle, K.Marek, *Abfallwirtschaft* , Springer Verlag, 1990.
/25/	Ahlstrom Pyroflow (company communication), ca. 1990
/26/	Report Dr.Graf AG, *Regelungen über die Abfallentsorgung in Zementwerken* , p.8, 1994
/27/	H.Belevi, *Was können Stoffflussstudien bei der Bewertung der thermischen Abfallbehandlung leisten?* VDI Bericht 1033, 1993
/28/	*. Sondergutachten* , 1990
/29/	Ct Environment, *RESH-Verwertungsanlage (CH)* , 2001
/30/	Kanton Basel-Land, Bericht: *Stoffliche Zusammensetzung von RESH* , Amt für Umweltschutz und Energie (CH).
/31/	Chyang, *Incineration of Biomass in a pilot plant fluidized bed combuster* , at IT3 Conference 1999, Orlando (USA).
/32/	N.Krzikalla, *Das Verhalten ausgewählter Spurenelemente in GuD-Kraftwerken mit integrierter Kohlevergasung* , Forschungszentrum Jülich , Jül-2756, 1993.
/33/	P.Monkhouse, *On-line diagnostic methods for metal species in industrial process gas* , Progress in Energy and Combustion Science 28 (2002) 331-381.
/34/	Biochemistry
/35/	S.F.Miller, B.G.Miller, C.M.Jawdy, *The occurance of inorganic elements in various biofuels and ist effect on combustion behavior* , The Energy Institut of Pennsylvania State University and Oak Ridge Nat. Lab. Tennessee, USA, WWW 8.7.02.
/36/	Römpp Chemie Lexikon, after Mason and Moore.
/37/	in NREL #3, from A.Bliek, *Mathematical Modelling of a Co-Current Fixed Bed Coal Gasifier* , Ph.D.Thesis, Twente University of Technology, Eidhoven, NL, 1984

INDEX